M000273078

Applied Mathematical Sciences
Volume 101

Editors
J.E. Marsden L. Sirovich

Advisors
S. Antman J.K. Hale P. Holmes
T. Kambe J. Keller K. Kirchgässner
B.J. Matkowsky C.S. Peskin

Springer
New York
Berlin
Heidelberg
Barcelona
Budapest
Hong Kong
London
Milan
Paris
Santa Clara
Singapore
Tokyo

Applied Mathematical Sciences

(continued following index)

Helena E. Nusse   James A. Yorke

Dynamics:
Numerical Explorations

Second, Revised and Enlarged Edition

Accompanying Computer Program *Dynamics 2*
Coauthored by Brian R. Hunt and Eric J. Kostelich

With 206 Illustrations, 16 in Color, and a 3¹/₂" DOS Diskette

Springer

Helena E. Nusse
University of Groningen
Faculty of Economics
Dept. of Econometrics / Section Mathematics
P.O. Box 800
NL-9700 AV Groningen
The Netherlands
and
Institute for Physical Science and Technology
University of Maryland
College Park, MD 20742
U.S.A.

James A. Yorke
Institute for Physical Science and Technology
University of Maryland
College Park, MD 20742
U.S.A.

Coauthors of Dynamics
Brian R. Hunt
University of Maryland
Institute for Physical Science and Technology
College Park, MD 20742
U.S.A.

Eric J. Kostelich
Department of Mathematics
Arizona State University
Box 871804
Tempe, AZ 85287-1804
U.S.A.

Editors
J.E. Marsden
Control and Dynamical Systems 107-81
California Institute of Technology
Pasadena, CA 91125
U.S.A.

L. Sirovich
Division of Applied Mathematics
Brown University
Providence, RI 02912
U.S.A.

Mathematics Subject Classification (1991): 49L20, 58G28, 70K15, 90C39

On the cover: Three Wada basins for the Pendulum differential equation. For more details, see Figure 7-9a.

Library of Congress Cataloging-in-Publication Data
Nusse, Helena Engelina, 1952–
    Dynamics : numerical explorations / Helena E. Nusse, James A.
  Yorke. — 2nd ed.
        p.   cm. — (Applied mathematical sciences : 101)
    Includes bibliographical references and index.
    ISBN 0-387-98264-7 (alk. paper)
    1. Dynamics. 2. Dynamics—Data processing. 3. Chaotic behavior
in systems—Data processing. I. Yorke, James A. II. Title.
III. Series: Applied mathematical sciences (Springer-Verlag New York
Inc.) ; v. 101.
    QA1.A647   vol. 101   1997
    [QA845]
    510 s—dc21
    [515´.352´078553042]   97-26972

Printed on acid-free paper.

© 1998, 1994 Springer-Verlag New York, Inc.
This work consists of a printed book and a diskette packaged with the book, both of which are protected by federal copyright law and international treaty. This work may not be translated or copied in whole or in part without the written permission of the publisher (Springer-Verlag New York, Inc., 175 Fifth Avenue, New York, NY 10010, USA), except for brief excerpts in connection with reviews or scholarly analysis. For copyright information regarding the diskette, please consult the printed information packaged with the diskette in the back of this publication. Use in connection with any form of information storage and retrieval, electronic adaptation, computer software, or by similar or dissimilar methodology now known or hereafter developed is forbidden.
The use of general descriptive names, trade names, trademarks, etc., in this publication, even if the former are not especially identified, is not to be taken as a sign that such names, as understood by the Trade Marks and Merchandise Marks Act, may accordingly be used freely by anyone. Where those designations appear in this book and where Springer-Verlag was aware of the trademark claim, the designations follow the capitalization style used by the manufacturer.
The publisher may be contacted directly regarding site licenses: Site Licenses Administration, Springer-Verlag New York, Inc., 175 Fifth Avenue, New York, NY 10010, USA; (212) 460-1500.

Production managed by Lesley Poliner; manufacturing supervised by Johanna Tschebull.
Camera-ready copy prepared by the authors using Chiwriter 4.
Printed and bound by R. R. Donnelley & Sons, Harrisonburg, VA.
Printed in the United States of America.

9 8 7 6 5 4 3 2 1

ISBN 0-387-98264-7 Springer-Verlag New York Berlin Heidelberg   SPIN 10629791

PREFACE

When investigating a dynamical system, it is of course important to be able to plot trajectories, but that is just the beginning. The Maryland Chaos Group developed an array of tools to help visualize the properties of dynamical systems including automatic method for plotting all "basins and attractors", computing "straddle trajectories", and for automatically searching for all periodic orbits of a specified period.

In the investigations of the Maryland Chaos Group, we found it useful to be able to combine these various basic tools with each other into a single package that grew with time so that each new study could benefit from the previous programming efforts. The earliest version of this program *Dynamics* was distributed in 1985, and it has grown continuously since then. The resulting program *Dynamics* requires either a Unix workstation running X11 graphics or an IBM PC compatible computer. Some basic tools in *Dynamics*, such as the computation of Lyapunov exponents and the use of Newton's method are standard. The method of computation of stable and unstable manifolds is superior to standard procedures. *Dynamics* is currently being used extensively in our research and it is being used in undergraduate courses.

Dynamics: Numerical Explorations provides an introduction to and overview of fundamental tools and numerical methods together with many simple examples. All the numerical methods described in this book are implemented in *Dynamics*. While the algorithms to implement these ideas are sometimes fairly sophisticated, they are elementary in that what they do can be understood by undergraduates. They should be available to everyone exploring dynamical systems. Many of the examples reveal patterns that are not fully understood and have surprises lurking just beyond the edges of the imagination. This package can be used by undergraduates, by graduate students, and by researchers in a variety of scientific disciplines.

Improving the program is a continuous process so there may be improvements in the program not reflected in this book. In such cases, these changes are self explanatory.

An overview of capabilities of Dynamics

This program will continue to evolve. It is a toolkit in which the tools are all available at any moment, enabling you to explore the system with much greater ease than if each tool was a separate program. These tools are elementary and should be available to everyone exploring dynamical systems. Some of the tools have not previously been available anywhere, indicating only that the ideas used in implementing them are not obvious. Nonetheless, the capabilities presented here are quite basic.

● The program iterates maps and solves differential equations. The program utilizes fixed and variable step size Runge-Kutta solvers for differential equations.

● Trajectories can be plotted and you can interactively store results and change initial conditions, parameters, and the scale of the screen.

● If desired, the user can split the screen into quadrants. The different quadrant windows can have different coordinates being plotted, with simultaneous plotting in all windows. Figure 1-1 on page 3 is an example of a screen divided into quadrants.

● The program features an array of simple commands. They can be invoked while a complicated process is being carried out. Some examples are: the screen can be cleared or refreshed; crosses can be plotted; the system can be paused or "single stepped" one point at a time; current positions can be stored; a trajectory can be reinitialized, to mention a few. The arrow keys may be used for drawing boxes, rescaling the screen, or choosing different initial points. The images of the small cross can be plotted, and as the user moves the small cross using the arrow keys the program continues to show the images of the small cross.

● The state of the program can be stored for later use. The program can create a file of parameters that have been set, and this file of values and settings can automatically be reinstalled later when restarting the program.

● Pictures created on the screen can be printed in resolution higher than that of the screen (960 dots wide by 960 vertically or the resolution 720 by 720 which is mainly used throughout this book), and these images can be stored on a disk (in a data compressed format). Pictures can be recalled from a disk and added to each other. Currently supported printers are the Epson printer MX-80 printer, Hewlett Packard printers, and printers compatible with these. It also supports PostScript printers including color PostScript printers. The PostScript printer support was added by Eric J. Kostelich, Tim Sauer and Brian Hunt.

An overview of advanced capabilities of Dynamics
With this program you can:
- Find fixed points and periodic orbits using Quasi-Newton's method and (when the system is 2-dimensional) find the eigenvalues and eigenvectors of the derivatives of the processes evaluated at the periodic orbits.
- Calculate Lyapunov exponents and the Lyapunov dimension of an attractor.
- Automatically plot all basins and attractors for 2-dimensional processes.
- Follow periodic orbits as a parameter is varied. Attracting and unstable orbits are plotted in different colors.
- Compute straddle orbits that are chaotic but do not lie on a chaotic attractor. For example, compute a bounded chaotic trajectory of the Henon map when almost all initial points diverge to infinity.
- Plot unstable and stable manifolds of periodic orbits.
- Create bifurcation diagrams showing how attractors change as a parameter is varied.
- Automatically find and plot the periodic orbits of a specified period.
- Create Chaos plots and Period plots in two dimensional parameter space.
- Zooming in on small regions. For example, zooming in toward a point on the common boundary of two basins of attraction.

Help for the novice
Help files are available on-line. There is a menu of help facilities and the program provides an on-line quick start tutorial for the beginner.

Examples and Exercises
Chapter 2 presents examples of pictures you can make simply. The required commands are printed in bold. In the Chapters 5 through 12, we present more examples for creating reliable pictures. You are invited to make the exercises to get familiar with the majority of the features of *Dynamics*. A few exercises are preceded with a '*', and are considered to be rather difficult.

Topics of discussion
Below many of the figures, we suggest a "topic of discussion" and pose a question concerning some feature of the figure. Sometimes there is a simple answer but more often there are a variety of correct possible answers, and sometimes there are obvious extensions. For such questions, the exchange of ideas in discussions is beneficial to all.

How does Dynamics 2 differ from the original?

The second edition of "*Dynamics: Numerical Explorations*" presents a version of *Dynamics* with changes that we hope will make it even easier to use. This version of the program is called *Dynamics* 2. All of the commands of *Dynamics* still work in *Dynamics* 2. The program *Dynamics* 2 is quite big. If the original source code of *Dynamics* was printed out, it would be nearly 40,000 lines or about 700 pages. For *Dynamics* 2, it is a third longer. This source code is included in the disk that comes with the book.

- *Dynamics* 2 is faster on PCs than *Dynamics*. On an Intel Pentium based computer for example, *Dynamics* 2 runs time-consuming processes (like plotting a trajectory for the forced pendulum differential equation) about 3 times as fast as *Dynamics*. Furthermore *Dynamics* 2 can use much more of your computer's memory, if your computer has four of more megabytes of RAM. Hence the pictures you create can be of higher resolution with more colors.

- *Dynamics* 2 has an Add-Your-Own-Equations facility by Marc Parmet. That means it will be unnecessary to have a compiler. When you provide a map or differential equation, the program can automatically compute partial derivatives of the process for Lyapunov exponents and Newton method for finding periodic orbits.

- *Dynamics* 2 supports color PostScript printers, so color pictures can be easily printed. In addition, depending on your computer's graphics card, you can create pictures with 256 colors on the screen, with full resolution (640 by 480 pixels).

- *Dynamics* 2 has a new menu system. When you type in a command, you are often prompted by an options menu that lists the various parameters you might want to set. This means the program is easier to learn and to use since the commands are easier to find.

- *Dynamics* 2 supports the mouse. If you wish, you can use the mouse for working through menus and selecting initial points of trajectories, drawing boxes, etc.

- *Dynamics* 2 has many additional minor capabilities. For example,
  - There is a "zoom" facility allowing you to automatically zoom in on small regions.
  - There is are features for creating Chaos plots and Period plots in two dimensional parameter space.
  - There is a help facility that gets help with a command using a single key stroke (the 'Tab'-key).

Compiler

Dynamics 2 was compiled using Microsoft Visual C++ Professional Edition, version 2.0 or later, together with Phar Lap's TNT DOS Extender SDK. This combination allows us to provide a program accessing available memory on PCs. (The version of TNT that is provided with Microsoft Visual C++ does not have sufficient capability to compile Dynamics). Dynamics 2 is a "32 bit" program which runs about three times as fast as the first edition of this program on computationally intensive projects. It is capable of using several megabytes of RAM. Usually 2 megabytes is quite sufficient. By comparison, *Smalldyn* is compiled with Microsoft Visual C++ Professional Edition version 1.5x and is a "16 bit", compiled using overlay features. It is restricted as to the amount of memory that it can access and it typically runs at roughly half the speed of *Dynamics 2*. It is designed for computers with only 1 Megabyte of RAM memory.

References

In the Chapters 5 through 12 we include a section entitled "References related to *Dynamics*". The purpose of this section is twofold. One purpose is to include references which establish the reliability of the numerical methods of the program. A second one is to illustrate how *Dynamics* is used or can be used.

At the end of this book we give a selection of references. In this list, we have obviously left out many important contributions to the field of dynamical systems. Consult, for example, the 4405 references in Shiraiwa (1985) and the 7157 references in Zhang (1991).

SmallDyn

Our diskette also includes *SmallDyn* 2, a small version of *Dynamics 2*. *SmallDyn* can be freely distributed provided it is complete (including the file *readme*) and it is unchanged, but the user must have access to this book to understand its capabilities. It has the format of *Dynamics 2* and accepts most of the commands, but it cannot access as much of the computer's memory and runs more slowly. It can run on older PC's with only 1 MByte of RAM. We recommend that students in a course "Nonlinear Dynamics" or "Dynamical Systems" should have access to several copies of this book.

SmallDyn 2 cannot be made part of a commercial product without the permission of Springer-Verlag and the authors. We distribute it freely on our web site   http://www.ipst.umd.edu/dynamics/ .

Animations

See our web site http://www.ipst.umd.edu/dynamics/ to view animations of dynamical systems that students have made with the help of *Dynamics 2*.

Disclaimer

While most of the routines have been tested during a period of years, they are not designed for commercial application. The authors and the publisher assume no responsibility for losses that might result from errors in this program. Comments, bugs, typos, and suggestions on the software package can be directed to H.E.N. (book) and J.A.Y. (program). However, we generally cannot give help individuals to use *Dynamics* or give help in adapting it to your needs.

Acknowledgment

We would like to thank the many people who have made useful comments on the program and this book. We would like to thank all the researchers of the Maryland Chaos Group for their comments. In particular, we would like to thank Kevin Duffy, Jason Gallas, Olaf Hammelburg, Brian Hunt, Hüseyin Koçak, Ajay Kochhar, Jacob Miller, Marc Parmet, Miguel Sanjuan, Tim Sauer, Paul Schure, and Guohui Yuan for their comments and suggestions.

The basic research reflected in this program has been supported in part by the Air Force Office of Scientific Research/Applied Mathematics, the Department of Energy (Mathematical, Information, and Computational Sciences Division, High Performance Computing and Communications Program), the Office of Naval Research/Defense Advanced Research Projects Agency/Applied and Computational Mathematics Program, the National Science Foundation (Division of Mathematical Sciences and Physics), and the W.M. Keck Foundation supported our Chaos Visualization Laboratory.

IBM is a registered trademark of International Business Machines, Visual C++ and MS-DOS are registered trademarks of Microsoft, Unix is a registered trademark of American Telegraph & Telephone company, and PostScript printer is a registered trademark of Adobe Systems Inc. TNT DOS-Extender and Pharlap are a trademark and registered trademark respectively of Pharlap Software Inc. Other brand and product names are included herein and are trademarks or registered trademarks of their holders. ChiWriter 4.20b was used for typesetting the text of this book.

Helena E. Nusse; H.E.Nusse@eco.rug.nl, nusse@ipst.umd.edu
James A. Yorke; yorke@ipst.umd.edu
July 1997, College Park, Maryland

Visit our World Wide Web site

http://www.ipst.umd.edu/dynamics/

for up-to-date information on *Dynamics*

CONTENTS

CHAPTER 1

GETTING THE PROGRAM RUNNING

1.1 THE DYNAMICS PROGRAM AND HARDWARE

Chaotic dynamics has a long history that can be traced back to the papers of Henri Poincaré and James Clerk Maxwell. That history is quite thin compared with the volumes written on how to study small perturbations of a stable equilibrium point. The field of chaotic dynamics did not come alive sufficiently to affect scientists until chaotic systems could be studied on small interactive computers with good graphics. Edward Lorenz's studies in 1963 were carried out on a desk-sized computer. One day after having previously made a long computer run he entered an intermediate result, a vector giving an intermediate state of the system. But the intermediate result had less precision than the true computer state. To his surprise he found this tiny change caused the system to evolve in quite a different manner. This was the first computer study illustrating sensitivity to initial data. Now with the advent of personal computers with good graphics, now that we know what to look for, it is easy to see chaotic trajectories.

When investigating a dynamical system, it is of course important to be able to plot trajectories, but that is just the beginning. An array of tools have been developed and combined in a single software package to help visualize what is happening in dynamical systems. Many of these tools such as plotting of "basins of attraction", computing "straddle trajectories", and automatically searching for all periodic orbits of a specified period are in this program. These tools are elementary in that what they do can be understood by undergraduates and they are fundamental. The algorithms to implement these ideas are sometimes fairly sophisticated. As an example, the computation of stable manifolds as presented in this program is superior to standard procedures. Some other basic tools such as the computation of Lyapunov exponents and the use of Quasi-Newton's method are rather standard.

Interactive exploration: hit a key at any time

The interactivity of *Dynamics* is reflected in the basic loop structure of the program:

compute a point
plot the point
check to see if a key has been hit

The program can go through this cycle hundreds of times per second, depending on the process. At any point you can pause or start a new procedure.

Hardware requirements and recommendations

This program requires either a Unix workstation running the X Window System or a PC running DOS or MS Windows 3.x or MS Windows 95. It does not run on a Macintosh. The PC must have at least two megabytes of "RAM" memory with an EGA, VGA, or super VGA graphics board. *Dynamics* needs less than two megabytes of disk storage. For those computers with less than 1 megabyte of "RAM" memory, we provide the program "*Smalldyn.exe*".

Recommended but not strictly needed are a color screen and a printer. Currently supported printers are the Epson MX-80 printer, Hewlett-Packard printers, PostScript printers, color PostScript printers, and printers compatible with these. Even without a printer, this should be an entertaining program.

Source code and compiler for Unix

The program comes with the full source code written in the language C (about 42,000 lines of programming). The same source code is used on both PC's and Unix workstations. To run *Dynamics* on a Unix computer, the X Window System is required. You must transfer the source code from a PC to your Unix workstation and then compile the program using your computer's C compiler. See Chapter 14 for details.

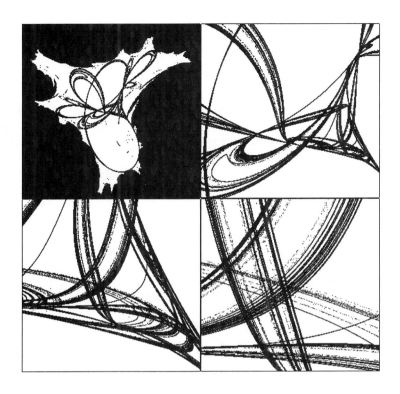

Figure 1-1: A screen of Dynamics

"Topics of discussion." *Many of the figures in this book are accompanied by one or more questions under the heading "Topic of discussion". The questions often ask for an explanation of some feature in the picture. A discussion of the questions benefits all because there will often be different correct interpretations and answers to the questions.*

Warning: the techniques useful in these investigations often have not been reached in the book when the questions arise. The techniques of Chapters 3 and 4 are especially helpful.

Installing Dynamics

First make a copy of the *Dynamics* disk with the files *install.exe*, *cfiles.exe* and *readme*. The program **dynamics.exe** calls upon the files

y.pic, **yprom.txt**, **ymenus.txt**, **yps.txt**, **yhelp.txt,** and **yalert.txt**.

and so the computer needs to know where to look for these. They should be in the active directory when you start the program. For example, if you are going to run the program from the directory C:\Dyn97, use the DOS commands "md \Dyn97" and "cd \Dyn97" to make that the active directory. Next, insert the *Dynamics* disk into the "A-drive" and use the DOS command "A:install" to install *Dynamics* in C:\Dyn97. The file *cfiles.exe* is for Unix users who want to change the program.

Smalldyn: small version of Dynamics

Dynamics is a big program which requires more than one megabyte of "RAM" memory. Does your computer have sufficient memory to run *Dynamics*? There is an alternative version of the program named **smalldyn.exe** which is also installed when *Dynamics* is installed. It requires the use of the files

yy.pic, **ymenus.txt**, **yalert.txt**, **yprom,txt**, **yhelp,txt**

It is for use only with computers having a hard disk. With the standard version, the entire program *dynamics.exe* is read from the disk into memory. When running the small version *smalldyn.exe*, the computer fetches only the parts of the program it needs at the moment. Each time a command is executed, some part of the program will be brought into memory from the disk. Hence, this small version requires a fast disk, as is usual with a hard disk. The advantage of this version is that it needs less memory. Hence, if your computer is not permitting you to run *Dynamics*, then you may find the small version helpful. *Smalldyn* will restrict you to using lower resolution pictures, or to use pictures with fewer than 16 colors and the program *Smalldyn* runs at half the speed of *Dynamics*. For this version, it is important to minimize the number of additional drivers and RAM resident programs in memory to allow as much space as possible for the program. Many computers have taken advantage of much of their memory by installing several files permanently in that memory, files such as printer drivers, "RAM resident" programs, and of course parts of the operating system. If *Smalldyn* will not run, it is possible that too much was installed on the computer. The solution is to create a new *autoexec.bat* file which installs less.

Helping Dynamics find the files

Before running *Dynamics* you can instruct your computer where to look for the *.txt, *.pic and *.dd files if they are not in the current directory. If, for example, the files are in directory C:\dyn97, then type

set DYNAMICS = c:\dyn97 *<Enter>*

This procedure works in MS-DOS and Unix, and can be helpful when the program is being used on a network. (*Smalldyn* will also look in C:\dyn97 for the files.)

Unix computers

To run *Dynamics* on a Unix computer, the X Window System is required. You must transfer the source code from a PC to your Unix workstation and then compile the program using your computer's C compiler. Some minor changes may be required for some Unix C compilers. It has been compiled and run successfully on a variety of computers including computers made by Hewlett-Packard, IBM (RS/6000), Sun, Digital Equipment Corporation, Silicon Graphics, as well as PCs running Linux, and others.

Read Chapter 14 if you have a Unix workstation. All our files' names are lower case in Unix. In MS-DOS the case does not matter. **Brian R. Hunt** and **Eric J. Kostelich** have put in a great deal of effort to make the MS-DOS source code compatible with Unix supporting version 11 of the X Window System.

Changes in the program

The menus in this program are subject to change. It is easy for the authors to make changes in the program. It is almost impossible to incorporate the final improvements in the program into the text. Read the file *readme* on the distribution disk for information about these changes in the program. Frequently, you will find that the menus may include processes or commands in addition to those we list since we emphasize only essential aspects of the program. We encourage you to experiment with the other commands and check the help files for information. In some cases the menus of *Dynamics* are more extensive than those of *Smalldyn*.

For the latest changes of *Dynamics*, see the **readme** file on the disk

---

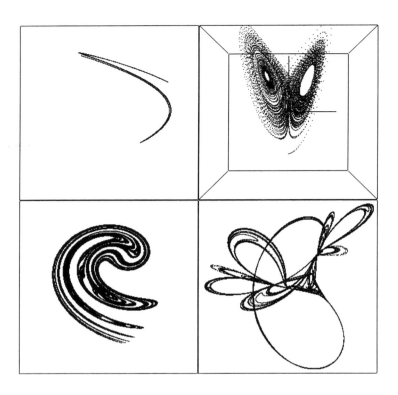

Figure 1-2: Another screen of Dynamics

This figure shows Trajectories of four different processes of Dynamics. The picture in the upper left window is a trajectory for the Henon map (H); see also Figure 2-1a. The picture in the upper right window is a trajectory for the Lorenz differential equations (L); see also Figure 2-16. The picture in the lower left window is a trajectory for the Ikeda map (I); see also Figure 3-1a. The picture in the lower right window is a trajectory for the Tinkerbell map (T); see also Figure 2-22.

1.2 GETTING STARTED WITH DYNAMICS

To get started, if the current directory does not contain the program *Dynamics*, change to the directory which contain it. If that is \dyn97, for example, then type

cd \dyn97 *<Enter>*

Next type

dynamics *<Enter>*

First, the title page of *Dynamics* appears on the screen for a few seconds

and then the **PROcess Menu**. This menu will include the following

---

```
╔═══════════════════════════════════════════╗
║                 MAP MENU 1                  ║
║   MORE  –  map menu 2                       ║
║   DE    –  Differential Equations menu      ║
║   OWN   –  enter your OWN process           ║
║                                             ║
║                  MAPS                       ║
║   C     –  complex Cubic map                ║
║   CAT   –  linear map on unit square        ║
║   CIRC  –  CIRCle map                       ║
║   CW    –  CobWeb map, a 1-dim map          ║
║   DR    –  Double Rotor map                 ║
║   H     –  Henon map                        ║
║   I     –  Ikeda map                        ║
║   KY    –  Kaplan/Yorke map                 ║
║   LOG   –  LOGistic map                     ║
║   Q     –  Quasiperiodicity map             ║
║   R     –  pulsed Rotor & standard map      ║
║   T     –  Tinkerbell map                   ║
║   TT    –  TenT map                         ║
║                                             ║
║   QX    –  Quit and eXit the program        ║
╚═══════════════════════════════════════════╝
```

Hit <*Enter*> and more maps of the PROcess Menu appear on the screen

```
╔═══════════════════════════════════════════╗
║                 MAP MENU 2                  ║
║                                             ║
║   CG    –  Chossat/Golubitsky symmetry map  ║
║   DDR   –  Degenerate Double Rotor map      ║
║   GB    –  Generalized Baker map            ║
║   GM    –  Gumowski/Mira map                ║
║   H2    –  Henon with 5th order polynomial  ║
║   HZ    –  Hitzl/Zele map                   ║
║   LPR   –  Lorenz system's Poincare Return map ║
║   N     –  Nordmark truncated map           ║
║   PL    –  Piecewise Linear map             ║
║   RR    –  Random Rotate map                ║
║   TRI   –  TRIangle map                     ║
║   YAK   –  2 competing populations          ║
║   Z     –  Zaslavskii rotation map          ║
╚═══════════════════════════════════════════╝
```

After selecting the Differential Equations menu, hit <*Enter*> and the differential equations of the PROcess Menu appear on the screen

```
┌─────────────────────────────────────────────────────────────────────────┐
│                   DIFFERENTIAL EQUATIONS MENU                             │
│                                                                           │
│   MORE –   map menu 1                                                     │
│                                                                           │
│   CC     –   Chua's Circuit                                               │
│   CF     –   Cylinder-Flow                                                │
│   D      –   forced double-well Duffing equation                         │
│   GN     –   GoodwiN equation                                             │
│   HAM    –   HAMiltonian system, h = (p^2 + q^2)/2 + u^2*v^2 = const      │
│   L      –   Lorenz system                                                │
│   LPR    –   Lorenz-Poincare-Return-map                                   │
│   LV     –   Lotka/Volterra equations                                     │
│   P      –   forced damped Pendulum equation                             │
│   PD     –   Parametric Duffing equation                                 │
│   ROS    –   ROeSsler equation                                           │
│   R3     –   Restricted-3-body-problem                                   │
│   SG     –   Samardzija/Geller odd-symmetry Lorenz-like system          │
│   VP     –   forced Van der Pol equation                                 │
│   VP2    –   2 forced coupled Van der Pol equations                     │
│                                                                           │
│   QX     –   Quit and eXit the program                                   │
└─────────────────────────────────────────────────────────────────────────┘
```

To select the Henon map (H) in the initial process menu (Map menu 1), either use the arrow keys to select this process and hit

< Enter >

or type (both upper case and lower case letters may be used)

H *< Enter >*

   The **Main Menu** appears on the screen in approximately the following format

```
┌──────────────────────────────────────────────────────────────┐
│  NumExplM – FileM – ParameterM – VectorM – ScreenM – Help     │
└──────────────────────────────────────────────────────────────┘
```

Being in the Main Menu, use the arrow keys to select NumExplM and hit <Enter> or type

NEM <*Enter*>

and the Numerical Explorations Menu appears on the screen in approximately the following format

```
NEM    - - NUMERICAL EXPLORATIONS MENU

T      -  plot Trajectory
DYN    -  quit & start new map or Differential Eqn.
OWN    -  quit & create OWN process
P      -  Pause the program: NOT paused
Q      -  Quit Dynamics program

       MENUS
BIFM   -  BIFurcation diagram Menu
BM     -  Basin of attraction Menu
DIM    -  DImension Menu
FOM    -  Follow (periodic) Orbit Menu
LM     -  Lyapunov (exponents) Menu
POM    -  Periodic Orbit Menu
STM    -  Straddle Trajectory Menu
UM     -  Unstable and stable manifold Menu
```

If you are in the Numerical Explorations Menu, either hit

<*Esc*>

or type

MM <*Enter*>

to return to the Main Menu.

Being in the Main Menu, use the arrow keys to select FileM and hit
<Enter> or type

FM *<Enter>*

and the **File Menu** appears on the screen in approximately the following
format

```
FM      - -    FILE MENU

FD      -   retrieve picture (From Disk)
PP      -   Print Picture
TD      -   save picture (To Disk)

DYN     -   quit & start new map or Differential Eqn.
OWN     -   quit & create OWN process
Q       -   Quit Dynamics program

            MENUS
SCM     -   Size of Core picture Menu
DM      -   Disk files Menu
```

If you are in the File Menu, either hit

<Esc>

or type

MM *<Enter>*

to return to the Main Menu.

Being in the Main Menu, use the arrow keys to select Help and hit
<Enter> or type

H *<Enter>*

and the **Help Menu** appears on the screen in approximately the following
format

```
┌─────────────────────────────────────────────────────────────────┐
│  H          – –  HELP MENU                                        │
│                                                                   │
│  INDEX    –  list all the commands whose descriptions             │
│                  have a word you choose                           │
│  MOUSE    –  what you can do with a MOUSE                          │
│  TUT      –  for the quick start TUTorial                         │
├───────────────────────────────────────────────────────────────────┤
│  Mouse is ON/OFF                                                  │
│  To cycle through the menus, hit  < & >  repeatedly.              │
│  To get an explanation of a high-lighted command, hit  < Tab >.  │
│  To get an explanation of a non-high-lighted command             │
│        like RN, hit '*' and type the command name:   *RN < Enter >│
│  To list all commands that begin with a letter or phrase          │
│        like "bif", enter the letter(s) and '*':  bif* < Enter >  │
│                                                                   │
│  To review the main features of the program, read                │
│  Chapter 2 of the Nusse/Yorke book                                │
│  "Dynamics: Numerical Explorations", Second Edition               │
└─────────────────────────────────────────────────────────────────┘
```

There are two windows in the Help Menu. The top window contains the commands and the bottom window contains information. Many menus do have this structure. If a menu has two windows, then the bottom window that provides some information, is referred to as the **Information window**.

Quick start tutorial

If you use the arrow keys to select the command TUT and hit < Enter >, you will be presented with a quick start tutorial. The purpose of the quick start tutorial is to get the novice working with *Dynamics* right away. Therefore, it will only deal with the simplest commands available in *Dynamics* and yet provide a view of the capabilities of this program. *Dynamics* provides a wide range of capabilities to the experienced user and yet is rather easy for the novice to use. The quick start tutorial will teach you the minimum set of skills that you need to start using *Dynamics*. If you exit from the tutorial and want to reenter at, say, page 9, type

9 <Enter>

Cycle through the menus

If you hit < & > repeatedly (do not hit < Enter >), you will cycle through the most important menus.

Enter a menu item or enter a command

There are two ways to select the menu item H. One approach is to use the "point and shoot" menu system. When a menu appears, the bottom line of the box that contains the menu presents an item from the menu. That item is high-lighted and enclosed by two asterisks. By hitting the arrow key, you can move from item to item. When the one you want is selected, hit <Enter> to implement this selection. If you have a mouse installed you can click on a command with the left mouse button to high-light an item. Clicking a highlighted item will implement it.

The second approach allows you to ignore which menu, if any, is on the screen. When you remember the command, this approach may be easier. Just type the command and hit <Enter>. When typing a command, you can use either upper or lower case letters.

For purposes of exposition in this book, we emphasize the latter approach, but you may choose occasionally to use the "point and shoot" approach. The phrase "enter a command" means type the command (or select the command from a menu on the screen using the arrow keys) and then hit the <Enter> key. Some computers have a key marked "Return" instead of "Enter" and others have a bent arrow pointing left.

Retrieving menus

Start the program (type *dynamics<Enter>*) and select the Henon map (type *H<Enter>* (both upper case and lower case letters may be used)). Now the Main Menu, a list of menus, appears on the screen. Once you have selected a process and are working with it, if no menu is on the screen, you can fetch the previous menu by hitting <*Enter*> (assuming you are not in the midst of typing a command). If a menu is on the screen, fetch its parent menu by hitting <*Esc*>. This <Esc> can be repeated until you arrive at the Main Menu MM. You can go directly to the Main Menu by entering command MM.

Throughout this book, as an aid to your memory, we capitalize the parts of descriptives that correspond to a command. Hence MM is the command for Main Menu.

Viewing the latest menu while plotting

Sometimes you want to see the last-called menu while the computer is plotting. If the computer is plotting, then whenever you hit <Enter>, the program pauses and the last called menu appears on the screen. In case you do not want to have this latest menu, just hit <space bar> and the menu is erased and the computer continues plotting.

Now hit

$$<Enter>$$

and you will see the last-called menu (for example, the Parameter Menu).
Now hit

$$<Esc>$$

and the parent menu will appear. Hit

$$<space\ bar>$$

and the trajectory resumes iterating again.

Help with the commands

To get help with the command that is high-lighted in a menu on the screen, just hit the <Tab> key. ("Hit" a key means gently depress and release the key.) For most commands, a description will appear on the screen. There is another way to get this information and can be used even if the command is not showing on the screen. To get help with the commands of the Main Menu or any of the other (sub)menus, hit * and enter the command. That is, hit the * key, then type the command, and then hit the <Enter> key. The result is that the relevant text from this book will be called to the screen. For example, to see what command B does, type

$$*B<Enter>$$

and the program prints out the description below. (Notice there is no space between * and B.) Sometimes, it is advisable to get an overview by hitting the * key, typing the command that retrieves the menu of concern and hitting <Enter>; for example, type

$$*MM<Enter>$$

and you will see the description of the "Main Menu" command MM.

Chapter 2 introduces some basic commands and provides detailed examples, and this could be a good point to go to Chapter 2 to try out the examples there.

INTERRUPTS

A few commands (called **interrupts**) are executed without hitting
< Enter > and are called by a single key stroke such as < . >, < Tab >, < Esc >,
< space bar >, the Function keys < F1 > - < F10 >, < End >, < Home >,
< PgUp >, < PgDn >, and the arrow keys. Descriptions of interrupts that are not
given here, can be found in the Appendix of this chapter. The "Interrupt
List" command **IL** fetches a list of these interrupts with brief
descriptions. Some interrupts can be used only while some routine like
plotting a trajectory is being carried out.

< · >

The key < · > (one dot) pauses the program *Dynamics* after plotting one
dot. Each time < · > is hit, the program computes one more dot and plots it
and pauses. The large cross appears at the current location of the
trajectory. Holding < · > down will produce a string of iterates; < space bar >
returns the program to normal continuous plotting of dots.

< space bar >

< space bar > ends the pause mode; see also command P (Pause) and
interrupts < · > and < Enter > which each initiate a pause mode.

The program often pauses a few seconds to allow you to read a message.
Hitting the space bar will often shorten the pause. Note that it will also
unpause the program if it is paused.

< Esc >

When < Esc > is hit, the current routine terminates or if a menu is on
the screen, it terminates the menu and retrieves the parent menu. Hit it
again and the program calls its parent menu. (On DEC stations there is no
< Esc > key, so use the key < F11 >.)

< Tab >

If a process is being run and you hit < Tab >, the program prints the
speed (in dots per second) and a selection of parameter values.

The speeds below are indications of speeds the user will get but
actual speed depends on various factors including the graphics board being
used and the windowing environment. Most applications of *Dynamics* do not
require high speeds.

Smalldyn runs at about half the speed of *Dynamics* on PC's, except that
an "Own" process runs about 8 times slower.

If a menu is on the screen, hitting <Tab> will fetch information about the high-lighted command with information similar to that given below.

dots per second		
process	H (Henon)	P (Pendulum)
PS/2 Model 50*	160	0.54
IBM PC*	260	2.5
DEC-ALPHA	16200	1360
Pentium 120**	9390	1920
Pentium 166**	12420	2780
* *Dynamics* (version 1), 1994 ** DOS window of Windows 95		

MS-DOS computers

"PS/2 Model 50" is an IBM computer without a mathematics coprocessor.

"IBM PC" means an IBM PC running at the speed 4.77 MHz, the speed of the original IBM PC, using the Intel 8087 math coprocessor.

"Pentium 120" denotes the Toshiba Tecra 700CT laptop computer based on a 120 MHz Intel Pentium chip.

"Pentium 166" denotes the Gateway 2000 P5-166 computer based on a 166 MHz Intel Pentium chip.

Computers running Unix running under the X Window System

"DEC-ALPHA" denotes a DECstation 3000/500 running at 175 MHz.

The Pendulum example above takes 30 Runge-Kutta steps per dot. It is computation intensive. While iterating with the Henon map however, the program spends almost no time on numerical computation and almost all its time on plotting. Those operations that are screen intensive such as the Henon map are primarily tests of how fast the computer can plot dots. All computers plot dots fast enough for most purposes.

To print out process parameters, hit <Tab>. If the process is running, it will also report the number of dots per second, so you can compare the speed of different computers on different processes. This timer for computing the speed is initialized whenever dot is set to 0, as for example when using commands T (Trajectory) or I (initialize).

USING THE MOUSE

The mouse can be used for a variety of purposes. It is not essential for Dynamics, but it can simplify procedures. This summary can be used as a reference. At this point in the book, you are only familiar with menus, and the other uses of the mouse will become clearer later in the book.

Double clicks

The program does **not** use double clicks.

Selecting items in menus *(left mouse button)*

If there is a menu on the screen, you can select an item in that menu using the left mouse button. If you click on a command that is not illuminated, it will become illuminated. If it is illuminated and you click on it, then that command will be executed.

Picking a new initial point of a trajectory *(left mouse button)*

If no menu is displayed on the screen, then the mouse can be used to set the initializer y1. (If a menu is displayed and you want it to vanish, just hit the <space bar>). Move the mouse and the mouse arrow will appear. Move it to the position on the screen that you select to be the new initial point. When you have selected a new initial point, click the left mouse button. If you are plotting a trajectory, this will reinitialize the trajectory and plotting will continue from this newly selected point.

If you are not using a mouse, the same can be accomplished by using the arrow keys to move the small cross, and then hitting command I (Initialize).

If no menu is displayed, you can use the mouse to set other vectors. Move the mouse arrow to the desired position. Depress the left button until a beep sounds. Release it and a menu of possibilities will appear.

Selecting a box for rescaling the screen *(right mouse button)*

See section 3.2 for a description of rectangular "boxes".

Position the mouse arrow at an appropriate position to be the corner of a new box. Then depress the right mouse button, hold it down, and move the mouse arrow to the diagonally opposite corner of the box you intend to create. Release the button, and the box is drawn on the screen. This draws the box *only on the screen*, not in core memory. If you want the box to be a permanent part of your picture, you must enter command B. Otherwise you can erase the box using command R. The right button is only for drawing boxes. You must still use a command like OW0 (Open Window 0) to use it to change scales.

TROUBLESHOOTING

If you move your mouse and no mouse arrow appears

On PCs, the mouse must be turned on to be used by DOS programs. It is often turned on by the computer's autoexec.bat file. If your computer comes with a mouse and routines for turning on your mouse but it is not automatically turned on when the computer is started, then you must look for the directory containing the corresponding files. They may be in the directory C:\MOUSE and in our computers the mouse can be turned on with the DOS command

C:\MOUSE\MOUSE

MS Windows 3.x

Dynamics can be run from a DOS window in MS Windows 3.x, provided the window is a full screen. If the DOS window is only part of the screen, the < Alt-Enter> will toggle to a full screen.

If you run Windows 3.x, you may see the mouse arrow, but Windows has its own Mouse routines. You may still have to turn on the mouse before starting *Dynamics*, if you want to use the mouse. The fact that you can use your mouse in Windows does not mean it is available for use in other programs.

If you are in a DOS window of MS Windows 3.x and the DOS mouse is off, you must exit Windows and start the mouse as described above. If you restart windows, then *Dynamics* will still have the mouse available to it.

MS Windows 95

If you run *Dynamics* in a DOS window of Windows 95, the mouse is automatically available. The window need not be a full screen.

Mouse and 256 colors on the screen

The mouse does **not** work in DOS when you are using 256 colors on the screen (command COL).

Figure 1-3: Another screen of Dynamics

This figure shows Trajectories of four different processes of Dynamics. The picture in the upper left window is a trajectory for the Henon map (H); see also Figure 2-4. The picture in the upper right window is a trajectory for the Quasiperiodicity map (Q); see also Figure 5-1b. The picture in the lower left window is a trajectory for the pulsed Rotor map (R); see also Figure 5-6a. The picture in the lower right window is a trajectory for the forced damped Pendulum differential equation (P); see also Figure 9-7.

APPENDIX: DESCRIPTION OF THE INTERRUPTS

Recall that "Enter a command" means type the command (using upper or lower case) and then hit the <Enter> key.

A few commands (called interrupts) are executed without hitting <Enter> such as <.>, <Tab>, <Esc>, <space bar>, and the arrow keys and Function keys. Enter the "Interrupt List" command IL:

IL <*Enter*>

and the following list will appear on the screen.

```
IL -- INTERRUPT LIST

                        <Esc> -- terminate current process
                        <Esc> -- get parent menu of current menu
                        <Tab> -- get process speed and parameters
CHANGE COLOR:           <F7> -- decrease color number by 1
                        <F8> -- increase color number by 1
                        <F9> -- choose color number
CHANGE PARAMETER:       <+> -- increase PRM by the amount PS
                        <-> -- decrease PRM by the amount PS
        <Home> or <Shift 3> -- halve PS
        <PgUp> or <Shift 4> -- double PS
CHANGE WINDOW:          <F1> -- upper left      <F2> -- upper right
                        <F3> -- lower left      <F4> -- lower right
                        <F10> -- whole screen
PAUSE MODE:             <·> -- single step process
                        <Enter> -- get previous menu and pause
                <space bar> -- undo the pause mode
SMALL CROSS:            <Arrow> -- turn on & move small cross
        <End> or <Shift 1> -- halve step size for small cross
        <PgDn> or <Shift 2> -- double step size for small cross
```

Now we describe the interrupts appearing in the Interrupt List, which are not described in Section 1.2.

<+>, <->, <Home>, <PgUp>

Hitting <+> and <-> increases or decreases respectively the value of the parameter PRM. PRM is a variable whose value is the name of a parameter. You can set the name using command PRM though initially it is

"RHO" (default of PRM is "RHO"). Hence, if you have not changed PRM, then hitting $<+>$ while running a process increases RHO by 1.0 and $<->$ decreases it by 1.0. Actually $<+>$ increases the parameter by the amount PS, which by default is 1.0, while $<->$ decreases PRM by the amount PS. This value can be changed using command PS (Parameter Step).

PS can also be changed while the program is plotting using either the $<PgUp>$ or $<Home>$ keys. The key $<PgUp>$ doubles PS and $<Home>$ halves it. Hitting $<+>$ or $<->$ sets DOT to 0 and reinitializes Lyapunov exponents. The interrupts $<Home>$ and $<PgUp>$ are **not listed in the menus**.

$<F1>$, $<F2>$, $<F3>$, $<F4>$, $<F10>$

The Function keys $<F1>$, $<F2>$, $<F3>$, and $<F4>$ can be used to switch back and forth between windows that have been opened. Use $<F10>$ to switch back to the whole screen.

Each window is a quadrant of the screen, situated as follows:

$$<F1> \qquad <F2>$$
$$<F3> \qquad <F4>$$

$<F7>$, $<F8>$, $<F9>$

The color numbers (or rather positions on the color table) on the screen are usually numbered 0 to 15 for color EGA and VGA and Unix/X11 graphics. The "Color Table" command CT displays the Color Table. The program has an active color number that is used in all plotting (and in refreshing the screen). Function key interrupt $<F9>$ is used for setting the value and the program will prompt the user for a color number; $<F8>$ increases that number while $<F7>$ decreases it.

$<End>$, $<PgDn>$

The arrow keys move the position of the vector y1 which is marked by the small cross; the small cross is turned on if it was off; in single stepping mode, the big cross at the position of y is plotted. Initially each move of the small cross is about 0.0015 of the screen width or height, but the size of these moves can be increased when the arrow keys are struck rapidly or are held down. After about 1 second of inaction, the step returns to the initial step size. This initial step size can be changed: it is doubled by hitting the $<PgDn>$ key and halved by hitting $<End>$, both of which are keys on the numeric key pad. Hitting either of these two will get the step size printed on the screen. The interrupts $<End>$ and $<PgDn>$ are **not listed in the menus**.

Note. Since some keyboards do not have certain keys, there are alternatives: use $<Shift\ 1>$ for $<End>$, $<Shift\ 2>$ for $<PgDn>$, $<Shift\ 3>$ for $<Home>$, and $<Shift\ 4>$ for $<PgUp>$, and $<F11>$ for $<Esc>$.

Figure 1-4: Computer art: Fractals and Dynamics

This figure shows a picture resembling the Sierpinski triangle in which the Henon attractor is super imposed on the white triangle; see Figure 2-4 for the Henon attractor and see Figure 4-3a for the Sierpinski triangle. Figure 4-3b is similar to this figure except that the Henon attractor is replaced by the Tinkerbell attractor.

See the caption of Figure 4-3a for an explanation of how to create a Sierpinski triangle; see the caption of Figure 4-3b for an explanation of how to create this picture.

1.3   QUESTIONS

Question: How do I get the program running?
Answer: See Section 1.1.

Question: How do I get the program running on a Unix computer?
Answer: See Section 1.1 and Chapter 14.

Question: How do I start and/or quit the program?
Answer: See Section 1.2.

Question: How do I investigate a new map or differential equation which is
           not in the process menu of *Dynamics*?
Answer: See Chapter 13.

Question: How do I plot a trajectory?
Answer: See Example 2-1a.

Question: How do I plot the Henon attractor?
Answer: See Example 2-4.

Question: How do I plot different 3-dim. views of the Lorenz attractor?
Answer: See Example 2-16.

Question: How do I plot a time series (a trajectory vs. time)?
Answer: See Example 2-2c.

Question: How do I plot the graph of an iterate of a one dimensional map?
Answer: See Example 2-3a.

Question: How do I plot a trajectory on a connected Julia set?
Answer: See Chapter 8.

Question: How do I plot the Mandelbrot set?
Answer: See Example 2-14.

Question: How do I plot the direction field of a differential equation?
Answer: See Example 2-6 and Section 4.3.

Question: How do I plot trajectories of a differential equation?
Answer: See Example 2-6 and Section 4.3.

Question: How do I find fixed points?
Answer: See Example 2-1k and Chapter 10.

Question: How do I compute Lyapunov exponents?
Answer: See Example 2-2b and Chapter 5.

Question: How do I plot Lyapunov exponents as a function of a parameter?
Answer: See Example 2-23 and Chapter 6.

Question: How do I plot a bifurcation diagram with parameter values printed?
Answer: See Example 2-8 and Chapter 6.

Question: How do I plot a bifurcation diagram on the screen?
Answer: See Example 2-7 and Chapter 6.

Question: How do I plot all the basins and attractors?
Answer: See Example 2-9 and Chapter 7.

Question: How do I plot the basin of infinity?
Answer: See Example 2-10 and Chapter 7.

Question: How do I find a periodic point of a specified period?
Answer: See Chapter 10.

Question: How do I search for many periodic orbits of a specified period?
Answer: See Example 2-1n, Example 2-11, and Chapter 10.

Question: How do I plot the unstable and stable manifolds of a fixed point?
Answer: See Example 2-17, Example 2-18, and Chapter 9.

Question: How do I compute the Lyapunov dimension of an attractor?
Answer: See Example 2-25 and Chapter 5.

Question: How do I compute the box-counting dimension of a picture?
Answer: See Example 2-25 and Chapter 12.

Question: How do I estimate the correlation dimension of an attractor?
Answer: See Chapter 12.

Question: How do I plot a trajectory on a basin boundary?
Answer: See Example 2-21, Example 2-22, and Chapter 8.

Question: How do I plot a bounded trajectory when almost all trajectories diverge?
Answer: See Example 2-19 and Chapter 8.

Figure 1-5: An attractor

This figure shows an attractor for the process "Van der Pol" of Dynamics. This is a periodically forced differential equation. See also Figure 3-4b which shows some consecutive pictures resulting from zooming in toward a point on the attractor.

Question: How do I zoom in on a point on a chaotic attractor?
Answer:　See Example 2-26 and Section 3.6.

Question: How do I choose initial conditions for plotting a trajectory?
Answer:　See Example 2-1i, Section 3.3, and Section 4.2.

Question: How do I print a picture that is on the screen?
Answer:　See Example 2-2 and Section 4.5.

Question: How do I store a screen picture on the disk?
Answer:　See Section 4.4 and Example 7-1.

Question: How do I draw boxes?
Answer:　See Example 2-1b and Section 3.2.

Question: How do I draw axes with tic marks?
Answer:　See Example 2-1h and Section 3.4.

Question: How do I change and/or set parameters?
Answer:　See Example 2-1n and Section 4.1.

Question: How do I change and/or set the scales?
Answer:　See Example 2-1q and Section 4.1.

Question: How do I change and/or set vectors?
Answer:　See Example 2-1p and Section 4.2.

Question: How do I create a figure with 4 windows containing consecutive
　　　　　blow-ups of a complicated picture?
Answer:　See Example 2-16 and Section 3.5.

Question: How do set I variables on the horizontal/vertical axis?
Answer:　See Example 2-14 and Section 4.1.

Question: How do I change and/or set colors?
Answer:　See Section 3.7.

Question: How do I erase colors?
Answer:　See Section 3.7.

Question: How do I superimpose two pictures?
Answer:　See Section 4.5.

CHAPTER 2

SAMPLES OF DYNAMICS:
PICTURES YOU CAN MAKE SIMPLY

2.1 INTRODUCTION

In this chapter, we give samples of capabilities of *Dynamics*. We begin with
an example that uses the Henon map. The Henon map has been selected because
(1) it exhibits many kinds of behavior and (2) computers can complete
computations for this map much faster than for a differential equation.

Given a pair (X,Y), applying the Henon map once gives a new pair using
the formula $(RHO - X*X + C1*Y, X)$. Throughout this book the symbol '*' in
maps and differential equations denotes multiplication. The original paper
by Hénon (1976) used a different formula, but Devaney and Nitecki (1979)
point out that this formula is nicer to deal with in order to analyze the
behavior as the process' parameters RHO and C1 are varied. In particular,
all periodic trajectories remain in a uniformly bounded region as the
parameters are varied. Our form can be obtained by a linear change of
variables from Hénon's original. The two differ only by a change in scale
of X and Y.

The program gives RHO the default parameter value of 2.12 and C1 the
value -0.3. The Jacobian determinant of the Henon map is -C1, so for this
C1 value the Jacobian determinant is positive. For this process with the
initial setting for the screen, X is plotted horizontally with a scale
running from -2.5 to 2.5, and Y is plotted vertically with the same scale.

We first present some elementary terms that we use throughout the
book. Let F be a continuous map from the n-dimensional phase space to
itself, and consider the corresponding discrete time process $x_{k+1} = F(x_k)$.
A point x in the phase space is called a **fixed point** of F if $F(x) = x$. For

---

each positive integer k, the **kth iterate of the map** F is the result of the procedure composing F with itself k times. We write F^k for the kth iterate of F. For each point x, the point $x_k = F^k(x)$ is called the **kth iterate of the point** x. The **trajectory** of any point x in the phase space is the finite or infinite sequence of consecutive iterates of x. A point x in the phase space is called a **periodic point** of period m if x is a fixed point of the mth iterate of F. Notice that a periodic point of period 2 is automatically also a periodic point of period 4, 6, 8, etc. The trajectory of a periodic point p is called a **periodic orbit** and consists of m different points, where m is the smallest period of the point p. A periodic orbit is called an **attracting periodic orbit** (or **locally stable periodic orbit**) if whenever an initial condition x_0 is chosen sufficiently close to the periodic orbit, the distance between the iterate x_k and the periodic orbit goes to zero as $k \rightarrow \infty$.

As discussed in Chapter 1, you should now start the program by typing

dynamics *< Enter >*

First the program's title page appears on the screen for a couple of seconds and then the PROcess Menu. We are interested in the line that says

| H | – | Henon map |

Recall from Chapter 1, that there are two ways to select the menu item H. One approach is to use the "point and shoot" menu system, and the second approach is to type the item and hit < Enter >.

Now type (both upper case and lower case letters may be used)

H *< Enter >*

and the Main Menu mentioned in Chapter 1 appears on the screen. As discussed in Chapter 1, once you have selected a process and are working with it, if no menu is on the screen, you can fetch the previous menu by hitting < Enter >. If a menu is on the screen, fetch its parent menu by hitting < Esc >. Select the "NumExplM" (Numerical Explorations Menu)

NEM *< Enter >*

Example 2-1a: Plot a trajectory
 Enter "Trajectory" command T:

$$T < Enter > < Enter >$$

that is, type in T (either upper case or lower case) and hit <Enter> (or <Return>) twice. There are two <Enter> keys above because after the first a menu of hints and options appears. By hitting the second <Enter> you are ignoring these options telling the program you do not want to use these options. The initial values for X and Y are 0.0 and 2.0 respectively, and the program will apply the process, which is the Henon map, to the point (X,Y) and then plot the resulting point. In other words, the program replaces the value of (X,Y) with the value (RHO - X*X + C1*Y, X), and it will do this repeatedly, that is it will "iterate" the map. A sequence of dots will appear in rapid succession, hundreds or thousands per second.
 While the program is plotting, the following instructions appear at the bottom right of the screen.

> Hit <.> for 1 more dot & pause.
> Hit <Enter> for previous menu.
> Hit <Esc> to stop process.

Example 2-1b: Draw a box
 The plotting of the trajectory can be paused at any time in order to perform other commands that enhance or change what the program is doing. For example, you can draw a box around the plotted trajectory by locating the "Box" command in the menu system and executing it.
 To locate the "Box" command, hit

$$< Enter >$$

to return the Numerical Explorations menu to the screen and hit <Esc> to get the Main Menu on the screen. Select the Screen Menu by typing

$$SM < Enter >$$

(or by typing *sm<Enter>*, or by highlighting the Screen Menu with the arrow keys and then hitting <Enter>). Within the Screen Menu, select the BoX Menu and enter the "Box" command:

$$BXM < Enter >$$

$$B < Enter >$$

---

2. **Samples of Dynamics** 29

and a box will be drawn around the screen which frames the trajectory.

This is an illustration of navigation through the menu system to find an appropriate command. For new and occasional users, this method will be the best one for locating and executing commands. On the other hand, if you had known ahead of time that "B" was the command for drawing a box, you could have avoided the menus entirely. Instead of calling up the Main Menu as was done above, you could have typed $B < Enter >$ (or $b < Enter >$). The program would have momentarily stopped plotting trajectory points, drawn the box, and then resumed the trajectory.

The BoX Menu (type $BXM < Enter >$ to see it) lists several alternatives to the simple "Box" command. Enter the "Box with tic marks" command B1:

$$\textbf{B1} < Enter >$$

and again a box is drawn, only now there are tic marks drawn around the edges of the box, suggesting how the box is scaled.

Retrieve the BoX Menu:
$$< Enter >$$

Enter the "Box with double tic marks" command B2:

$$\textbf{B2} < Enter >$$

results in a second set of tic marks, smaller than before, spaced 1/5 as far apart.

Hit
$$< Esc >$$

and the run is terminated and the BoX Menu appears (since it was the last menu you called). Hit this key two more times and the Main Menu appears on the screen.

Example 2-1c: Viewing the Parameter Menu

The Parameter Menu PM is listed in the Main menu. To view the parameter settings, fetch the Parameter Menu:

$$PM < Enter >$$

The program gives RHO the default parameter value of 2.12 and C1 the value -0.3. For this process, the initial setting for the screen is that X is plotted horizontally with a scale running from -2.5 to 2.5, and Y is plotted vertically with the same scale.

Hit $< Esc >$ to return to the Main Menu.

Example 2-1d: Refresh the screen and continue plotting
Select in the Main Menu the Numerical Explorations Menu and enter again the "Trajectory" command T:

$$T < Enter > < Enter >$$

(or equivalently, use arrow keys to select NumExplM from the Main Menu and hit < Enter >, and then use arrow keys to select command T from the Numerical Explorations Menu and hit < Enter >) and the previous picture returns to the screen and the program continues plotting the trajectory. The picture can be returned to the screen because the program has a separate copy of the picture in core memory, the "core" copy of the picture, in addition to the screen picture. The core copy of the picture usually has higher resolution than the screen. The core copy is 720 dots wide by 720 dots high if there is sufficient memory available, and when the picture is sent to the printer, this higher resolution picture is transmitted. Unix versions have a different resolution. The core picture does not include the words on the screen so pictures can be refreshed (that is, cleaned up) using the "Refresh" command R. Call the Screen Menu

$$SM < Enter >$$

and select the Refresh command:

$$R < Enter >$$

to get a refreshed picture without all the text that was also on the screen. Hence, the pictures sent to the printer are not muddied up with extraneous text such as menus.

Example 2-1e: Clear the screen and continue plotting
Assuming that a trajectory of the Henon map is still being plotted and that a box with double tic marks has been drawn around the screen, enter the "Clear window" command C:

$$C < Enter >$$

(or equivalently hit < Enter > to retrieve the Screen Menu, use arrow keys to select command C from the Screen Menu, and hit < Enter >). The screen clears and the computer continues plotting the trajectory from where it left off.

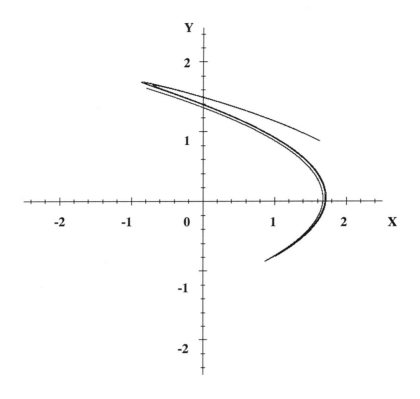

Figure 2-1a: Trajectory and axes with tic marks
This figure shows a trajectory of the Henon map (H)

$$(X, Y) \;\rightarrow\; (2.12 - X*X - 0.3*Y, \; X)$$

Difficult topic of discussion. *As with the trajectory of the Tinkerbell map in Figure 1-1, the picture of the trajectory is a result of many iterates of the map. Points appear one at a time. The first few look randomly positioned but as more appear, a pattern emerges. Why do the plots of chaotic attractors (see the next section for the definition) seen throughout this book appear to consist of curves? In the lower right window of Figure 1-1, there are still many dots that have not yet been connected into curves. Can you explain this phenomenon?*

Easier topic of discussion. *How many periodic orbits with period PR ($1 \leq PR \leq 12$) can you find in this picture? Clear the screen (command C) and use command RPK (Random Periodic orbit search) for a variety of values of period PR. What periods are missing?*

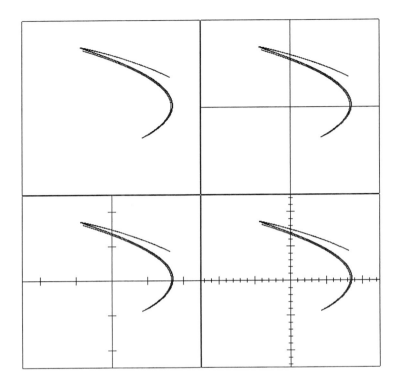

Figure 2-1b: Trajectory and axes with tic marks

The map parameters are as in the previous figure. The X Scale is from -2.5 (left) to 2.5 (right) and the Y Scale is from -2.5 (bottom) to 2.5 (top). All four windows show the same trajectory. To split the screen into windows as shown above, use the Open Window commands OW, OW1, ..., as explained in Chapter 3. See also Example 2-16.

In the upper left window no axes are drawn, while they are drawn in the upper right window using the "X AXis" command XAX and the "Y AXis" command YAX. In the lower left window, they are drawn with one set of tic marks using commands XAX1 and YAX1, and in the lower right window with two sets of tic marks using commands XAX2 and YAX2.

Example 2-1f: Single stepping through a trajectory
Sometimes the computer plots too fast for what you want to see or do. To see the process iterated at a slower rate, hit $< \cdot >$ and the process will pause. Hit $< \cdot >$ again and the program will iterate the map once and plot the resulting point. A large cross is plotted at the current (X,Y) point. Repeatedly hitting $< \cdot >$ you will see how the trajectory moves through the chaotic attractor. Hence, these processes can be walked through as slowly as you want. When no trajectory of the Henon map is plotted, enter the "Trajectory" command T again. Clear the screen:

$$C < Enter >$$

and immediately hit the period key $< \cdot >$ several times:

. . . .

and you will see the process single step, producing one dot for each ' · ', and pausing thereafter. While the program is in this "pause after one dot" mode, you can still enter commands to reinitialize or clear, etc., and while the program is paused, a cross will appear on the screen at the current (X,Y) point.
Now hit the

$$< space \ bar >$$

and the trajectory resumes iterating again as fast as it can.

Example 2-1g: Plot a cross at current position
Enter the "Kruis" command K (Kruis means cross in Dutch):

$$K < Enter >$$

and the computer will show the position of each dot it plots with a cross. This command is in the cross Menu KM, which is in the Screen Menu SM. This feature is a toggle. A "toggle" is a command that turns some feature on or off. If it is "ON", entering the command turns it "OFF" and vice versa. Hence, when the command K is entered again, the feature turns off and no crosses are plotted anymore. Turn this feature off:

$$K < Enter >$$

Clear the screen and core memory:

$$C < Enter >$$

Example 2-1h: Draw axes and print picture
Fetch the Screen Menu SM:

SM < Enter >

Fetch the AXes Menu AXM:

AXM < Enter >

All the commands involving axes are collected in this menu.
Enter the "X AXis" command XAX and "Y AXis" command YAX:

XAX *< Enter >*   **YAX** *< Enter >*

and the X and Y axes will be drawn. Hit

< Enter >

and the AXes Menu will reappear.
Enter the "X AXis with 1 set of tic marks" command XAX1:

XAX1 *< Enter >*

and retrieve the AXes menu:
< Enter >

and enter the "Y AXis with 1 set of tic marks" command YAX1:

YAX1 *< Enter >*

results in drawing both axes, only now there are tic marks drawn on the axes, suggesting how the axes are scaled.
Retrieve the AXes Menu:
< Enter >

Enter the "X AXis with 2 sets of tic marks" command XAX2:

XAX2 *< Enter >*

retrieve the menu again:
< Enter >

Enter the "Y AXis with 2 sets of tic marks" command YAX2:

YAX2 *< Enter >*

results in drawing both axes with a second set of tic marks, smaller than before, spaced 1/5 as far apart.

The resulting picture is similar to Figures 2-1a and 2-1b. To get the picture printed, make sure the printer is turned on, and enter the "Print Picture" command PP:

$$\textbf{PP} < Enter >$$

and select printer and hit $< Enter >$ twice. After the picture is printed, plotting continues. Hit

$$< Esc >$$

and the run is terminated and you are suddenly faced with the AXes Menu.

Note. If you have a problem printing the picture, you may wish to consult "Troubleshooting" in Section 4.6.

Example 2-1i: Initializing

The current state of the trajectory is a vector called y (or sometimes called y0). When the trajectory is paused, a large cross appears at that point. Another vector, the "initialization" vector, denoted y1, has certain special uses. When the trajectory is paused (or when the vector y1 is changed using the arrow keys) a small cross appears at its location. It is the initial condition for y.

Enter the "Initialize" command I:

$$\textbf{I} < Enter >$$

which replaces y by the position of y1. After the command I has been entered, the trajectory will be restarted from the small cross whose position is given by the initialization vector y1. The command I is in the Kruis (cross) Menu KM, which is in the Screen Menu.

A note that novices should skip. We call y the "state" of the system. The dimension of y depends on the process, and currently is permitted to be as high as 21. The coordinates of a vector are indicated by brackets [ ] so that y[0] and y[1] are the coordinates of y. Coordinates are numbered starting with 0 as the beginning coordinate. For the trajectory you have been plotting for the Henon map, y[0] is X the horizontal coordinate and y[1] is Y the vertical coordinate. The coordinates of the initialization vector y1 are y1[0] and y1[1].

Note. If you have a mouse, you may wish to select a new initial condition by pointing the arrow at the desired position and click the left button for initialization; see also "Picking a new initial point of a trajectory" in Chapter 1.

Example 2-1j: Viewing the Y Vectors
The values of the coordinates of the vectors y and y1 and some of the other storage vectors can be seen by entering the "Y Vectors" command YV:

$$YV < Enter >$$

Refresh the screen to get rid of the text:

$$R < Enter >$$

Example 2-1k: Find a fixed point
Recall from Example 2-1i above that the position of the initialization vector y1 is indicated by a small cross. Enter the "Initialize plus Iterate" command II:

$$II < Enter >$$

and use the arrow keys to move the small cross (or move the mouse and click the left button). First move the small cross downward. The first iterate of this point (under the Henon map) will also move. It is marked by the large cross. Then move the small cross to the right. Continue moving the small cross (in all possible directions) and try to bring the large cross near the small cross. When you succeed, you have found an approximate fixed point of the Henon map. Enter the "Y Vectors" command YV:

$$YV < Enter >$$

to see which fixed point has been located. For the default parameter values RHO = 2.12 and C1 = -0.3, the Henon map has two fixed points which are approximately (0.94,0.94) and (-2.24,-2.24).
Now apply Quasi-Newton method Q1 (see Chapter 10):

$$Q1 < Enter >$$

once or twice, and mark the fixed point with a permanent cross:

$$KK < Enter >$$

The commands II and KK are in the Kruis (cross) Menu KM, which is in the Screen Menu.
Hit $< Esc >$ and Clear the screen and core memory:

$$C < Enter > .$$

Example 2-1l: Find a period 2 orbit

Enter the "Maximum Iterates" command MI:

$$MI < Enter >$$

After the reply type in

$$2 < Enter >$$

to set MI to be 2.

Enter the "Initialize plus Iterate" command II:

$$II < Enter >$$

and use the arrow keys to move the small cross. The routine II will now iterate the map twice (since MI = 2) starting from the small cross. The resulting second iterate of this point is marked by the large cross. Move the small cross (in all possible directions) until the position of the large cross is near the position of the small cross. If this is the case, you have found an approximate period 2 point of the Henon map. Enter the "Y Vectors" command YV:

$$YV < Enter >$$

to see which period 2 point has been located. For the default parameter values RHO = 2.12 and C1 = -0.3, the Henon map has two period 2 points (or equivalently, one period 2 orbit) which are approximately (1.57,-0.27) and (-0.27,1.57). Hit $< Esc >$ and Clear the screen and core memory:

$$C < Enter > .$$

Note. Another way to find a period 2 orbit is set PR to 2 and apply the Quasi-Newton method, see Chapter 10. Still another way is to use either of the commands RP or RPK to Randomly seek all Periodic orbits of a given period (see Example 2-1m).

Example 2-1m: Search for all periodic points of period 5

Commands RP and RPK allow you to search for all the periodic points of period 5. Command RPK differs from RP in that a cross is drawn at each period 5 point while RP merely plots a point.

Set the PeRiod to be 5:

$$PR < Enter > \quad 5 < Enter >$$

and search at Random for the Periodic points with period 5:

$$RPK < Enter >$$

The commands PR, RP, and RPK are in the Periodic Orbit Menu POM, which is in the Numerical Explorations Menu NEM. A resulting picture is given in Figure 2-1c.

Hit *<Esc>* and Clear the screen and core memory:

$$C < Enter > \ .$$

Note. For more detailed information on searching for all periodic points, see Example 2-11 and Chapter 10.

Example 2-1n: Change RHO

Enter "Trajectory" command T:

$$T < Enter > \ < Enter >$$

and a trajectory will be plotted. The parameter RHO of the Henon map can be changed from its default value of 2.12. Set RHO to be 2. To do this, enter command RHO (and after the reply) enter 2:

$$\textbf{RHO} < Enter > \quad 2 < Enter >$$

(or 2. or 2.0 instead of 2) and if you have a color screen immediately hit Function key 'F7' (without <Enter>)

$$< F7 >$$

to change the color used for plotting. A second trajectory is plotted.

Now set RHO to be the value 1.9:

$$\textbf{RHO} < Enter > \quad 1.9 < Enter >$$

and immediately hit *<F7>* to change the color. A third trajectory is plotted. To view this third trajectory separately, Clear the screen and core memory:

$$C < Enter >$$

The current trajectory is a period 4 orbit. See Figure 2-1d.

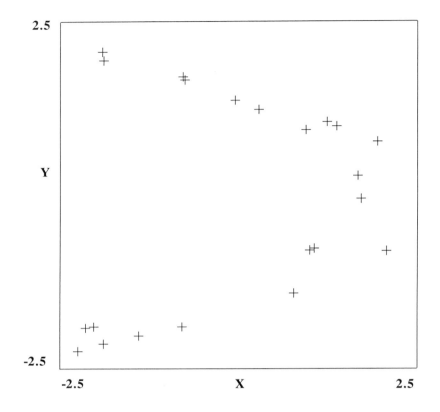

Figure 2-1c: Periodic points with period 5

The map parameters and scale are as in Figure 2-1a. The figure shows period 5 points.

The routine RPK found 6 distinct orbits of period 5. These periodic orbits include 2 fixed points.

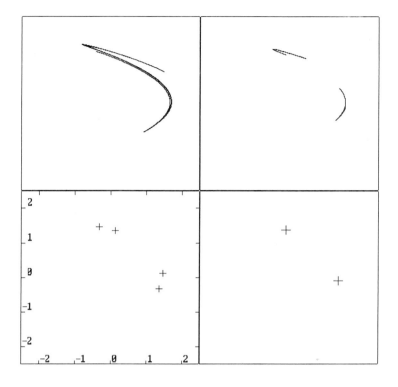

Figure 2-1d: Trajectories
This figure shows trajectories of the Henon map (H)
$$(X, Y) \rightarrow (RHO - X*X - 0.3*Y, X)$$
   In the upper left window RHO is 2.1; in the upper right window RHO is 2.0; in both these upper windows the trajectory is chaotic. In the lower left window RHO is 1.9, and the position of an attracting period 4 orbit is marked by crosses. In the lower right window RHO is 1.8, and the position of an attracting period 2 orbit is marked by crosses. In all four windows, the variable X is plotted horizontally and the variable Y is plotted vertically, and the X Scale is from -2.5 (left) to 2.5 (right) and the Y Scale is from -2.5 (bottom) to 2.5 (top).

   Topic of discussion. *Find the bifurcation value of RHO between 1.8 and 1.9 which denotes the transition between the lower two figures. What are the eigenvalues (see the Quasi-Newton command Q1) of the period 2 orbit at the transition? How are these eigenvalues reflected in the above pictures? Where is the (unstable) period 2 orbit in the lower left window? See the warning on Figure 1-1.*

Example 2-1o: Plotting permanent crosses

Assuming the computer is still plotting, draw a permanent cross at each point of the trajectory:

$$KKK <Enter> <Enter>$$

which results in 4 permanent crosses.

Enter the command KKK again to turn the cross plotting off, so that points instead of crosses will be plotted in the next examples.

Hit <Esc> and Clear the screen and core memory: **C** <Enter> .

Example 2-1p: Set storage vector y1 and initialize

Set RHO to be 1.4, set C1 to be 0.3, Initialize, and enter command T:

$$RHO <Enter> \quad 1.4 <Enter>$$

$$C1 <Enter> \quad 0.3 <Enter>$$

$$I <Enter>$$

$$T <Enter> <Enter>$$

and a warning "Trajectory is too far from the screen: PROCESS PAUSED" appears on the screen with a menu (Recover Menu) of possible corrections.

```
RM    – –    RECOVER MENU

CENT –  move small cross to CENTer of screen; this is
            the position of the initialization vector
I      –  reInitalize trajectory to small cross position
PM     –  to reset parameters
SD     –  increase Screen Diameters SD (=  1.00 diameters)
            so the trajectory can go farther from the screen
SV1    –  Set value of initialization Vector,
            whose position is marked by the small cross
```

Now set y1 to be (0,0). Enter the "Set Vector" command SV:

$$SV <Enter>$$

After the reply, enter the "Set Vector y1" command:

$$SV1 <Enter>$$

Select SV10 from the menu on the screen

$$SV10 <Enter> \quad 0 <Enter>$$

to set coordinate #0 of y1 to 0, enter

$$SV11 <Enter> \quad 0 <Enter>$$

to set the coordinate #1 of y1 equal to 0. Notice that if you are using the menu, you should now respond with OK twice by hitting $<Enter>$ twice.
Now initialize, and plot a trajectory:

$$I <Enter> \quad T <Enter> <Enter>$$

and (if appropriate) hit $<space\ bar>$ to unpause the routine.

Example 2-1q: Change X Scale or Y Scale
In Example 2-1p above, the horizontal or X Scale runs from -2.5 to 2.5. To change it to -2 to 2, fetch the Parameter Menu:

$$PM <Enter>$$

and enter the command XS:

$$XS <Enter> \quad -2\ 2 <Enter>$$

with a space between -2 and 2. You can similar change the Y Scale to run from -2 to 2 using the command YS:

$$YS <Enter> \quad -2\ 2 <Enter>$$

Clear the screen and core memory:

$$C <Enter> \quad .$$

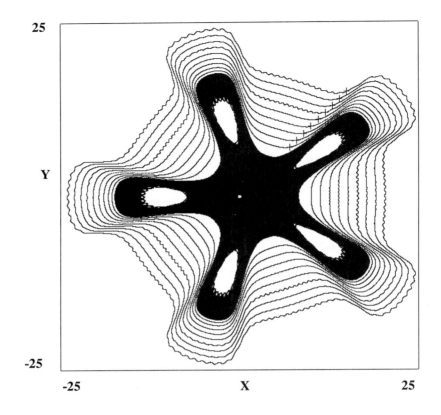

25

Y

-25

-25   X   25

Figure 2-1e: Trajectories of the Gumowski/Mira map
The figure shows ten trajectories of the the Gumowski/Mira map (GM)
$$(X,Y) \;\rightarrow\; (Y + F(X), \; -X + F(Xnew)$$
where $F(u) = 0.3u + 1.4u^2/(1+u^2)$, so $C1 = 1$, $C2 = 0$ and $RHO = 0.3$ (default values); see also Figure 4-2.

The initial conditions for the ten trajectories are: (1,1), (7,7), (8,8), (9,9), (10,10), (11,11), (12,12), (13,13), (14,14) and (15,15). All these initial conditions are marked with a cross. The trajectory of (1,1) seems to travel troughout a chaotic attractor, while all the other trajectories seem to fill up a closed curve.

Topic of discussion. *Try other initial conditions and find out what the corresponding trajectories look like.*

2.2 COMPLEX PICTURES THAT ARE SIMPLE TO MAKE

The examples below are simple to make. We specify the required commands, though in some cases, we must defer detailed explanation of these commands to the later chapter. As explained in Example 2-1b, you do not have to navigate through the menu system for locating and executing commands. From now on, to carry out the examples it is sufficient to enter the bold commands. However, new and occasional users are recommended to carry out the examples in the way they are presented.

When making complicated pictures, you may make errors, ruining the picture. When making a complex picture, we regularly store the picture after each major step using the To Disk command TD. Then the From Disk command FD will retrieve the picture, erasing what you have. Or you may use the "Add From Disk" command AFD which adds the old picture onto the screen, keeping new points you have created.

All the examples below are presented independently, although some of them are clearly related. The majority of these examples use the Henon map. If while running an example you want to quit and exit to the menu, just hit the escape key <Esc>. The examples are selected in such a way that the default values of the program can be used extensively.

When an example in this book tells you to type **dynamics** <*Enter*>, you are expected to start the program. If the program is already running, you still can type DYNAMICS <*Enter*> (or DYN <*Enter*>). Then the program will return to the PROcess Menu and reset the parameter values to the default values without the delay of terminating the program. Hence, you do not have to exit the program while going through a number of examples.

Example 2-2a: Chaotic attractor

The trajectory in Example 2-1a fills out a set and the later iterates of the initial point seem to run throughout the set. A set A is called an **attractor** if whenever an initial point x_0 is chosen sufficiently close to the set A, the distance between the kth iterate x_k and the set A goes to zero as k $\rightarrow$ ∞. Entering command C clears the screen, but the successive iterates fill out the set again; hence the trajectory certainly appears to be covering the same set repeatedly. Of course, the points plotted on the screen are plotted with limited precision. The fact that the trajectory hits the same point on the screen twice only means that two of the points of the trajectory were close to each other. In this case the attractor is a complicated set. In any case, typical trajectories in the attractor keep retracing the attractor on the screen. See Chapter 6 for a more detailed discussion of "attractor".

As we iterate a point, the trajectory repeatedly goes over the same set. It does this in an irregular nonperiodic manner, in the following sense. If we choose 2 initial points close together on the attractor, the kth iterate of one would unlikely to be near the kth iterate of the other if k is large. We say the attractor is **sensitive to initial data**, when a small change in initial data usually causes big changes in where the kth iterate will be in the attractor. When there are small changes in the initial point, the attractor does not change, but the order in which the attractor is traced out does change. When the attractor is sensitive to initial data, we say the attractor is a **chaotic attractor**. In Chapter 5 we give a criterion for a trajectory to be chaotic, namely we will say that the trajectory is **chaotic** if all the iterates of the initial point are different and "one of its Lyapunov exponents is positive". In Chapter 6 we use this criterion for determining whether an attractor is chaotic.

The purpose of this example is to get a picture of the chaotic attractor of the Henon map (H)

$$(X,Y) \;\to\; (2.12 - X*X - 0.3*Y, \; X).$$

Start the program by typing

dynamics < *Enter* >

followed by

H < *Enter* >

to get the Main Menu for the Henon map. The parameters RHO and C1 are by default 2.12 and -0.3.

1. Since we are interested in the attractor, we do not want to plot some initial part of the trajectory that starts with the default value. We say we **pre-iterate** the trajectory, that is, iterate without plotting. Fetch the Parameter Menu (which is in the Main menu):

PM < *Enter* >

and get the When and What to plot Menu:

WWM < *Enter* >

Enter the "PreIterates" command PI:

PI < *Enter* >

After the reply type in the number of preiterates, that is, the value of PI:

$$100 <Enter>$$

2. Enter the "Trajectory" command T:

$$T <Enter> <Enter>$$

and you will see that the plotted trajectory is similar to the one in Figure 2-1a.

3. After 10 seconds or more, get information on line by hitting

$$<Tab>$$

(do not hit <Enter>) and get the values of several variables printed on the screen. You can see how fast your computer is, and you may compare it with the table in Chapter 1.

4. When you would like to print the attractor on paper, make sure the printer is on, fetch the File Menu (which is in the Main Menu):

$$FM <Enter>$$

enter the "Print Picture" command PP:

$$PP <Enter>$$

select the printer you have and hit <Enter> twice.

Wait until the picture is printed, and hit

$$<Esc>$$

a few times to return to the Main Menu.

Note. If you have a problem printing the picture, you may wish to consult "Troubleshooting" in Section 4.6.

Example 2-2b: Computing Lyapunov exponents

The Lyapunov exponents and Lyapunov numbers of a trajectory are numbers that measure an average rate of expansion or contraction of trajectories extremely close to the trajectory; see Chapter 5 for more details. The Lyapunov exponents are the natural logarithms of the Lyapunov numbers. Every few iterates, the results of the computation will be printed on the screen, becoming more accurate as more iterates are included in the average. If a trajectory has a Lyapunov exponent that is greater than zero, then typical initial points chosen near the trajectory will move away, that is, there is sensitivity to initial data. A trajectory which has a positive Lyapunov exponent, and is not asymptotically periodic, is called a **chaotic trajectory**; see Chapter 5.

Enter the "Clear current window" command C to clear the screen and core memory:

$$C < Enter >$$

and the picture will be cleared when plotting is restarted, or if you are not running the program yet, type *dynamics* $< Enter >$ $H < Enter >$ to get the Main Menu for the Henon map. This example deals with computing the Lyapunov exponents of a trajectory.

1. Get the Lyapunov Menu LM (which is in the Numerical Explorations Menu):

$$LM < Enter >$$

2. The allowable number L of Lyapunov exponents to be computed is 0, 1 or 2 since this process is two dimensional, that is, the phase space is two dimensional, and the default value is L = 0. Set the number of Lyapunov exponents to be computed to 2:

$$L < Enter > \quad 2 < Enter >$$

3. Enter the "Trajectory" command T:

$$T < Enter > < Enter >$$

and plotting begins. The computer is considerably slowed by the printing, much more so than by the computation. Every 50 dots (iterates of the map) the results are printed on the screen, for example:

```
dot = 14350; Lyapunov exponents --
0.392939      -1.596912
              Lyapunov numbers --
1.481327      0.202521
Lyapunov Attractor Dimension estimate = 1.24606
```

When command T2 is entered (*T2 <Enter>*), the text level is set to 2 instead of the more verbose default value of 3. This reduces the amount of text appearing on the screen in a variety of situations. Entering command T2, you would be able to obtain the results of the computation of the Lyapunov exponents up to the present by entering "List Lyapunov exponents" command LL. Refresh the screen to get a refreshed picture without all the text that was also on the screen:

R *< Enter >*

Now set the number of exponents to 0:

L *< Enter >*   **0** *< Enter >*

and hit *< Esc >* to return to the Lyapunov Menu.

Note. The approximate Lyapunov exponents computed in Example 2-2b, can also be plotted as a function of the time. See Chapter 5 for more information.

Example 2-2c: Plotting trajectory versus time
Enter the "Clear window" command C to clear the screen and core memory:

C *< Enter >*

and the picture will be cleared when plotting is restarted, or if you are not running the program yet, type *dynamics < Enter > H < Enter >* to get the Main Menu for the Henon map.

1. We are interested in plotting a trajectory on the attractor versus "time". To have the time (the number of the iterate) on the horizontal axis, the "Plot Time" toggle PT has to be turned on. A "toggle" is a command that turns some feature on or off. If it is "ON", entering the command turns it "OFF" and vice versa. To see the current status of PT, fetch the Parameter Menu:

PM < Enter >

and then get the When and What to plot Menu:

$$WWM < Enter >$$

This menu has an entry

PT	- T Plots Time horizontally if ON; now OFF

Now turn the "Plot Time" toggle on:

$$\textbf{PT} < Enter >$$

2. To have the trajectory on the attractor, we want to plot the iterates 101 through 200. Either we can set the number of PreIterates PI to be 100 (as in Example 2-2a) and the X Scale to run from 0 to 100, or we can set the PreIterates to be 0 and the X Scale to run from 100 through 200. We choose the latter option.
Set PI to be 0 (the default value):

$$\textbf{PI} < Enter > \quad \textbf{0} < Enter >$$

and set the X Scale to run from 100 to 200:

$$\textbf{XS} < Enter > \quad \textbf{100 200} < Enter >$$

3. To have a better insight into the behavior of the trajectory, connect consecutive dots with a straight line. The command in the program for connecting consecutive dots is in the When and What to plot Menu. Retrieve the When and What to plot Menu:

$$< Enter >$$

(since it was the last menu you called) and turn the "CONnect" toggle on:

$$\textbf{CON} < Enter >$$

4. Initialize:

$$\textbf{I} < Enter >$$

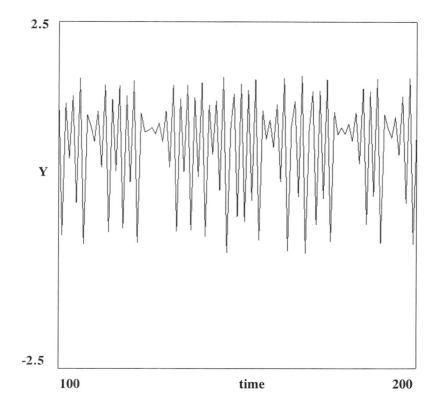

Figure 2-2: Trajectory versus time
This picture shows a trajectory versus time of the Henon map (H)
$$(X, Y) \;\rightarrow\; (2.12 - X*X - 0.3*Y, \; X)$$
The initial condition of the trajectory is (0,2).

Topic of discussion. *The sequence of values appears to be somewhat random despite its deterministic origin. If we continued the sequence forever, would there necessarily be a maximum and minimum for the values? See also Example 2-5 and Figure 2-5.*

5. Plot the Trajectory:

$$\mathbf{T} <Enter> <Enter>$$

and you will see that the plotted trajectory is similar to the one in Figure 2-2.

6. When you would like to print the attractor on paper, make sure the printer is on, fetch the File Menu (which is in the Main Menu):

$$FM <Enter>$$

enter the "Print Picture" command PP:

$$\mathbf{PP} <Enter>$$

select the printer you have and hit *<Enter>* twice.

Wait until the picture is printed, and hit

$$<Esc>$$

a few times to return to the Main Menu.

Note. If you have a problem printing the picture, you may wish to consult "Troubleshooting" in Section 4.6.

Example 2-3a: Graph of iterate of one dimensional map
The purpose of this example is to get a picture of the graph of the third iterate of the LOGistic map (LOG)

$$X \rightarrow 3.83*X*(1-X)$$

Start the program and enter command LOG:

dynamics *< Enter >*   **LOG** *< Enter >*

to get the Main Menu for the LOGistic map.
1. Fetch the Parameter Menu and set the parameter RHO to be 3.83:

PM < Enter >

RHO *< Enter >*   **3.83** *< Enter >*

2. Since we are interested in the graph of the third iterate, we want to plot the third iterate of a certain number of points and connect consecutive points. We say we "Iterate Per Plot" 3 times, that is, plot each third iterate, and "CONnect" the resulting consecutive dots. You can retrieve the Parameter Menu as follows:

< Enter >

(since it was the last menu you called). Retrieve now the When and What to plot Menu:

WWM < Enter >

Set the "Iterates Per Plot" to be 3:

IPP *< Enter >*   **3** *< Enter >*

3. To see the current status of CON, retrieve the When and What to plot Menu:

< Enter >

Enter the "CONnect consecutive dots" command to turn the toggle "CON" on:

CON *< Enter >*

4. The periodic points with period 3 are just the points at which the graph of the third iterate intersects the diagonal line "y = x". Plot the diagonal line "y = x":

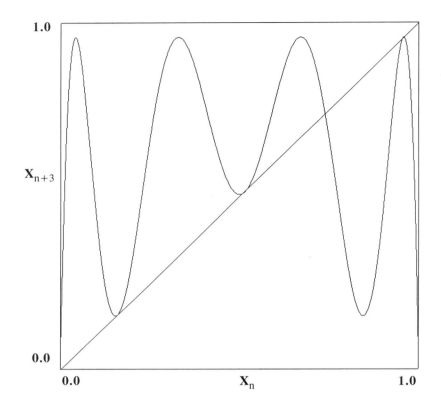

Figure 2-3a: Graph of iterate of map
This figure shows the third iterate of the LOGistic map (LOG)
$$X \;\rightarrow\; 3.83*X*(1-X)$$

Topic of discussion. *The logistic map has a number of periodic points. In particular, there are periodic points of period 3 and fixed points. How many of each of these does the map have? What can you say about other periodic points? There are points above where the graph is nearly tangent to the diagonal line. What does that signify?*

AB < *Enter* > < *Enter* >

5. Plot the graph of the third iterate:

G < *Enter* >

and you will see that the graph is similar to one in Figure 2-3a.
Now reset the "Iterates Per Plot" to the default value 1:

IPP < *Enter* >   1 < *Enter* >

If the command G was entered now, it would plot a parabola since the logistic map is a quadratic function; see the graph in the upper left window of Figure 2-3d.

Example 2-3b: Cobweb plot of a trajectory

The **cobweb method** for plotting a trajectory x_0, x_1, x_2, ... of a one dimensional map f is illustrated in Figure 2-3b. It is the following: (a) Draw a vertical segment along $x = x_0$ up to the graph of f intersecting at x_1, $x_1 = f(x_0)$; (b) Draw a horizontal segment from x_1 to the line "$y = x$"; the abscissa of this point of intersection is x_1; and (c) Repeat (a) and (b) to obtain x_{n+1} from x_n for any positive integer n.

If you are continuing from Example 2-3a, enter the "Clear window" command C to clear the screen and core memory:

C < *Enter* >

and hit < *Esc* > a few times to return to the Main Menu of the LOGistic map. If you start here with the LOGistic map, type *dynamics* < *Enter* > *LOG* < *Enter* > to get the Main Menu for the LOGistic map.

The purpose of this LOGistic map example is to illustrate a picture of the trajectory that starts at $x(0) = 0.4$ with parameter RHO as in Example 2-3a. We assume that RHO has been set to be 3.83.

1. To have a better insight into the behavior of the trajectory, connect consecutive dots with a straight line. Retrieve the Parameter Menu:

PM < *Enter* >

and get the When and What to plot Menu:

WWM < *Enter* >

to see the current status of CON. If it is on, you may hit < *Esc* > ; and if it is off, turn the "CONnect" toggle on:

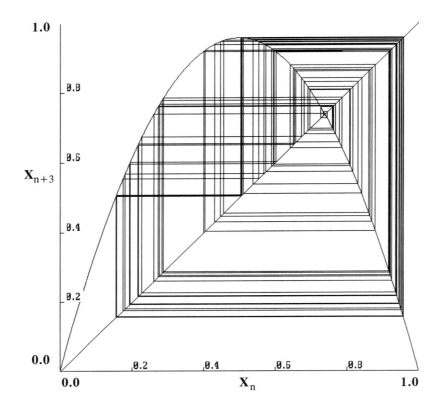

Figure 2-3b: Cobweb plot of a trajectory

This figure shows a trajectory x_0, x_1, x_2, ... obtained by the cobweb method using the graph of the one dimensional LOGistic map LOG. The parameter RHO is as in Figure 2-3a. The initial condition of the trajectory is 0.4.

The procedure consists of the following steps. a. Draw a vertical segment along $x_0 = 0.4$ up to the graph of LOG intersecting at x_1, the first iterate of x_0. b. Draw a horizontal segment from x_1 to the diagonal line "y = x"; the abscissa of this point of intersection is x1. c. Repeat the steps (a) and (b) to obtain x_{n+1} from x_n for any positive integer n.

Topic of discussion. *The sequence of values appears to converge to a period 3 orbit. What can you say about trajectories that start at other initial conditions? Are there trajectories that stay away from the stable period 3 orbit? See also the figures 2-3a and 2-3c.*

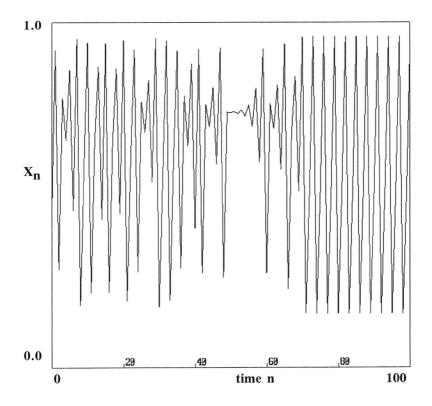

Figure 2-3c: Trajectory versus time
This figure shows a trajectory versus time of the LOGistic map (LOG)
$$X \rightarrow 3.83*X*(1-X)$$
The initial condition of the trajectory is 0.4. Note that the parameter value and initial condition are the same as in Figure 2-3b.

Topic of discussion. *Using only this graph and the map, identify where there is a fixed point. What can you conclude about the map near the fixed point?*

CON<Enter>

2. We want to plot the trajectory that starts at 0.4. Iterating this initial condition yields a trajectory that eventually gets attracted to a period 3 orbit. Fetch the Vector Menu (which is in the Main Menu):

VM<Enter>

and view the Y Vectors:

YV<Enter>

Set Vector y1 to be 0.4:

SV<Enter>

SV1 *<Enter>*

SV10 *<Enter>*   **0.4** *<Enter>*

3. Initialize:

I *<Enter>*

4. The cobweb method uses the diagonal line "y = x". Plot the diagonal line "y = x":

AB *<Enter> <Enter>*

5. Plot the graph of (the first iterate of) the map:

G *<Enter>*

6. For one dimensional maps, the trajectory x_0, x_1, x_2, ... is plotted by the routine T as follows. It plots (x_0,x_0), (x_0,x_1), (x_1,x_1), (x_1,x_2), etc. Notice that every other point is on the diagonal line. Plot the Trajectory:

T *<Enter> <Enter>*

and you will see that the resulting picture is similar to Figure 2-3b.

7. Draw the X Axis with tic marks:

XAX1 *<Enter>*

and draw the Y AXis with tic marks:

$$YAX1 < Enter >$$

After a while, hit

$$< Esc >$$

a few times to return to the Main Menu.

Note. For Unix workstations, commands XAX1 and YAX1 will not print the numerical values of the tic marks.

Example 2-3c: Plotting trajectory versus time

The purpose of this example is to plot the trajectory of Example 2-3b versus "time". Enter the "Clear window" command C to clear the screen and core memory:

$$C < Enter >$$

If you are not running the program yet, type *dynamics < Enter > LOG < Enter >* to get the Main Menu for the LOGistic map.

We assume that (a) RHO has been set to be 3.83, (b) the toggle "CON" has been turned on, and (c) the vector y1 has been set to be 0.4 (see Example 2-3b for the appropriate commands).

1. Trajectory T will plot the time (the number of the iterate) on the horizontal axis if the "Plot Time" toggle PT has to be turned on. See also Example 2-2c. To see the current status of PT, fetch the Parameter Menu:

$$PM < Enter >$$

and then get the When and What to plot Menu:

$$WWM < Enter >$$

Turn the "Plot Time" toggle on:

$$PT < Enter >$$

2. We want to plot iterates 1 through 100. Retrieve the When and What to plot Menu WWM:

$$< Enter >$$

Get the Parameter Menu:

$<Esc>$

Set the X Scale to run from 0 to 100:

XS $<Enter>$ **0 100** $<Enter>$

3. Initialize:

I $<Enter>$

4. Plot the Trajectory:

T $<Enter>$ $<Enter>$

and you will see that the plotted trajectory is similar to the one in Figure 2-3c. This command automatically sets DOT (that is, time) to 0.

5. Draw the X Axis with tic marks:

XAX1 $<Enter>$

6. When you would like to print the attractor on paper, make sure the printer is on, and enter the "Print Picture" command PP:

PP $<Enter>$

select the printer you have and hit $<Enter>$ twice.

Wait until the picture is printed, and hit

$<Esc>$

a few times to return to the Main Menu.

Note. If you have a problem printing the picture, you may wish to consult "Troubleshooting" in Section 4.6.

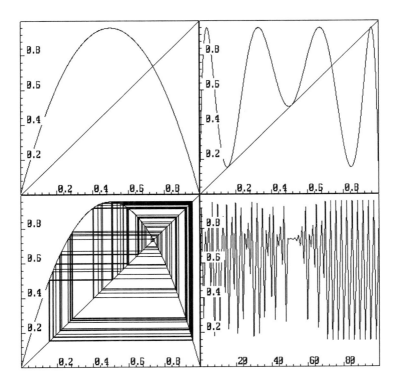

Figure 2-3d: Different scalings in windows

This figure shows iterates and trajectories of the LOGistic map (LOG)

$$X \rightarrow 3.83*X*(1-X)$$

The upper left window shows the first iterate of the map LOG. The upper right window contains the third iterate of this map and is similar to Figure 2-3a. The lower left window displays a trajectory x_0, x_1, x_2, ... obtained by the cobweb method using the graph of this map. This figure is similar to Figure 2-3b. The lower right window shows a trajectory versus time of this map. This figure is similar to Figure 2-3c. Notice that the scalings in the windows are different.

Topic of discussion. *The "time series" in the lower right window can be obtained directly from the figure in the lower left window. Describe this procedure.*

Example 2-4: The Henon attractor

The purpose of this example is to generate a picture of the famous **Henon attractor** which is the attractor of the Henon map (H)

$$(X,Y) \quad \rightarrow \quad (1.4 - X*X + 0.3*Y, \; X)$$

Type *dynamics* <*Enter*> followed by *H* <*Enter*> to get the Main Menu for the Henon map.

1. Set the parameter RHO to be 1.4:

RHO <*Enter*>   **1.4** <*Enter*>

2. Set the parameter C1 to be 0.3:

C1 <*Enter*>   **0.3** <*Enter*>

3. Since we are interested in the attractor, we want to neglect some initial part of the trajectory that starts with the default value, say we neglect the first 100 iterates. Set the number of "PreIterates" PI to 100:

PI <*Enter*>   **100** <*Enter*>

4. Since the trajectory of the initial point (0,2) is not in the basin of the attractor (it is diverging to infinity), we want to change the vector y1 which equals (0,2). Therefore, set y1 to be (0,0). Fetch the Vector Menu:

VM <*Enter*>

and get the Y Vectors:

YV <*Enter*>

Set Vector y1:

SV <*Enter*>

SV1 <*Enter*>

After the reply, type in

SV10 <*Enter*> **0** <*Enter*>

to set coordinate #0 of y1 to 0, and type

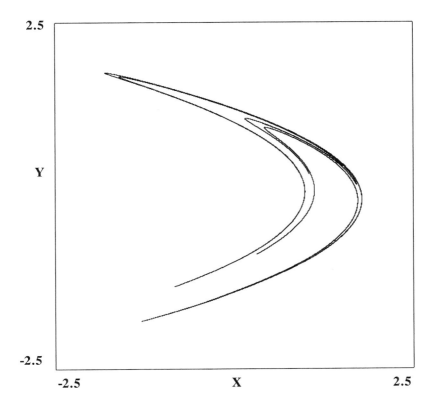

Figure 2-4: Henon attractor

This picture shows the attractor of the Henon map (H)

$$(X, Y) \;\rightarrow\; (1.4 - X*X + 0.3*Y, \; X)$$

This attractor is called a "strange attractor" because it appears to be a Cantor-like set of curves, and it is called a "chaotic attractor" because it contains a trajectory which has a positive Lyapunov exponent.

Topic of discussion. *Hénon may have chosen RHO = 1.4 and C1 = 0.3 because of the beauty of the attractor. What do the attractors look like when the Jacobian C1 is chosen significantly larger than 0.3 for various RHO values? (There are no attractors if │C1│ > 1.) See also Figure 5-3.*

The lower left part of the attractor seems to end in a sharp spike. Is it really? Find which point maps to the tip and which point in turn maps to that point. You may wish to use command II and possibly V (inVerse).

$$\textbf{SV11} <\textit{Enter}> \ \textbf{0} <\textit{Enter}>$$

to set the coordinate #1 of y1 equal to 0. Notice that if you are using the menu, you should now respond with OK twice by hitting $<\textit{Enter}>$ twice.

    5. Initialize:

$$\textbf{I} <\textit{Enter}>$$

    6. Plot the Trajectory:

$$\textbf{T} <\textit{Enter}> \ <\textit{Enter}>$$

and observe that the plotted trajectory is similar to the one shown in Figure 2-4. The resulting picture is an attractor, the Henon attractor, and it is an example of a so-called **strange attractor**. See also Example 2-5.

    7. To have the picture surrounded with a box, enter the "Box" command:

$$\textbf{B} <\textit{Enter}>$$

    8. Make sure the printer is on and send the picture to the printer:

$$\textbf{PP} <\textit{Enter}>$$

select the printer you have and hit $<\textit{Enter}>$ twice.

Wait until the picture is printed, and hit

$$<\textit{Esc}>$$

a few times to return to the Main Menu.

Note. If you have a problem printing the picture, you may wish to consult "Troubleshooting" in Section 4.6.

Note. For Unix computers, the picture will be sent to a binary file which must be sent in binary mode to a printer.

Example 2-5: Demonstration that the Henon map has an attractor: computing the first iterate of a quadrilateral

In this example, the purpose is to plot a quadrilateral and its first iterate for the Henon map (H)

$$(X,Y) \rightarrow (1.4 - X*X + 0.3*Y, X)$$

such that the first iterate of the quadrilateral is inside the quadrilateral and surrounds the chaotic attractor obtained in Example 2-4.

Hénon (1976) gave the coordinates of the 4 points of a quadrilateral having the desired property that it is mapped into itself when the map H is applied to it once. Based on this information, we consider the quadrilateral defined by the four points A = (-1.862,1.96), B = (1.848,0.6267), C = (1.743,-0.6533), and D = (-1.484,-2.3333). Using the feature of opening windows, we will plot the attractor in window W1 (the upper left window), the quadrilateral ABCD in window W2 (the upper right window), the quadrilateral ABCD and its first iterate in window W3 (the lower left window), and the superposition of the the quadrilateral ABCD and its first iterate, and the chaotic attractor in window W4 (the lower right window).

We assume you are faced with the Main Menu for the Henon map. If not, type *dynamics<Enter>* followed by *H<Enter>* to get this Main Menu.

1. Fetch the Parameter Menu:

PM<Enter>

and set the parameters RHO to be 1.4 and C1 to be 0.3:

RHO*<Enter>* **1.4***<Enter>*

C1 *<Enter>* **0.3** *<Enter>*

2. We are interested in the attractor, and do not want to plot the first 100 points of the trajectory (see also Example 2-4). Set the number of PreIterates to be 100:

PI*<Enter>* **100** *<Enter>*

3. Fetch the Vector Menu and change the initial condition y1 (see also Example 2-4):

VM<Enter>

CENT*<Enter>*

4. We want to store the coordinates of the points A, B, C, and D in the storage vectors ya, yb, yc, and yd, respectively. Fetch the Vector Menu and get the Y Vectors:

VM < *Enter* >

YV < *Enter* >

Set Vector ya to be (-1.862,1.96):

SV < *Enter* >

SVA < *Enter* >

SVA0 < *Enter* >   **-1.862** < *Enter* >

SVA1 < *Enter* >   **1.96** < *Enter* >

Set Vector yb to be (1.848,0.6267):

SVB < *Enter* >

SVB0 < *Enter* >   **1.848** < *Enter* >

SVB1 < *Enter* >   **0.6267** < *Enter* >

Set Vector yc to be (1.743,-0.6533):

SVC < *Enter* >

SVC0 < *Enter* >   **1.743** < *Enter* >

SVC1 < *Enter* >   **-0.6533** < *Enter* >

Set Vector yd to be (-1.484,-2.3333):

SVD < *Enter* >

SVD0 < *Enter* >   **-1.484** < *Enter* >

SVD1 < *Enter* >   **-2.3333** < *Enter* >

Notice that if you are using the menu, you should now respond with OK twice by hitting < *Enter* > twice.

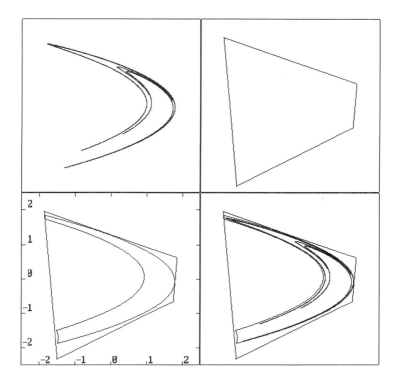

Figure 2-5: A quadrilateral and its first iterate for the Henon map
This figure shows a quadrilateral and its first iterate surrounding the chaotic attractor of the Henon map (H)

$$(X, Y) \rightarrow (1.4 - X*X + 0.3*Y, X)$$

The upper left window shows the chaotic attractor. In the upper right window, the quadrilateral ya-yb-yc-yd is drawn, where ya = (-1.862,1.96), yb = (1.848,0.6267), yc = (1.743,-0.6533), and yd = (-1.484,-2.3333). In the lower left window, the quadrilateral and its first iterate are shown, and the first iterate is contained in the quadrilateral. (This should be verified.) The superposition of the three foregoing pictures is shown in the lower right window. The first iterate of the quadrilateral surrounds the chaotic attractor. In all four windows, the variable X is plotted horizontally and the variable Y is plotted vertically, and the X Scale is from -2.5 (left) to 2.5 (right) and the Y Scale is from -2.5 (bottom) to 2.5 (top).

Topic of discussion. What does this construction imply about the existence of an attractor?

5. Fetch the Window Menu and Open window W1:

WM<Enter>

OW<Enter>

OW1 *<Enter> <Enter>*

6. Initialize, and plot the Trajectory:

I *<Enter>*

T *<Enter> <Enter>*

7. After a while, terminate routine T (hit *<Esc>*) and Open window W2:

<Esc>

OW<Enter>

OW2 *<Enter> <Enter>*

Fetch the Unstable and stable manifold Menu:

UM<Enter>

and plot the quadrilateral ya-yb-yc-yd:

AB *<Enter> <Enter>*

8. Open window W3:

OW<Enter>

OW3 *<Enter> <Enter>*

and plot the quadrilateral ya-yb-yc-yd and its Iterate:

ABI *<Enter> <Enter>*

---

9. Open window W4:

$$OW<Enter>$$

$$\textbf{OW4}<Enter><Enter>$$

Plot the quadrilateral ya-yb-yc-yd and its Iterate:

$$\textbf{ABI}<Enter><Enter>$$

and Initialize and plot the Trajectory:

$$\textbf{I}<Enter>\quad\textbf{T}<Enter><Enter>$$

The resulting picture is given in Figure 2-5.

10. When you would like to print the attractor on paper, make sure the printer is on, and enter the "Print Picture" command PP:

$$\textbf{PP}<Enter>$$

select the printer you have and hit $<Enter>$ twice.

Wait until the picture is printed, and hit

$$<Esc>$$

a few times to return to the Main Menu.

Note. If you have a problem printing the picture, you may wish to consult "Troubleshooting" in Section 4.6.

Example 2-6: Plotting direction field and trajectories

The solution of a system of 2 first order differential equations has velocity $(\frac{dX}{dt}, \frac{dY}{dt})$. The **direction field** of a system of 2 first order differential equations shows the direction of the vector $(\frac{dX}{dt}, \frac{dY}{dt})$ but does not reflect the length of that vector. In this book X' denotes $\frac{dX}{dt}$, and Y' denotes $\frac{dY}{dt}$. The purpose of this example is to plot some trajectories together with the direction field of the Lotka/Volterra differential equations (LV)

$$X' = (C1 + C3*X + C5*Y)*X, \quad Y' = (C2 + C4*Y + C6*X)*Y$$

where the parameter values are given by C1 = 0.5, C2 = 0.5, C3 = -0.0005, C4 = -0.0005, C5 = -0.00025, and C6 = -0.00025.

Start the program, fetch the Differential Equations Menu, and enter LV to get the Lotka/Volterra equations:

dynamics *< Enter >*

DE *< Enter >*

LV *< Enter >*

and the Main Menu for the Lotka/Volterra equations appears on the screen. We use the default values of the parameters and the scales.

1. Fetch the Differential Equations Menu (which is in the Main Menu):

DEM < Enter >

and plot the Direction Field:

DF *< Enter > < Enter >*

A resulting picture is given in Figure 2-6a.

---

2. For plotting different trajectories, we want to mark the initial conditions by a small cross. Clear the screen and fetch the cross Menu:

C *< Enter >*

KM < Enter >

and set the Scale KKS of the permanent cross to be 5 pixels by 5 pixels:

KKS *< Enter >*   **5 5** *< Enter >*

3. To get continuous trajectories, connect the consecutive computed dots. Fetch the When and What to plot Menu and CONnect consecutive dots:

WWM < Enter >

CON *< Enter >*

4. Pause the program:

P *< Enter >*

5. Plot the Trajectory:

T *< Enter > < Enter >*

6. Initialize and draw a permanent cross at y:

I *< Enter >*   **KK** *< Enter >*

7. Hit the *<* space bar *>* to un-pause: *< space bar >*

8. Use the arrow keys to move the small cross to a new initial condition for a trajectory, and pause the program:

P *< Enter >*

9. Repeat the steps 6 through 8. A resulting picture is given in Figure 2-6b.

Note. Instead of going through steps 6-8 repeatedly, recall that an easy way to initialize is to move the mouse arrow to the desired position and then click the left mouse button. To have a cross drawn at this position of y1, just enter command **KK1**.

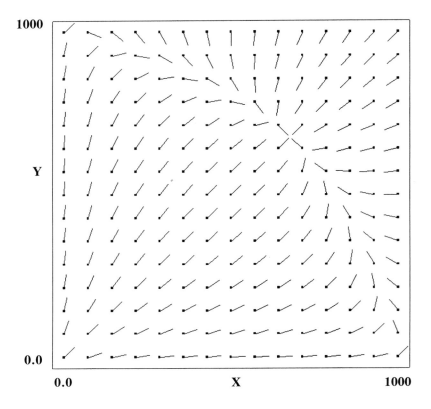

Figure 2-6a: Direction field

This figure shows a direction field for the Lotka/Volterra differential equations

$$X' = (C1 + C3*X + C5*Y)*X, \quad Y' = (C2 + C4*Y + C6*X)*Y$$

where $C1 = 0.5$, $C2 = 0.5$, $C3 = -0.0005$, $C4 = -0.0005$, $C5 = -0.00025$, and $C6 = -0.00025$.

Topic of discussion. *This figure suggests there are "at least" four equilibria. What are they?*

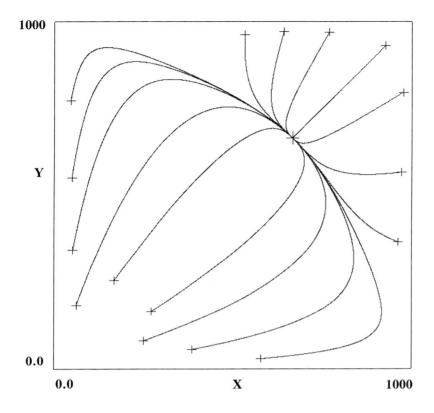

Figure 2-6b: Trajectories of an ODE

This figure shows some trajectories for the Lotka/Volterra differential equations, with parameters as in Figure 2-6a. The initial value of each trajectory is indicated with a small cross.

Topic of discussion. *There is a stable equilibrium where the trajectories converge. What does this picture suggest about the eigenvalues and eigenvectors of that equilibrium?*

Example 2-7: Bifurcation diagram for the scalar quadratic map

A **bifurcation diagram** displays the long term values of one of the variables (like X below) as some parameter is varied (RHO in the case below). See Chapter 6 for more details. The purpose is to produce a bifurcation diagram of the scalar map (H)

$$X \; \rightarrow \; RHO - X*X$$

where RHO varies between 1.2 and 2. Type *dynamics* $<Enter>$ followed by $H<Enter>$ to get the Main Menu for the Henon map.

1. In order to have the quadratic map, fetch the Parameter Menu and set the parameter C1 to be 0:

$$PM<Enter>$$

$$C1<Enter> \quad 0<Enter>$$

2. Enter the "BIFurcation diagram Menu" command BIFM:

$$BIFM<Enter>$$

(The menu BIFM is in the Numerical Explorations Menu.) The command BIFR in the BIFurcation diagram Menu BIFM is for specifying range of the bifurcation parameter RHO. The default setting is from NOT SET to NOT SET.

In order to make the parameter RHO vary between 1.2 and 2, set the "BIFurcation Range":

$$BIFR<Enter> \quad 1.2 \; 2<Enter>$$

with a space between 1.2 and 2.

3. Retrieve the BIFurcation diagram Menu:

$$<Enter>$$

and enter the "BIFurcation Screen" command BIFS:

$$BIFS<Enter> <Enter>$$

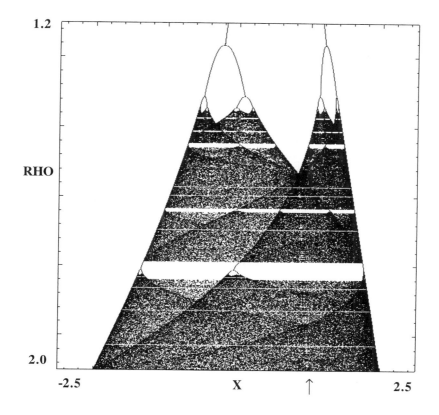

Figure 2-7: Bifurcation diagram for the scalar quadratic map
*This figure shows a bifurcation diagram of the scalar map (the Henon map
(H) with C1 = 0)*

$$X \rightarrow RHO - X*X$$

where parameter RHO plotted vertically varies between 1.2 and 2.

Topic of discussion. *Why are there curves marking the boundaries
between dark and very dark regions?*

Difficult topic of discussion. *In the dark region, there is a string
of white dots along a line rising above the arrow. What does this nearly
straight line mark? The hard question is: why do the white dots exist?*

When the plotting of the diagram has been completed, the message

```
Enter command: _____   or
hit <Enter> (fetch menu) or
H<Enter> (Help)
```

appears on the screen. The resulting picture will be similar to Figure 2-7.

4. To reproduce this figure on paper do the following. Make sure the printer is on and send the picture to the printer:

PP<Enter>

and select the printer. Return to the BIFurcation diagram Menu by hitting

<Esc>

Note. To get a high quality picture like Figure 2-7, set BIFV to be 720 before invoking command BIFS. See Chapter 6 for more details.

Example 2-8: Bifurcation diagram with bubbles
The purpose is to produce a bifurcation diagram of the Henon map (H)

$$(X,Y) \; \rightarrow \; (1.25 - X*X + C1*Y, \; X)$$

where C1 varies between 0.1 and 0.35. Type *dynamics<Enter>* followed by
H<Enter> to get the Main Menu for the Henon map.
1. Fetch the Parameter Menu and set the parameter RHO to be 1.25:

PM<Enter>

RHO<*Enter*> **1.25**<*Enter*>

2. Fetch the BIFurcation diagram Menu:

BIFM<Enter>

The command BIFR in the BIFurcation diagram Menu BIFM is for specifying
range of the bifurcation parameter RHO. The default setting is from NOT SET
to NOT SET.

3. In order to guarantee the parameter C1 is the PaRaMeter that will
be varied, set PRM to be C1:

PRM<*Enter*>

C1<*Enter*>

4. The parameter C1 must vary between 0.1 and 0.35. Retrieve the menu
BIFM:

<*Enter*>

and set the "BIFurcation Range":

BIFR<*Enter*> **0.1 0.35**<*Enter*>

with a space between 0.1 and 0.35.

5. Make sure the printer is on, and create the BIFurcation diagram on
the printer:

BIF<*Enter*> <*Enter*>

and select the printer.

When the plotting of the diagram has been completed, the standard message appears on the screen. The resulting picture will be similar to Figure 2-8. The bifurcation diagram in this example exhibits so-called bubbles, a feature which is not found in the bifurcation diagram of the quadratic map in Example 2-7.

Note. You can make the same picture without the parameter values printed by turning the toggle BIFP off. Furthermore, you can vary the length of the diagram by using command BIFV; see Chapter 6 for more details.

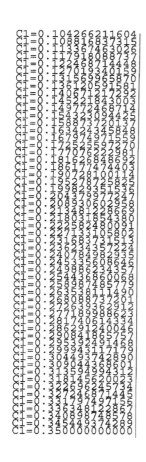
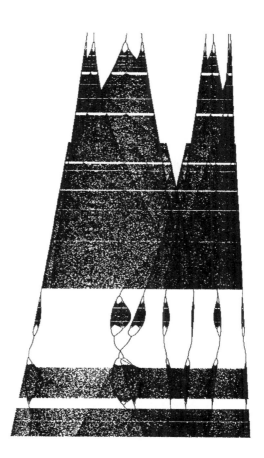

Figure 2-8: Bifurcation diagram with bubbles

This figure shows a bifurcation diagram with bubbles of the Henon map

$$(X, Y) \;\rightarrow\; (1.25 - X*X + C1*Y, \; X)$$

where parameter C1 range BIFR is from 0.1 to 0.35, and the variable X is plotted horizontally. The X Scale is from -2.5 (left) to 2.5 (right)

This projection can make curves seem to cross (as in the lower center) while in fact they do not cross in the true 3-dimensional (C1,X,Y) situation. The strange-looking oval shapes in the lower part of the figure are called bubbles.

The diagram was produced using command BIF. The settings are:
y1 = (0,1), BIFI = 0, BIFP = ON, BIFPI = 200, BIFD = 700, and BIFV = 880.

Topic of discussion. *What is the cause of the plateau, that is, the sudden broadening, at C1 ≈ 0.17?*

Example 2-9: All the Basins and Attractors

The **basin of an attractor** is the set of points whose iterates tend to that attractor; see Chapter 7 for more details. In this example we want to plot all the basins and attractors for the Henon map (H)

$$(X,Y) \rightarrow (1 - X*X + 0.475*Y, X)$$

Type *dynamics < Enter >* followed by *H < Enter >* to get the Main Menu for the Henon map.

1. Fetch the Parameter Menu:

$$PM < Enter >$$

Set the parameter RHO to be 1:

$$\textbf{RHO} < Enter > \quad 1 < Enter >$$

and set the parameter C1 to be 0.475:

$$\textbf{C1} < Enter > \quad \textbf{0.475} < Enter >$$

2. Fetch the Basin of attraction Menu (which is in the Numerical Explorations menu):

$$BM < Enter >$$

A method for plotting basins of attraction, which demands less skill and patience of the user than the routine BAS, is the method for plotting all the Basins and Attractors without setting any vectors (command BA). The "Basins and Attractors Precision" number BAP is used when plotting the basins and attractors without setting any vectors. Higher values produce more accurate pictures but result in slower computations.

Enter the plot the "Basins and Attractors" command BA:

$$\textbf{BA} < Enter > \, < Enter >$$

A 100 by 100 grid of points will be tested. The colors used to color the grid boxes indicate something that is in the box. Initially all grid boxes are uncolored but as the routine BA proceeds, all grid boxes will eventually be colored. The BA routine selects an uncolored grid box and tests the box by examining the trajectory that starts at the center of the grid box. The trajectory may or may not pass through grid boxes that have previously been colored, but what it encounters will determine how the initial grid box is colored.

First, pixels are colored dark blue, indicating that the trajectories of the centers of these boxes diverge. After a while, some boxes are colored light blue followed by boxes which are colored green, purple, and red, respectively. The dark blue region is the basin of infinity. There are 2 other attractors, a period 8 orbit (red) and a 2-piece chaotic attractor (green). The basin of the period 8 orbit is colored purple, and the basin of the chaotic attractor is colored light blue. See Chapter 7 for a more detailed explanation.

3. If you print the current picture on paper, you just get a solid black plot. The program has the feature to erase a color (see Section 3.7 for colors and color numbers). The basin of the period 8 attractor is colored with color number 5.

Fetch the Color Menu:

$$CM < Enter >$$

and then get the Erase color Menu:

$$EM < Enter >$$

and Erase color number 5:

$$\textbf{E5} < Enter >$$

The resulting picture is similar to Figure 2-9a. Make sure the printer is on and Print the Picture:

$$\textbf{PP} < Enter >$$

You may want to compare the picture with Figure 7-2, in which the union of the basin of infinity and the basin of the period 8 attractor is plotted using the routine BAS. See also Figure 2-9b for a feature of printing all three basins.

Note. The resolution of the picture that appears on the screen, is low. By first invoking a command like RH (High Resolution), the command BAS generates a picture of high resolution.

Note. Another interesting example of BA is:
The Henon map with RHO = 1.25 and C1 = -0.3. There is a 3-piece chaotic attractor and a fixed point attractor and a period-4 attractor.

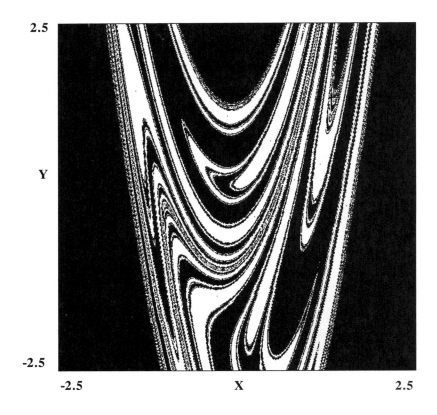

2.5

Y

-2.5

-2.5   X   2.5

Figure 2-9a: All the basins of attraction and attractors
The black area is the union of two basins of the Henon map (H)
$$(X, Y) \; \longrightarrow \; (1 - X*X + 0.475*Y, \; X)$$
*It shows points whose trajectories diverge (to infinity) and points that
are attracted to the chaotic attractor. There are three attractors for this
choice of parameters. The routine BA plots all attractors and their basins.
It has plotted the basins of infinity, of a chaotic attractor and of a
stable period 8 orbit. The white area is the basin of attraction of the
period 8 attractor, and is obtained by erasing the color of the basin of
the period 8 attractor. We used command E5 to Erase color 5. The isolated
dots (in the white region) are on the period 8 attractor. Its basin has
been erased and so is white.*

Topic of discussion. *Compare this figure with Figure 7-2. What are the
similarities, and what are the differences?*

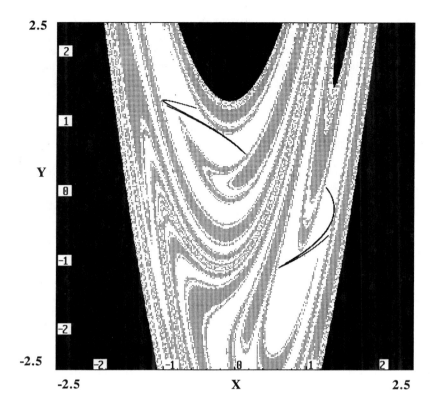

Figure 2-9b: Shading of basins when there are more than 2 attractors
The black and grey areas are two basins of the Henon map with parameter values as in Figure 2-9a. Recall, there are three attractors for this choice of parameters. The routine BA has plotted the basins of infinity, of a chaotic attractor and of a stable period 8 orbit. The white area is the basin of attraction of the chaotic attractor. The two hair pin shaped "curves" in the white region are actually a single chaotic attractor, and they are not in fact curves, but more complicated shapes. We used command E3 (with PAT = 0) to Erase the plotted region using color 3. The grey area is the basin of attraction of the period 8 attractor, and is obtained by shading the color of the basin of the period 8 attractor using the Erase command E5 with PATtern PAT = 6. The isolated dots (in the grey region) are on the period 8 attractor.

Topic of discussion. *Compare this figure with Figure 2-9a. What are the similarities, and what are the differences?*

In Example 2-9 all the Basins and Attractors were plotted by using the routine BA. The routine BAS provides another method for plotting BASins; see Chapter 7. In the next example we use the simplest form of the routine BAS to plot the basin of infinity. A more advanced use of this routine for plotting a basin will be applied in Example 2-20. In order to be able to use the routine for plotting a Basin boundary Straddle Trajectory (that is, a trajectory on a basin boundary), you must be familiar with the more advanced use of the routine BAS. For Basin boundary Straddle Trajectories, see the Examples 2-21 and 2-22.

Example 2-10: Metamorphoses in the Henon map's basin of infinity
The purpose of this example is to show that the basin of infinity of the Henon map (H)

$$(X,Y) \rightarrow (RHO - X*X - 0.3*Y, X)$$

depends on the parameter RHO. The **basin of infinity** is the set of points that go to infinity under iteration of the map. For the present we will consider the *basin of infinity* to be the set of points whose trajectory exits a large box enclosing the screen area within 60 iterates. See Chapter 7 for a discussion of this computation.

We consider four values for the parameter RHO, namely,
(a) RHO = 1.31; (b) RHO = 1.32; (c) RHO = 1.395; (d) RHO = 1.405
Type *dynamics < Enter >* followed by *H < Enter >* to get the Main Menu for the Henon map. The parameter C1 is by default -0.3, so it does not need to be reset.

The purpose is to plot the four basins in one picture. We want to determine the basin of infinity of the Henon map when RHO = 1.31, and plot it in the upper left window. Similarly, we want to plot the basin of infinity for the cases (b), (c), and (d) in the upper right window, the lower left window, and the lower right window, respectively.

1. Retrieve the Window Menu WM:

WM < Enter >

All the commands involving the opening of new windows and the switching between windows that are open, are collected in this menu.

2. Open Window W1:

OW < Enter >

OW1 *< Enter > < Enter >*

3. Set the parameter RHO to be 1.31:

RHO *<Enter>* **1.31** *<Enter>*

4. Fetch the Basin of attraction Menu:

BM <Enter>

All the commands involving plotting of basins of attraction are collected in this menu. Notice that storage vector y2 is NOT SET.

5. Enter the command BAS:

BAS *<Enter> <Enter>*

and the computer starts computing the basin in the upper left window of the screen. A 100 by 100 grid of points will be tested. (In creating the picture we have used a higher resolution of 720 by 720. See the Basin Resolution Menu BRM.) The program determines if the trajectory through each of those points goes to infinity (or more accurately, goes more than SD diameters from the screen within 60 iterates -- see the "Screen Diameter" command SD) in which case a small box about the point will be plotted. Then the next grid point is tested. The resolution of the picture that appears on the screen, is low. Invoking a command like RH followed by the command BAS, generates a picture of High Resolution similar to the one that is presented in Figure 2-10.

6. When the computations for the basins have been completed, the message

```
Enter command: _____ or
hit <Enter> (fetch menu) or
H<Enter> (Help)
```

appears on the screen, fetch the Window Menu, and Open Window W2:

WM <Enter>

OW<Enter>

OW2 *<Enter> <Enter>*

7. Set the parameter RHO to be 1.32:

$$\textbf{RHO} <Enter> \quad \textbf{1.32} <Enter>$$

8. Plot the basin by entering command BAS:

$$\textbf{BAS} <Enter> <Enter>$$

and the computer starts computing the basin in the upper right window of the screen.

9. When the plotting of the second basin has been finished, Open Window W3:

$$OW<Enter>$$

$$\textbf{OW3} <Enter> <Enter>$$

Set the parameter RHO to be 1.395:

$$\textbf{RHO}<Enter> \quad \textbf{1.395} <Enter>$$

Plot the basin of infinity:

$$\textbf{BAS} <Enter> <Enter>$$

and the computer starts computing the basin in the lower left window of the screen.

10. When the third basin has been plotted, Open Window W4:

$$OW<Enter>$$

$$\textbf{OW4} <Enter> <Enter>$$

Set the parameter RHO to be 1.405:

$$\textbf{RHO}<Enter> \quad \textbf{1.405} <Enter>$$

Plot the basin of infinity:

$$\textbf{BAS}<Enter> <Enter>$$

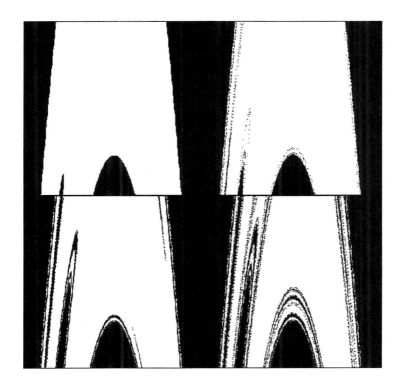

Figure 2-10: The basin of attraction of infinity

This figure shows the basin of attraction of the attractor at infinity of the Henon map (H)

$$(X, Y) \rightarrow (RHO - X*X - 0.3*Y, X)$$

The basin of attraction of the attractor at infinity is the set of points that go to infinity under iteration of the map. In all four windows, the variable X is plotted horizontally, and the variable Y is plotted vertically. The X Scale runs from -2.5 (left) to 2.5 (right) and the Y Scale is from -2.5 (bottom) to 2.5 (top).

In the upper left window RHO is 1.31; in upper right window RHO is 1.32; in the lower left window RHO is 1.395; in lower right window RHO is 1.405. Notice the major change that occurred in changing RHO from 1.395 to 1.405. There is a sudden discontinuity at RHO ≈ 1.396. There is a similar discontinuity between 1.31 and 1.32. Such discontinuities in boundaries are called **metamorphoses**.

Topic of discussion. *In what sense are these sudden changes in the basins discontinuous, and in what sense are they continuous?*

11. To Print the Picture on paper, make sure the printer is on and enter the command PP:

$$PP < Enter >$$

and select the printer.

Comparing the four basins of infinity, one sees clearly that they are different. For example, the basins in the lower left window (RHO = 1.395) and the lower right window (RHO = 1.405) are rather different. For more details on this phenomenon of metamorphoses, see C. Grebogi, E. Ott and J.A. Yorke (1987).

In making basin pictures like this, one might ask if there are additional points whose trajectories are going to infinity, that is, points whose trajectories take more than 60 iterates to leave the big box mentioned above. Example 2-20 studies how to check that. In that case, the extra care produces very few extra points whose trajectories are going to infinity. It is nonetheless important to be able to check the reliability of a picture.

Note. In this example we check for only 60 iterates to see if a trajectory diverges, that is, the Maximum number of Checks MC is 60. In Example 2-20 we set MC = 1000 and yet by use of a more sophisticated approach, the computation time remains about the same.

Note. While BA can be used for creating this example, we demonstrate here another technique by using BAS.

Example 2-11: Search for all periodic points with period 10

The purpose of this example is to search for all periodic points with period 10 of the Henon map (H):

$$(X,Y) \;\rightarrow\; (2.12 - X*X - 0.3*Y, \; X).$$

The attractor for this map was discussed in Example 2-2a.

Start the program by typing *dynamics* <*Enter*> H <*Enter*> to get the Main Menu of the Henon map. The parameters RHO and C1 are by default 2.12 and -0.3.

1. Fetch the Screen Menu:

SM <*Enter*>

and then get the BoX Menu:

BXM <*Enter*>

and draw the Box and a double set of tic marks:

B2 <*Enter*>

2. Fetch the Numerical Explorations Menu:

NEM <*Enter*>

and then get the Periodic Orbit Menu:

POM <*Enter*>

and set the PeRiod to be 10:

PR <*Enter*> **10** <*Enter*>

3. Retrieve the Periodic Orbit Menu POM:

<*Enter*>

and search at Random for the Periodic points with period 10:

RPK <*Enter*> <*Enter*>

A resulting picture of the periodic points is given in Figure 2-11.

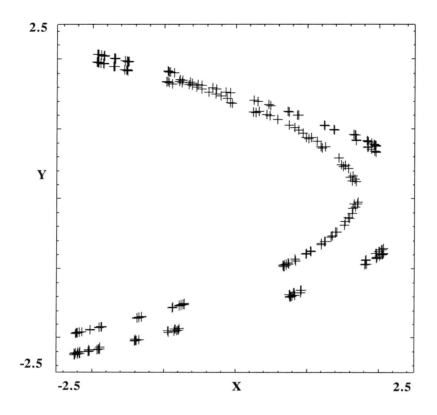

Figure 2-11: Periodic points with period 10
This figure shows period 10 orbits of the Henon map (H)

$$(X, Y) \rightarrow (2.12 - X*X - 0.3*Y, X)$$

The routine RPK found 30 distinct orbits of period 10. These periodic orbits may include fixed points and periodic orbits with period 2 or 5, since these orbits also have period 10. Many of these period 10 points are **not** *on the attractor.*

Topic of discussion. *What can be said about the location of period 10 points which are not on the attractor?*

---

90   Dynamics: Numerical Explorations

Example 2-12: search for all fixed points and periodic points with period 2, and mark them with permanent crosses of different sizes

The purpose of this example is to search for all fixed points and periodic points with period 2 of the time-2π map of the forced damped Pendulum differential equation (P)

$$X''(t) + 0.2*X'(t) + \sin X(t) = 2.5*\cos(t)$$

Start the program and enter command P:

dynamics $<Enter>$ **P** $<Enter>$

to get the Main Menu of the forced damped Pendulum equation. The parameters RHO, C1, C2, C3, and PHI are by default 2.5, 0.2, 1, 0, and 1, respectively.

1. Fetch the Kruis (cross) Menu KM (which is in the Screen menu):

KM $<Enter>$

and set the size of the permanent crosses at the fixed points 8 by 8 (the default is 6.4 by 6.4):

KKS $<Enter>$ **8 8** $<Enter>$

2. Fetch the Periodic Orbit Menu:

POM $<Enter>$

and search at Random for the Periodic points with period 1 (fixed points):

RPK $<Enter>$ $<Enter>$

The routine chooses random initial points for the Quasi-Newton method and then tries up to 50 Quasi-Newton steps. Eventually we can expect it to find periodic orbits of the specified period, but we do not know at any point if they all have been found.

3. After a while, there are 8 fixed points found. Hit

$<Esc>$

to return to the Periodic Orbit Menu. Set the PeRiod to be 2:

PR $<Enter>$ **2** $<Enter>$

4. Call the Kruis (cross) Menu KM:

$$KM <Enter>$$

and set the size of the permanent crosses at the period 2 points 5 by 5:

$$\textbf{KKS} <Enter> \quad \textbf{5 5} <Enter>$$

5. To plot a permanent cross at each point of the trajectory, retrieve the Kruis (cross) Menu:

$$<Enter>$$

and enter command KKK:

$$\textbf{KKK} <Enter> <Enter>$$

6. Get the Periodic Orbit Menu:

$$POM <Enter>$$

and search at Random for the Periodic points with period 2 which are not fixed points:

$$\textbf{RP} <Enter>$$

$$\textbf{RPX} <Enter> <Enter>$$

7. Fetch the AXes Menu:

$$AXM <Enter>$$

Draw the AXes with a single set of tic marks:

$$\textbf{XAX1} <Enter>$$

$$\textbf{YAX1} <Enter>$$

After a while, hit

$$<Esc>$$

to return to the AXes Menu. A resulting picture is given in Figure 2-12.

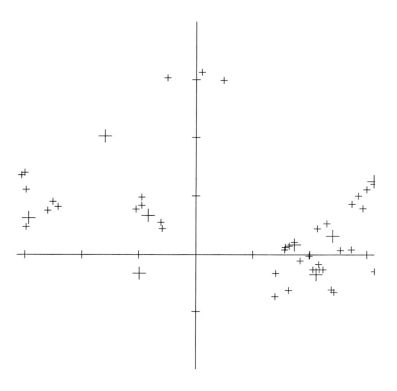

Figure 2-12: Fixed points and period 2 points

This figure shows the period 2 orbits and fixed points of the time-2π map of the forced damped Pendulum (P)

$$X''(t) + 0.2*X'(t) + sin\ X(t) = 2.5*cos(t)$$

The variable X is plotted horizontally, and the variable X' is plotted vertically. The X Scale is from -3.141593 (left) to 3.141593 (right) and the Y Scale is from -2 (bottom) to 4 (top).

Command KKK results in a permanent cross at each point that is plotted. Command KKS is used to specify the size of the crosses. The larger crosses indicate the positions of the fixed points, while the small crosses indicate the positions of the period 2 points. The routine RP found 8 distinct fixed points, and 40 distinct points of period 2 which are not fixed points. The computations for plotting these fixed points and period 2 points may take some hours.

Topic of discussion. *Is there a region ($|Y| \leq$ constant) which is mapped into itself? Compare this with Figure 2-5 for the Henon map.*

Example 2-13: Following periodic orbits as a parameter is varied

In this example we illustrate following periodic orbits of the Ikeda (Ikeda/Hammel/Jones/Maloney laser) map (I)

$$Z \rightarrow RHO + 0.9*Z*exp\{i[0.4 - 6/(1 + |Z*Z|)]\}$$

where $Z = X + iY$, and the parameter RHO will be varied. Type

dynamics *< Enter >*   **I** *< Enter >*

to get the Main Menu for the Ikeda map. The objective of this example is to follow a periodic orbit with period PR = 5.

1. Enter the "Parameter Menu" command PM:

PM < Enter >

to verify that the values above are the default values for the parameters.

2. Get the Periodic Orbit Menu:

POM < Enter >

and set the period PR to be 5. Enter the "PeRiod" command PR:

PR *< Enter >*   **5** *< Enter >*

3. Fetch the Follow Orbit Menu (which is in the Numerical Explorations Menu):

FOM < Enter >

All the commands involving following periodic orbits are collected in this menu.

4. Enter the "Follow Orbit" command FO:

FO *< Enter > < Enter >*

In the lower left corner of the screen, the color codes for attractor repellor, flip saddle and regular saddle appear; white color indicates no periodic orbit.

---

5. To reduce printing on the screen, enter the "Text level 2" command:

T2 *< Enter >*

6. After a little while, clear the screen:

C *< Enter >*

to get rid of the nonperiodic part. (Refresh the screen *R< Enter >* to get the information that was in the lower left corner, back on the screen.) The resulting plotted picture will be similar to the one in Figure 2-13a.

Note. If you enter command FO when RHO is different from the default value, then you may wind up with a different family of period 5 orbits.

After the period 5 is plotted in the phase space while RHO is varied, we can get another view on following periodic orbits as follows.
Hit

< Esc >

to return to the Following Orbit Menu and clear the screen:

C *< Enter >* .

7. Set the X COordinate to be RHO:

XCO *< Enter >*   **RHO** *< Enter >*

8. Set the X Scale to run from 0.9 to 1.9:

XS *< Enter >*   **0.9 1.9** *< Enter >*

9. Enter the "Follow Orbit" command FO:

FO *< Enter >*

The resulting plotted picture will be similar to the one in Figure 2-13b.

Note. The periodic orbit followed, may depend on the initialization vector y1.

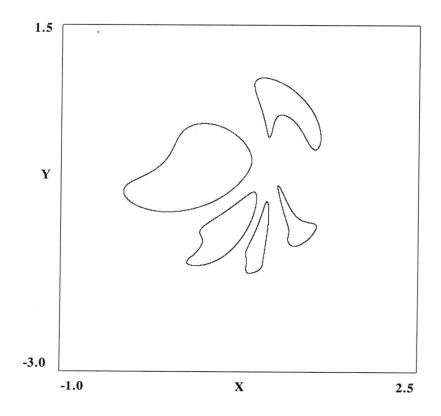

Figure 2-13a: A family of periodic orbits

The figure shows a family of period 5 periodic orbits of the Ikeda (Ikeda/Hammel/Jones/Maloney laser) map (I)

$$Z \rightarrow RHO + 0.9*Z*exp\{i[0.4 - 6/(1 + |Z*Z|)]\}$$

where $Z = X + iY$. This family is created by varying RHO.

The period PR was set to be 5, the initialization vector y1 was set to (0,1) and y0 was set equal to y1 (command I); thereafter the "Follow Orbit" command FO was invoked. Color screens will show different colors for the period 5 attractors, regular saddles and flip saddles (see Chapter 11).

Notice that this picture shows the orbit for a range of values of RHO. For a single value of RHO, there would be 5 discrete points. Compare with Figure 2-13b.

Topic of discussion. *On color screens, the above plot is red, blue and green. Find period 10 orbits bifurcating from points where the red and blue curves meet.*

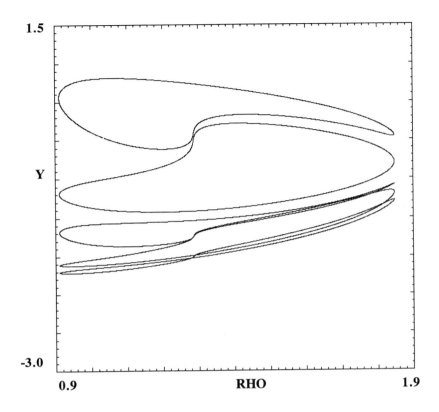

Figure 2-13b: Following a family of periodic orbits

This family of periodic orbits is the same as in Figure 2-13a, only here the parameter RHO is plotted horizontally. The X COordinate is RHO, and the "Follow Orbit" command FO was invoked. On a color screen, the color indicates which periodic points are stable and which are saddles. See Chapter 11.

Topic of discussion. *As the family is followed, RHO occasionally changes from an increasing to a decreasing variable, and vice versa. Describe the changes in the types of orbits seen near the points where RHO changes direction.*

Example 2-14: The Mandelbrot set

The purpose of this example (using process C) is to generate a set in the complex plane which is known as the "Mandelbrot set". The **Mandelbrot set** is the set of parameter values (C5,C6) such that the trajectory through (0,0) is bounded under iteration of the complex quadratic map.

Start the program and enter C to get the Cubic map:

$$\textbf{dynamics} < Enter > \quad \textbf{C} < Enter >$$

and the Main Menu for the Cubic map appears on the screen.

The Cubic map is a map in the complex plane

$$Z \;\rightarrow\; RHO*Z*Z*Z \;+\; (C1+iC2)*Z*Z \;+\; (C3+iC4)*Z \;+\; C5+iC6$$

with $Z = X + iY$ where $X = y[0]$, $Y = y[1]$.

The default values of the parameters are:

C1 = 1.00000000   C2 = 0.00000000
C3 = 0.00000000   C4 = 0.00000000
C5 = 0.25000000   C6 = 0.00000000
RHO = 0.00000000

With the default values RHO = 0, C2 = C3 = C4 = 0 the Cubic map in the complex plane reduces to the **complex quadratic map**:

$$Z \;\rightarrow\; Z*Z \;+\; C5+iC6$$

The Mandelbrot set is the set of complex numbers C5+iC6 for which the trajectory through 0+i0 remains bounded, under iteration of the complex quadratic polynomial map. For the Cubic map C, the initial point y1 has coordinates (0,0), so the initial point is as needed.

1. Change the horizontal coordinate to be the parameter C5:

$$\textbf{XCO} < Enter > \quad \textbf{C5} < Enter >$$

2. Change the vertical coordinate to be the parameter and C6:

$$\textbf{YCO} < Enter > \quad \textbf{C6} < Enter >$$

3. Enter command BAS:

$$\textbf{BAS} < Enter > \, < Enter >$$

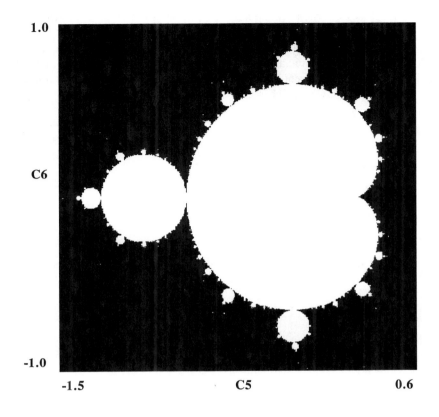

1.0

C6

-1.0

-1.5   C5   0.6

Figure 2-14: The Mandelbrot set

The black area is the set of parameter values (C5, C6) such that the trajectory through (0,0) goes to infinity under iteration of the complex quadratic map (the complex Cubic map (C) with RHO = 0)

$$Z \;\rightarrow\; Z*Z \;+\; C5 + iC6$$

The white area is the set of parameter values (C5, C6) such that the trajectory through (0,0) is bounded under iteration of the complex quadratic map. This set is called the Mandelbrot set.

On a color screen, the black region is plotted in a variety of colors, different colors for different escape or exit times.

Before creating this picture, the resolution command RH was invoked, and MC was set to be 1000.

Topic of discussion. *Change RHO to 1 in the map C and make a picture corresponding to this picture, and compare the results.*

and the black region is the Mandelbrot set, or as good as an approximation as we can get with a 100 by 100 grid. The horizontal and vertical scales have been selected so that the picture has a reasonable scale. See figure 2-14 for the resulting picture with a high resolution. Different pictures can be made using different parameters. For example, the same parameters can be varied but RHO can be set to 1.

Example 2-15: All the Basins and Attractors

The purpose of this example is to plot all the basins and attractors for the complex quadratic map (the complex Cubic map (C) with RHO = 0):

$$Z \rightarrow Z*Z + 0.32 + i0.043$$

where $Z = X + iY$. See also Example 2-9.

Type *dynamics* <*Enter*> followed by *C* <*Enter*> to get the Main Menu for the Cubic map.

    1. Fetch the Parameter Menu:

PM <*Enter*>

Set the parameter C5 to be 0.32:

C5 <*Enter*>   **0.32** <*Enter*>

and set the parameter C6 to be 0.043:

C6 <*Enter*>   **0.043** <*Enter*>

    2. Retrieve the Parameter Menu

<*Enter*>

Set the X Scale to run from -1 to 1:

XS <*Enter*>   **-1 1** <*Enter*>

and set the Y Scale to run from -1.35 to 1.35

YS <*Enter*>   **-1.35 1.35** <*Enter*>

    3. Fetch the Basin of attraction Menu (see also Example 2-9):

BM <*Enter*>

Enter the "Basins and Attractors" command and plot the Basins and Attractors:

BA <*Enter*> <*Enter*>

A 100 by 100 grid of points will be tested. First, pixels are colored dark blue, indicating that the trajectories of the centers of these boxes diverge. After a while, some boxes are colored light blue followed by boxes which are colored green. The dark blue region is the basin of infinity. There is one other attractor, a period 11 orbit (green); its basin is colored light blue. See Chapter 7 for more details.

5. If you print the current picture on paper, you just get a solid black plot. The basin of infinity is colored with color 1 (see Section 3.7 for colors and color numbers).

Fetch the Color Menu:

$$CM < Enter >$$

Get the Erase color Menu:

$$EM < Enter >$$

and Erase color 1:

$$E1 < Enter >$$

The resulting picture is similar to Figure 2-15. Make sure the printer is on and Print the Picture:

$$PP < Enter >$$

and select the printer you have. You may want to compare the picture with Figure 7-3, in which the basin of infinity is plotted using the routine BAS.

Note. The resolution of the picture that appears on the screen, is low. By first invoking a command like RH (High Resolution), the command BA generates a picture of high resolution.

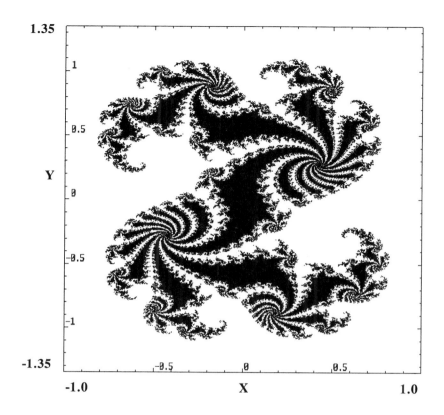

Figure 2-15: All the basins of attraction

The black area is the basin of attraction of a stable period 11 orbit of the complex quadratic map (the complex Cubic map (C) with RHO = 0)

$$Z \rightarrow Z*Z + 0.32 + i0.043$$

where Z = X + iY. The routine BA has plotted the basins of infinity and of a stable period 11 orbit. The white area is the basin of infinity, and is obtained by erasing the color of the basin of infinity (color 1).

The printed scales shown inside the boundary of the figure, were made using command B2.

Note. These scales cannot be made in Unix.

Topic of discussion. *Compare this figure with Figure 7-3 which was made using command BAS; see also Figure 3-3b. Aside from the reversing of colors, what are the differences? Why is the figure invariant under 180 degrees rotation? Does this symmetry also hold for the stable period 11 orbit?*

Example 2-16: 3-Dimensional views on the Lorenz attractor
   The purpose of this example is to rotate the Lorenz attractor around X, Y and Z axes, that is, to rotate the attractor of the Lorenz differential equations (L):

$$X' = 10*(Y-X), \quad Y' = 28*X - Y - X*Z, \quad Z' = X*Y - 2.66667*Z$$

Start the program and enter L to get the Lorenz equations:

dynamics *< Enter >*   **L** *< Enter >*

and the Main Menu for the Lorenz equations appears on the screen. We use the default values of the parameters and the scales.

   1. Fetch the Window Menu:

WM < Enter >

and Open window W1:

OW< Enter >

OW1 *< Enter > < Enter >*

   2. Plot the Trajectory:

T *< Enter > < Enter >*

the resulting picture is similar to the upper left window in Figure 2-16.

   3. After a while, hit

< Esc >

and Open window W2:

OW< Enter >

OW2 *< Enter > < Enter >*

Fetch the AXes Menu:

AXM < Enter >

Get the 3-dimensional AXes Menu:

AX3M < Enter >

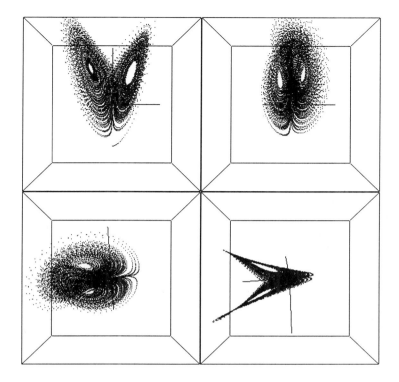

Figure 2-16: 3-Dimensional views on the Lorenz attractor

The chaotic attractor of the Lorenz differential equations (L)

$$X' = 10*(Y-X), \quad Y' = 28*X - Y - X*Z, \quad Z' = X*Y - 2.66667*Z$$

In all four windows, the variable X is plotted horizontally, and the variable Z is plotted vertically, and the X Scale is from -30 (left) to 30 (right) and the Y Scale is from -10 (bottom) to 40 (top).

The upper left window shows the chaotic attractor in the default situation. In the upper right window, the Y (i.e., the vertical) axis is the rotation axis, and the attractor has been rotated 125 degrees around the Y axis. In the lower left window and lower right window show the attractor rotated around the Y axis, the Z (i.e., the perpendicular-to-the-screen) axis and the X (i.e., the horizontal) axis.

Topic of discussion. *Explain the left-right symmetry of the upper left window using the equations.*

4. Make the Y axis the rotation Axis:

$$AY <Enter>$$

and plot the Trajectory:

$$T <Enter> <Enter>$$

5. While plotting, hit

$$< + >$$

and the computing of the trajectory starts anew, since it will be rotated with an ANGLE of 5 degrees around the Y Axis. Hit this key, say 24, more times, and the emerging picture will be similar to the upper right window in Figure 2-16.

6. After a while, hit
$$<Esc>$$

and Open window W3:

$$OW<Enter>$$

$$OW3 <Enter> <Enter>$$

Fetch the AXes Menu:

$$AXM <Enter>$$

Get the 3-dimensional AXes Menu:

$$AX3M <Enter>$$

7. Make the Z axis the rotation Axis:

$$AZ <Enter>$$

and plot the Trajectory:

$$T <Enter> <Enter>$$

8. While plotting, hit

$$< + >$$

and the computing of the trajectory starts anew, since it will be rotated with an ANGLE of 5 degrees around the Z Axis. Hit this key a few more times, until you are satisfied with the resulting picture.

9. After a while, hit

\<Esc\>

and Open window W4:

OW\<Enter\>

OW4 *\<Enter\> \<Enter\>*

Fetch the AXes Menu:

AXM \<Enter\>

Get the 3-dimensional AXes Menu:

AX3M \<Enter\>

10. Make the X axis the rotation Axis:

AX *\<Enter\>*

and plot the Trajectory:

T *\<Enter\> \<Enter\>*

11. While plotting, hit

$$< + >$$

and the computing of the trajectory starts anew, since it will be rotated with an ANGLE of 5 degrees around the X Axis. Hit this key a few more times, until you are satisfied with the picture.

Example 2-17: Unstable manifold of the fixed point for the Henon map
The **stable manifold** of a fixed point p of the Henon map is the set of points whose trajectories tend to p under forward iteration of the map; the **unstable manifold** of p is the set of points whose trajectories tend to p under backward iteration of the map. See Chapter 9 for more details. In this example, we want to produce the unstable manifold of the fixed point that lies on the attractor of the Henon map (H)

$$(X,Y) \;\rightarrow\; (1.4 - X*X + 0.3*Y, \; X).$$

Type *dynamics* <*Enter*> *H*<*Enter*> to get the Main Menu for the Henon map.

1. Set the parameter RHO to be 1.4:

RHO <*Enter*>   **1.4** <*Enter*>

2. Set the parameter C1 to be 0.3:

C1 <*Enter*>   **0.3** <*Enter*>

3. While we are interested in the unstable manifold of the fixed point on the attractor, it is not necessary to determine that fixed point first. The program will first automatically look for a fixed point. Fetch the Unstable and stable manifold Menu:

UM <*Enter*>

4. Enter the "Unstable manifold Left" command UL:

UL <*Enter*> <*Enter*>

and watch the screen. After a while type in

UR <*Enter*> <*Enter*>

Do you also see a picture which resembles the Henon attractor? The resulting picture will be similar to Figure 2-17. It is widely believed, but still unproven, that the closure of the unstable manifold of the fixed point of the Henon attractor equals the Henon attractor. For a discussion, see for example C. Simó (1979). One of the obstacles in proving or disproving such a conjecture is that there might well be stable periodic points with a high period in the closure of the unstable manifold of the fixed point of the Henon attractor.

Hit <*Esc*> to terminate the UR process and return to the Unstable and stable manifold Menu, and hit it two more times to return to the Main Menu.

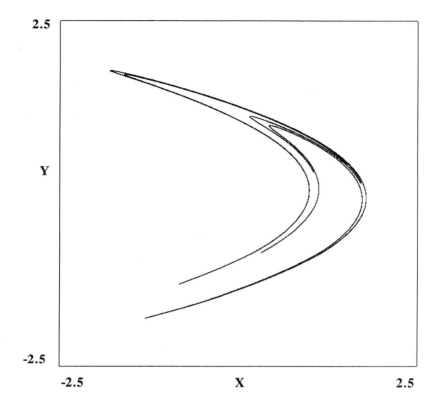

Figure 2-17: Unstable manifold of a fixed point of the Henon attractor
This figure shows the unstable manifold of the fixed point (appr. (0.88,0.88)) on the chaotic attractor of the Henon map (H)
$$(X, Y) \rightarrow (1.4 - X*X + 0.3*Y, X)$$

Topic of discussion. *Figure 2-4 is virtually identical to this figure. It has not been proved that the unstable manifold equals the attractor, but still we must ask why there is any similarity between these figures.*

Example 2-18: Stable and unstable manifolds

In this example, we want to produce both the stable and the unstable manifold of the fixed point that lies on the attractor of the Henon map (H)

$$(X,Y) \rightarrow (2.12 - X*X - 0.3*Y, X).$$

Type *dynamics* <*Enter*> followed by *H* <*Enter*> to get the Main Menu for the Henon map. The default values for the parameters RHO and C1 are 2.12 and -0.3 respectively.

1. Fetch the Unstable and stable manifold Menu and plot the Unstable manifold, Left side:

UM <*Enter*>

UL <*Enter*> <*Enter*>

and the program will first seek out a fixed point (a periodic point of period PR, which now is equal to 1), starting its search from y1, whose position on the screen is marked by the small cross. Initially y1 is at (0.0,2.0) for the Henon map. From that point, (and each point in a substantial region), the program will find a fixed point that is a saddle, and will proceed to compute the left side of its unstable manifold. As it plots the unstable manifold, the program prints out more information than is generally desirable. The "Text level 2" command T2 can be used to reduce the print being sent to the screen.

On color screens, when the function keys <F7> and <F8> are hit, the color being used for plotting changes. This may make it easier to see what the program has plotted compared to what it is plotting. The "Color Table" command CT displays the color table, that is, the available palette of colors. The key <F7> shifts the color one position to the left while <F8> increases color number (or equivalently, the color table position), shifting to the color one position to the right. Changing the color enables the user to distinguish new points from old points. The resulting picture will be similar to Figure 2-1a.

Hit <*Esc*> to return to the menu and clear the screen: **C** <*Enter*>

2. Retrieve the Unstable and stable manifold Menu UM:

<*Enter*>

and plot the Stable manifold, Left side:

SL <*Enter*> <*Enter*>

to compute and plot the stable manifold of the fixed point. The "Stable manifold, Right side" command SR gives the right side of the stable manifold. The resulting picture will be similar to Figure 2-18.

Hit <*Esc*> and return to the menu and clear the screen **C** <*Enter*>.

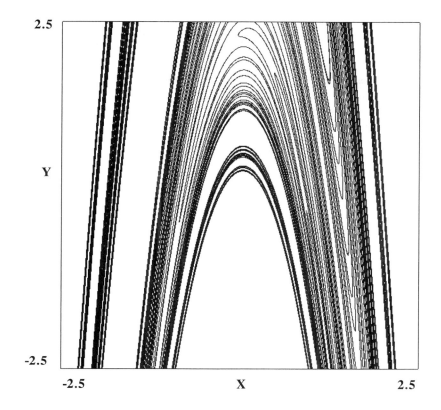

Figure 2-18: Stable manifold of a fixed point

This figure shows a part of the stable manifold of a fixed point on the attractor of the Henon map (H)

$$(X, Y) \longrightarrow (2.12 - X*X - 0.3*Y, X)$$

The fixed point used is approximately (0.9445, 0.9445).

Topic of discussion. *Where does the stable manifold in this figure lie in relation to the basin of the chaotic attractor (Figure 2-20)? Clearly we show only a part of the stable manifold. Is the entire stable manifold a bounded set?*

Example 2-19a: Plotting a bounded chaotic trajectory when almost all trajectories are unbounded (Saddle Straddle Trajectory)

A set is **invariant** if the process maps it onto itself: iterates of initial points in the set are in the set, and every point in the set is the iterate of some other point in the set. A **chaotic saddle** is a compact, invariant set which is not an attractor and contains a chaotic trajectory. The purpose of this example is to produce a numerical chaotic trajectory that is on a chaotic saddle of the Henon map (H)

$$(X,Y) \rightarrow (4.2 - X*X + 0.3*Y, X).$$

For this case, almost all trajectories are going to infinity, yet a complicated, bounded, invariant set exists. This procedure allows us to investigate the dynamics on this invariant set.

Type *dynamics* < *Enter* > followed by *H* < *Enter* > to get the Main Menu for the Henon map.

1. Set the parameter RHO to be 4.2:

RHO < *Enter* >   **4.2** < *Enter* >

2. Set the parameter C1 to be 0.3:

C1 < *Enter* >   **0.3** < *Enter* >

3. Since we are interested in a Saddle Straddle Trajectory on the chaotic saddle, we want the first few points of the trajectory to remain unplotted, say 10. Set the PreIterates to 10:

PI < *Enter* >   **10** < *Enter* >

4. From the work by Devaney and Nitecki (1979), we know that the box [-3,3] × [-3,3] includes the chaotic saddle. Therefore, we want to change the scale, since it is by default [-2.5,2.5] × [-2.5,2.5]. Set the X Scale to run from -3 to 3:

XS < *Enter* >   **-3 3** < *Enter* >

Thereafter, set the Y Scale to run from -3 to 3:

YS < *Enter* >   **-3 3** < *Enter* >

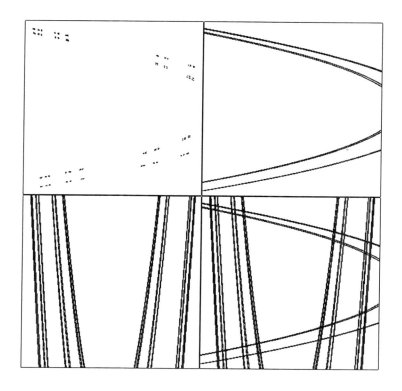

Figure 2-19: A Saddle Straddle Trajectory and manifolds

This figure shows a Saddle Straddle Trajectory and manifolds of a fixed point of the Henon map (H)

$$(X, Y) \;\rightarrow\; (4.2 - X*X + 0.3*Y, \; X)$$

Almost all trajectories diverge for this Henon map. The upper left window shows a Saddle Straddle Trajectory on a chaotic saddle (that is, a compact, invariant set which is not an attractor and contains a chaotic trajectory). The unstable manifold of the fixed point (appr. (1.729,1.729)) on the chaotic saddle is plotted in the upper right window, while the part of the stable manifold of the same fixed point is plotted in the lower left window. The lower right window W4 contains a superposition of the pictures in the windows W1, W2, and W3. In all the four windows, the variable X is plotted horizontally and the variable Y is plotted vertically, and the X Scale is from -3 (left) to 3 (right) and the Y Scale is from -3 (bottom) to 3 (top).

Topic of discussion. *Describe the dynamics of the trajectories in the upper left window. Find where the plotted points are mapped under H.*

5. Call the Straddle Trajectory Menu STM:

STM <Enter>

Unlike a regular trajectory that is initiated with a single vector, a Saddle Straddle Trajectory requires two vectors, ya and yb. The SST routine moves ya and yb so that they are close together and follow along with the SST trajectory. We use the default values for ya and yb. Chapter 8 describes the procedure for choosing ya and yb. Enter the "Saddle Straddle Trajectory" command SST:

SST *<Enter> <Enter>*

and the plotted trajectory will be similar to Figure 2-19 (upper left window). The resulting picture is a **Saddle Straddle Trajectory**.

6. Hit *<Esc>* a few times to return to the Main Menu.

For more applications using this method to generate Saddle Straddle Trajectories, see H.E. Nusse and J.A. Yorke (1989).

Example 2-19b: The unstable manifold of a fixed point

The purpose of this example is to produce a part of the unstable manifold of the fixed point that is on the chaotic saddle of the Henon map (H) with parameters and scales as in Example 2-19a.

Clear the screen and core memory $\mathbf{C} < Enter >$. If the program is not running yet, type *dynamics* $< Enter >$ followed by $H < Enter >$ to get the Main Menu for the Henon map.

We assume that RHO has been set to be 4.2, C1 has been set to be 0.3, the X Scale has been set to run from -3 to 3, and the Y Scale has been set to run from -3 to 3 (see Example 2-19a for the commands).

1. Since our objective is to plot the unstable manifold of the fixed point on the chaotic saddle, we want to approximate the fixed point on the chaotic saddle by using the Newton-like method. Enter the command Q5 to iterate the Quasi-Newton method 5 times starting at y1:

$$\mathbf{Q5} < Enter >$$

Very rapidly the program produces an approximation of the fixed point which is approximately (1.729, 1.729). Actually, the manifold routines automatically seek a fixed point, but this allows us to ascertain which fixed point is found.

2. Fetch the Unstable and stable manifold Menu:

$$UM < Enter >$$

Plot the Unstable manifold, Left side:

$$\mathbf{UL} < Enter > \ < Enter >$$

and after a while, plot the Unstable manifold, Right side:

$$\mathbf{UR} < Enter > \ < Enter >$$

The computer computes the part of the unstable manifold in which we are interested. The resulting plotted picture will be similar to the one in the upper right window of Figure 2-19.

3. Hit $< Esc >$ a few times to return to the Main Menu.

Example 2-19c: The stable manifold of a fixed point

The purpose of this example is to produce a part of the stable manifold of the fixed point that is on the chaotic saddle of the Henon map (H) with parameters and scales as in Example 2-19a.

Clear the screen and core memory $C < Enter >$. If the program is not running yet, type *dynamics* $< Enter >$ followed by $H < Enter >$ to get the Main Menu for the Henon map.

We assume that RHO has been set to be 4.2, C1 has been set to be 0.3, the X Scale has been set to run from -3 to 3, and the Y Scale has been set to run from -3 to 3 (see Example 2-19a for the commands).

1. Since our objective is to plot the stable manifold of the fixed point on the chaotic saddle, we want to approximate the fixed point on the chaotic saddle by using the Newton-like method. Enter the command Q5 to iterate the Quasi-Newton method 5 times starting at y1:

$$Q5 < Enter >$$

and very rapidly the program produces an approximation of the fixed point which is approximately (1.729, 1.729).
Note. Q1, Q2, ..., Q999 are permitted.

2. Fetch the Unstable and stable manifold Menu:

$$UM < Enter >$$

Plot the Stable manifold, Left side:

$$SL < Enter > \ < Enter >$$

and after a while plot the Stable manifold, Right side:

$$SR < Enter > \ < Enter >$$

The resulting plotted picture will be similar to the one in the lower left window of Figure 2-19.

3. Hit $< Esc >$ a few times to return to the Main Menu.

The intersection of the unstable manifold of the fixed point on the chaotic saddle and the stable manifold of this fixed point (of which the relevant parts are presented in the upper right window and lower left window of Figure 2-19, respectively) is the chaotic saddle shown in upper left window of this figure.

Example 2-19d: Saddle Straddle Trajectory, and manifolds

The purpose of this example is to show how to create the three pictures of the Examples 2-19a, 2-19b, and 2-19c in one figure.

We assume you have cleared the screen and core memory ($C<Enter>$), and you are faced with the Main Menu for the Henon map. If not, type *dynamics*$<Enter>$ followed by $H<Enter>$ to get this Main Menu.

1. Fetch the Window Menu

$$WM<Enter>$$

and Open window W1:

$$OW<Enter>$$

$$\mathbf{OW1}<Enter><Enter>$$

2. Carry out the steps in Example 2-19a.

3. Open window W2:

$$OW<Enter>$$

$$\mathbf{OW2}<Enter><Enter>$$

4. Carry out the steps in Example 2-19b.

5. Open window W3:

$$OW<Enter>$$

$$\mathbf{OW3}<Enter><Enter>$$

6. Carry out the steps in Example 2-19c.

7. To have a superposition of the Saddle Straddle Trajectory, the unstable manifold, and the stable manifold in window W4, all the computations have to be redone. Open window W4:

$$OW<Enter>$$

$$\mathbf{OW4}<Enter><Enter>$$

8. Carry out the steps in the Examples 2-19a, 2-19b, and 2-19c. The resulting picture is similar to Figure 2-19.

Note. If you have made a picture in a window, there is no command in the program to copy this picture to another window.

Example 2-20: The basin of attraction of infinity (moderately difficult)

The purpose of this example is to generate a picture of the basin of attraction of the attractor at infinity of the Henon map (H)

$$(X,Y) \rightarrow (2.12 - X*X - 0.3*Y, X).$$

The parameters RHO and C1 are by default 2.12 and -0.3. In this example, a small box will be plotted if the the trajectory starting in its center exits the enlarged screen area within 1000 iterates, and it will not be plotted if the trajectory of its center converges to the chaotic attractor within 1000 iterates. See Chapter 7 for a discussion of this computation.

Computing the basin of infinity discussed in Example 2-10 for a 720 by 720 grid and MC = 60 (default) but now utilizing window W0 (which is the whole screen), we find that if we enter the "CouNT" command CNT, we get that 203028 pixels being 39.16% of the total number are not colored. Hence, 60.84% of the trajectories of the centers of the small boxes (pixels) are diverging. The goal of this example is to find out if there is a significant difference when we increase MC to 1000. We could simply set MC to be 1000 and follow the procedure in Example 2-10, but then the trajectories that do not diverge will be iterated 1000 times, and the total computation time will increase by about a factor of 10. The method below avoids this difficulty.

Type *dynamics* <*Enter*> followed by *H*<*Enter*> to get the Main Menu for the Henon map. For the default set of parameters of the map, there are two attractors, namely the chaotic attractor that is outlined by the trajectory (see also Example 2-2a), and the "attractor at infinity". To plot the basin of infinity, that is to block out the region of initial points of which the trajectories go to infinity, carry out the following steps.

1. Plot the trajectory to find out where the Henon attractor is:

T <*Enter*> <*Enter*>

2. Move the small cross (the location of y1) to a representative point of the attractor (either by the mouse or the arrow keys).

3. If an initial point comes close to any of the storage vectors y5, y6, or y7, the routine BAS stops iterating it and the initial box is not plotted. The phrase "close to" means within a RAdius RA which has a default value of 0.1. These vectors did not enter into the calculation in the previous example because they were not set (see Chapter 7 for more details). To specify the storage vector y5, replace y5 by y1 (that is, replace the coordinates of y5 by those of y1) by entering "Replace Vector" command RV followed by entering '5' and '1':

RV <*Enter*>   **5**<*Enter*>   **1**<*Enter*>

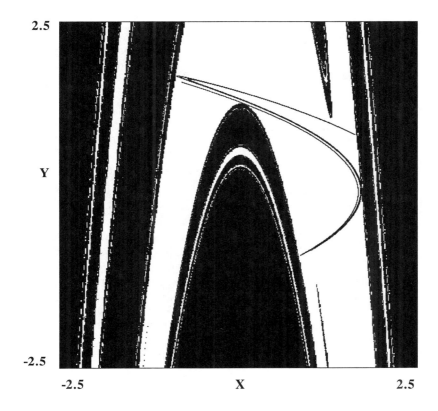

2.5

Y

-2.5

-2.5   X   2.5

Figure 2-20: The basin of attraction of infinity and a chaotic attractor
This figure shows the basin of attraction of the attractor at infinity and the basin of the chaotic attractor for the Henon map (H)
$$(X, Y) \;\longrightarrow\; (2.12 - X*X - 0.3*Y, \; X)$$
The black area is the basin of attraction of the attractor at infinity, that is, the set of points whose trajectories diverge. The white area is the basin of attraction of the chaotic attractor; the chaotic attractor is superimposed.

Topic of discussion. *As RHO is increased to the* **crisis** *value (RHO ≈ 2.1247), the attractor collides with the basin boundary simultaneously in a countable set of points which include one periodic orbit. Find that orbit (which has period 3), describe the collision set, and describe the dynamics on the collision set.*

Also consider the case C1 = 0.3 and RHO ≈ 1.4269.

4. The default value for RA is 0.1 (see the "Radius of Attraction" command RA). An extra test can be made to be sure that the disk of radius RA centered at the point y5 is in the basin of the chaotic attractor. Set MI to say 10 using the "Maximum number of Iterates" command MI:

$$MI < Enter > \quad 10 < Enter >$$

Enter now the "circle and its Iterates" command OI

$$OI < Enter >$$

The routine OI now plots a circle of radius RA about y1 and the first 10 iterates of points on the circle. These iterates should be spread out along the attractor. If it seems that some of the trajectories starting on the circle are diverging, that is, leaving the vicinity of the attractor, then either RA should be made smaller, or preferably, y5 should be replaced by another point near the attractor, for retesting. (y1 = y5 = (0.38,1.23) works fine.) The circle drawing routine is terminated by hitting $<Esc>$ and then hitting $<Esc>$ again will return the program to the Main Menu.

5. Clear the screen:
$$C < Enter >$$

6. Enter the "Basin of attraction Menu" command BM:
$$BM < Enter >$$
Notice that storage vector y2 is NOT SET, while y5 is set.

7. Set the "Maximum Check" MC to be 1000:

$$MC < Enter > \quad 1000 < Enter >$$

8. Retrieve the Basin of attraction Menu BM and enter the "BASin" command BAS:
$$< Enter >$$

$$BAS < Enter > < Enter >$$

which should cause the basin of infinity to be plotted. A 100 by 100 grid of points will be tested. The routine determines if the trajectory through each of those points diverges, that is, goes more than SD diameters from the screen within 1000 iterates, in which case a small box at the point will be plotted. Then the next grid point is tested. The resulting picture is similar to Figure 2-20.

9. Fetch the Vector Menu

$$VM < Enter >$$

Get the Y Vectors:

$$\mathbf{YV} < Enter >$$

and verify that ye is NOT SET. This means that the trajectory of the center of each small box that has been tested either diverges to infinity, or converges to the chaotic attractor within 1000 iterates.

10. Fetch the Color Menu:

$$CM < Enter >$$

and enter the "CouNT" command:

$$\mathbf{CNT} < Enter >$$

and the reply (for a 720 by 720 resolution) is that 39.16% of the pixels are not colored. Hence, a fraction of 60.84% of the trajectories of the centers of the small boxes (pixels) are diverging. Compare this with result for the default setting (MC = 60, and y5 is NOT SET) mentioned at the beginning of this example. Hence, in this example MC = 60 is a fairly good value. See Chapter 7 for more details.

11. To Print the Picture on paper, make sure the printer is on, select the printer you have, and enter the command PP ($PP < Enter >$). After the picture is printed, hit $< Esc >$ to return to the Color Menu.

Note. The resolution of the picture that appears on the screen, is low. By invoking a command like RH (High Resolution) followed by the command BAS, generates a picture of high resolution. The computations for plotting this basin take more time than for the basin of Figure 2-10.

If y5 is not set or if it is not near the attractor, then a warning will be printed that a trajectory has been iterated MC times (which is the maximum number of iterates), provided MC (Maximum Check) is set to be at least 100. The default value of MC is 60, and in the default case such warnings will not be printed on the screen. If MC is set to be at least 100, and if a point has been iterated MC times without going to an attractor, then the final point of that trajectory will be stored in ye for later investigation. The vector ye might be on an attractor that you did not know existed.

Note. Command TD stores the picture on disk for later use. Use root name H for simplicity.

Example 2-21: A trajectory on a basin boundary (moderately difficult)

There are two basins of attraction with the default parameter values of the Henon map, the basin of the chaotic attractor and the basin of the attractor at infinity. In this example, we want to produce a numerical trajectory that is on the common boundary of these two basins (see also Examples 2-2a and 2-20). In fact, we will produce a trajectory that neither goes to infinity nor enters a disk of radius 0.3 centered at some appropriate point on the chaotic attractor, and every point of the trajectory will lie on the boundary between the two basins. This trajectory we are looking for is called a **Basin boundary Straddle Trajectory**. See Chapter 8 for a more detailed explanation. If you use T, you cannot get a trajectory on the common boundary of the two basins. Therefore, we have a special routine, that routine is BST.

Type *dynamics* < *Enter* > followed by *H* < *Enter* > to get the Main Menu for the Henon map.

1. Plot the trajectory:

$$\textbf{T} < Enter > \ < Enter >$$

2. Two basins must be specified in order for the trajectory to lie on the boundary. The points whose orbits diverge are one basin. We now specify the other. We must specify that any point on the attractor should be avoided by the trajectory. The BST trajectory will avoid storage vectors y2 through y7 if they are set. We set y5 to be a point on the attractor.

While the program is still plotting, we must specify a point on the attractor: use the arrow keys to move the small cross (y1) to a point on the attractor. (Alternatively, move the mouse and the mouse arrow will appear. Move it to a point on the attractor and click the left button. This sets y1 to be this point on the attractor.)

Fetch the Vector Menu:

$$VM < Enter >$$

and then Replace the coordinates of Vector y5 (which is NOT SET) by those of this point:

$$\textbf{RV} < Enter > \quad 5 < Enter > \quad 1 < Enter >$$

The vector y1 is unaffected by this. Get the Y Vectors:

$$\textbf{YV} < Enter >$$

and verify that y5 equals y1 (we have y5 $\approx$ (1.12,0.82)). The routine BST automatically checks to see if trajectories of points diverge. It would be best to check that all points coming within RAdius RA of this point are in the basin of the chaotic attractor.

3. Since we have specified y5, the Basin boundary Straddle Trajectory will stay away from that point. We must specify how big the avoidance RAdius is. The default value is 0.1. The routine assumes any point whose trajectory comes within RA of y5 as being in a basin. While the default value would be acceptable, the value 0.3 makes the routine run slightly more quickly.

Fetch the Basin of attraction Menu and set the radius RA to be 0.3:

$$BM < Enter >$$

$$RA < Enter > \quad 0.3 < Enter >$$

4. To check that the disk with radius 0.3 centered at y5 is in the basin of the chaotic attractor, first either use command FD to read in From Disk the picture you made in Example 2-20 or plot the basins using command BA (see Example 2-9).

Assuming you have the basins on the screen, now enter the circle command:

$$O < Enter > \ < Enter >$$

and verify that the disk centered at y1 (which equals y5) is in the basin of the chaotic attractor. If it is not, move the small cross using the arrow keys and repeat the procedure. Hit $< Esc >$ to terminate the routine O.

Clear the screen:

$$C < Enter >$$

5. Call the Straddle Trajectory Menu STM (which is in the Numerical Explorations Menu):

$$STM < Enter >$$

Unlike a regular trajectory that is initiated with a single vector, a Basin boundary Straddle Trajectory requires two vectors, ya and yb.

It is necessary to choose ya in one basin and yb in the other. The BST routine moves ya and yb so that they are close together and follow along with the BST trajectory. There are two basins, the basin of the chaotic attractor and the basin of infinity. Next, set ya to be a point whose trajectory is in one basin and yb is in the other. For simplicity, Replace Vector yb by y5:

$$RV < Enter >$$

$$B < Enter > \quad 5 < Enter >$$

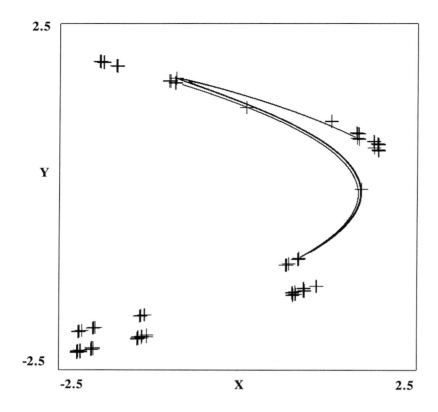

Figure 2-21: A Basin boundary Straddle Trajectory

This figure shows a Basin boundary Straddle Trajectory of the Henon map (H)

$$(X, Y) \rightarrow (2.12 - X*X - 0.3*Y, X)$$

The Basin boundary Straddle Trajectory, which is on the common boundary of the basin of the chaotic attractor and the basin of the attractor at infinity, is marked with crosses. The permanent cross command KKK was entered before plotting the Basin boundary Straddle Trajectory. The chaotic attractor is superimposed.

Topic of discussion. *The basin of infinity is very close to the attractor at some points. See Figure 2-20. In this case, do any of the crosses lie on the attractor? Appearance can be deceiving.*

The point y5 must in principle be chosen with some care (see step 4 above). In particular, the ball of radius RA centered at y5 must lie in a single basin for this routine to be guaranteed to work. If some of the points in that ball were initial points of trajectories diverging, the procedure could fail. In addition, this ball must cross the attractor because we want it chosen so that typical trajectories on the Henon attractor pass through the ball repeatedly.

6. Move the small cross using the arrow keys to a point whose trajectory diverges. For this case, see also the basin of infinity in Figure 2-20. To check that y1 (the position of the small cross) is in the basin of infinity, enter the "Where and When" command WW:

$$WW < Enter >$$

The program will report "when and where" the trajectory through y1 goes. For example, if y1 = (-2.5,-2.5) the output for WW is

> Trajectory through small cross diverges in 2 iterates

You can check Figure 2-20 for a plot of the basin of infinity, and in some cases one might resort to command BA to plot the basins. If y1 diverges, set ya to be that point:

$$RV < Enter >$$

$$A < Enter > \quad 1 < Enter >$$

7. Fetch the Kruis (cross) Menu:

$$KM < Enter >$$

and enter the "Kruis" command K:

$$K < Enter >$$

so that the trajectory will be highly visible; a cross will be plotted as each point is plotted and erased as the next point is plotted.

8. Retrieve the Straddle Trajectory Menu:

$$STM < Enter >$$

Now enter the "Basin boundary Straddle Trajectory" command BST:

$$BST < Enter > \ < Enter >$$

The attractor of the map has one piece; the boundary of the basin of this chaotic attractor is fractal; and the BST gives a complicated trajectory going rapidly to a Cantor set, that is, a totally disconnected closed set, none of whose points are isolated. The plotted trajectory will be similar to the one in Figure 2-21. The resulting picture is a Basin boundary Straddle Trajectory.

9. To get the resulting picture on paper, make sure the printer is on and enter the command PP:

PP *< Enter >*

and select the printer.

Hit *< Esc >* to return to the Straddle Trajectory Menu.

Example 2-22: A BST trajectory for the Tinkerbell map

The purpose of this example is to produce a Basin boundary Straddle Trajectory, which is on the common boundary of the basin of the chaotic attractor and the basin of infinity of the Tinkerbell map (T)

$$X \rightarrow X*X - Y*Y + 0.9*X - 0.6013*Y$$

$$Y \rightarrow 2*X*Y + 2.0*X + 0.5*Y$$

Type

dynamics <Enter>

T <Enter>

to get the Main Menu for the Tinkerbell map.

1. If you wish to verify that the parameters C1, C2, C3, and C4 have the default values, you can fetch the Parameter Menu:

PM <Enter>

to see their values. When you call the Parameter Menu, you will see at the top of your screen

$$X \rightarrow X*X - Y*Y + C1*X + C2*Y$$

$$Y \rightarrow 2*X*Y + C3*X + C4*Y$$

For example, C1 should be 0.9 and C2 should be -0.6013.

2. Before calling BAS, we used BRM the Basin Resolution Menu to obtain a higher resolution for BAS. The basin picture in Figure 2-22 was created with a resolution of 720 by 720. Plot the basin of infinity:

BAS <Enter> <Enter>

Now enter command CC to Clear the screen and the Core copy:

CC <Enter>

3. Enter command T to plot a trajectory:

T <Enter> <Enter>

See figure 2-22 upper right window for a resulting picture of the trajectory.

4. Move the small cross (y1) using the arrow keys or the mouse to a point on the attractor. This point should be chosen so that points in the ball of radius RA centered at y1 all go to the attractor. Recall that from basin calculations (see the upper left window of Figure 2-22) that inside the region surrounded by the trajectory, there are some small regions of points whose trajectories go to infinity. Hit <*Esc*> to terminate routine T.

5. Fetch the Vector Menu and Replace vector y5 by y1:

$$VM <Enter>$$

$$RV <Enter> \quad 5 <Enter> \quad 1 <Enter>$$

6. Verify that the disk with RAdius 0.1 centered at y5 is in the basin of the chaotic attractor (see also Example 2-21), and clear the screen.

7. The vector ya has to be in the basin of the attractor, replace ya by y1:

$$RV <Enter> \quad A <Enter> \quad 1 <Enter>$$

Next, move the small cross (y1) using the arrow keys or the mouse to a point on the edge of the screen or beyond and test it (enter command WW) to be sure that trajectory diverges. Then replace yb by that point y1:

$$RV <Enter> \quad B <Enter> \quad 1 <Enter>$$

Hence, one of the points ya and yb diverges while the other approaches y5.

8. Enter command K (to plot temporary crosses along the trajectory):

$$K <Enter>$$

9. Fetch the Straddle Trajectory Menu and plot the Basin boundary Straddle Trajectory:

$$STM <Enter>$$

$$BST <Enter> <Enter>$$

It may help to clear the screen (command CC) or change the color number (function keys <F7> or <F8>) to increase the visibility of the trajectory or use command KKK to plot a permanent cross at each point. Commands K and KKK are toggles, so entering them again turns them off. See figure 2-22 for a resulting picture of (a part of) the Basin boundary Straddle Trajectory. If the command CC was not invoked before entering command BST, then the picture that will appear on the screen is a superposition of the trajectory (upper right) and straddle trajectory (lower left window) of figure 2-22.

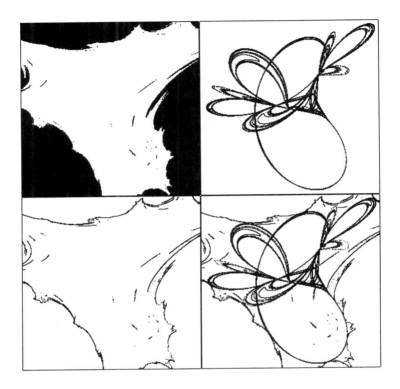

Figure 2-22: Basin, chaotic attractor, and straddle trajectory

This figure shows the basin of infinity, a chaotic attractor, and a Basin boundary Straddle Trajectory of the Tinkerbell map (T)

$$(X, Y) \rightarrow (X*X - Y*Y + 0.9*X - 0.6013*Y, \ 2*X*Y + 2*X + 0.5*Y)$$

There are two attractors: a bounded chaotic attractor and a point at infinity. In the upper left window, the black area is the basin of the attractor at infinity. The white area contains a chaotic attractor which is shown in the upper right window. The lower left window contains a Basin boundary Straddle Trajectory on the common boundary of the basin of the chaotic attractor and the basin of the attractor infinity. The lower right window contains both the chaotic attractor and a Basin boundary Straddle Trajectory.

In all the four windows, the variable X is plotted horizontally and the variable Y is plotted vertically. The X Scale is from -1.3 (left) to 0.5 (right) and the Y Scale is from 0.6 (bottom) to -1.6 (top). This choice of the Y Scale illustrates that the first number in a scale can be larger than the second.

Example 2-23: Lyapunov exponent bifurcation diagram

The purpose of this example is to plot the Lyapunov exponent bifurcation diagram for the CIRCle map (CIRC)

$$X \rightarrow RHO + X + \sin(X) \mod 2\pi$$

where RHO varies between 0 and 6.28319. The Lyapunov exponent is plotted horizontally, and the parameter RHO is plotted vertically. The X Scale is from -2 to 0.5.

Caution: the computations for plotting Lyapunov bifurcation diagrams (and in particular this example) will take several hours on some PC's.

Start the program and enter CIRC to get the CIRCle map:

dynamics $<Enter>$ **CIRC** $<Enter>$

and the Main Menu for the CIRCle map appears on the screen.

1. Fetch the Lyapunov Menu:

$LM<Enter>$

Set the number of Lyapunov exponents to be computed to be 1:

L $<Enter>$ 1 $<Enter>$

2. Call the Parameter Menu:

$PM<Enter>$

Set the parameter C1 to be 1:

C1 $<Enter>$ 1 $<Enter>$

Set the X Scale to range from -6 to 1:

XS $<Enter>$ **-6 1** $<Enter>$

3. The Lyapunov exponents may assume huge negative values. Therefore, the SD must be set to some big number. Retrieve the Parameter menu and set the Screen Diameter to be 1000:

$<Enter>$

SD $<Enter>$ **1000** $<Enter>$

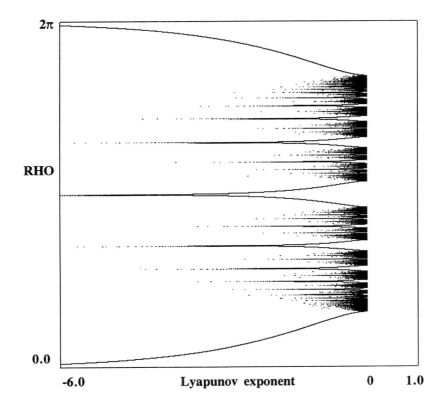

Figure 2-23: A Lyapunov exponent bifurcation diagram
This figure shows a Lyapunov exponent bifurcation diagram for the CIRCle map (CIRC)

$$X \rightarrow RHO + X + sin(X) \quad mod \ 2\pi$$

The diagram was printed using command BIFS. The settings are:
BIFI = 0, BIFPI = 5000, BIFD = 500, and BIFV = 200000.

There is a considerable fine structure in this picture which is why 200,000 values were used instead of the default value, which is 480 for computers with VGA graphics board. This picture with this detail takes hours on the fastest PC's. Lyapunov exponents are calculated during all BIFPI and BIFD iterates, and the trajectories are initialized with each new RHO value.

Topic of discussion. *Notice that the derivative of this map is 0 at one value of X and is positive elsewhere. What causes the prominent daggers in this figure?*

---

4. Get the BIFurcation diagram Menu:

BIFM < Enter >

The "PaRaMeter" command PRM does not need to be invoked, since PRM is RHO by default. Set the BIFurcation parameter Range of RHO to vary from 0 to 6.28319:

BIFR *< Enter >*   **0 6.28319** *< Enter >*

5. Retrieve the BIFurcation Menu:

< Enter >

Set the length of the transient time interval of the exponents to be 100:

BIFPI *< Enter >*   **100** *< Enter >*

6. Retrieve the BIFurcation Menu:

< Enter >

Set the the number of dots to be plotted to be 10:

BIFD *< Enter >*   **10** *< Enter >*

7. Retrieve the BIFurcation Menu and set the number of BIFurcation Values to be 600:

< Enter >

BIFV *< Enter >*   **600**

8. Plot the Lyapunov exponent bifurcation diagram:

BIFS *< Enter > < Enter >*

Compare the resulting picture with Figure 2-23.

Note. As with many calculations, it will be unclear how the resulting figure depends on parameters like BIFPI, BIFD, and BIFV, so you are advised to try other values to see if the picture changes significantly. Compare your picture with figure 2-23, which was obtained using much higher values for BIFPI and BIFD.

Example 2-24: Chaotic parameters

The purpose of this example is to generate a set in the parameter space such that each point in this set the trajectory through y1 is chaotic (that is, it has a positive Lyapunov exponent) for the CIRCle map (CIRC)

$$X \rightarrow RHO + X + C1*\sin(X) \mod 2\pi$$

where RHO varies between 0 and 6.28319, and C1 varies between 0 and 8.

Caution: the computations for plotting chaotic parameters will take many hours on some PC's.

Start the program and enter CIRC to get the CIRCle map:

dynamics *< Enter >*   **CIRC** *< Enter >*

and the Main Menu for the CIRCle map appears on the screen.

1. Fetch the Lyapunov Menu:

LM < Enter >

Set the number of Lyapunov exponents to be computed to be 1:

L *< Enter >*   **1** *< Enter >*

2. Call the Parameter Menu:

PM < Enter >

Set the X COordinate to be RHO:

XCO *< Enter >*   **RHO** *< Enter >*

Set the Y COordinate to be C1:

YCO *< Enter >*   **C1** *< Enter >*

3. The horizontal and vertical scales are by default from 0 to 6.28319, so only the vertical one has to be changed. Retrieve the Parameter Menu:

< Enter >

Set the Y Scale to range from 0 to 8:

YS *< Enter >*   **0 8** *< Enter >*

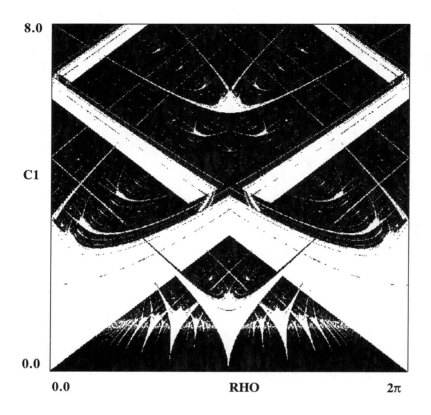

Figure 2-24a: Arnol'd tongues

The black area is the set of parameter values (RHO,C1) for which the trajectory through y1 = (0,0) has a nonnegative Lyapunov exponent under iteration of the CIRCle map (CIRC)

$$X \rightarrow RHO + X + C1*sin(X) \quad mod \ 2\pi$$

In fact, points are plotted only when the approximate Lyapunov exponent of 1000 iterates of (0,0) is at least -0.002. (We used the "Basin Lyapunov Minimum value" command BLM and set it to be -0.002, and set MC to be 1000.) **Arnol'd tongues** *are connected regions in the above diagram where a periodic orbit is attracting. This picture shows Arnol'd tongues (the white "triangles" in the lower part of the figure).*

Topic of discussion. *Find the period of the stable periodic orbit for each of several white regions for which C1 < 1.*

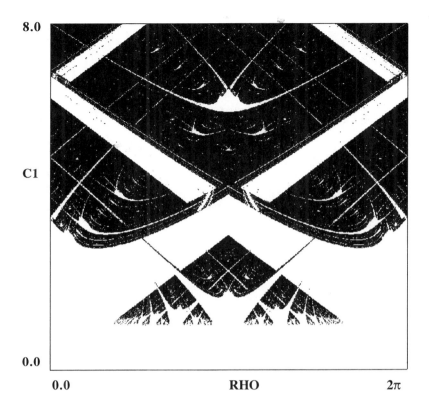

8.0

C1

0.0

0.0   RHO   2π

Figure 2-24b: The chaotic parameter set

The black area is the set of parameter values (RHO,C1) for which the trajectory through y1 = (0,0) has a positive Lyapunov exponent under iteration of the CIRCle map (CIRC)

$$X \rightarrow RHO + X + C1*sin(X) \quad mod \ 2\pi$$

It is called a **Chaos plot***. In fact, points are plotted only when the approximate Lyapunov exponent of 1000 iterates of (0,0) is at least 0.01. (We used the "Basin Lyapunov Minimum value" command BLM and set it to be 0.01, and set MC to be 1000.)*

 On a color screen, the black region is plotted in a variety of colors, different colors for different magnitudes of the approximate Lyapunov exponent.

 Topic of discussion. *Determine the dynamical meaning of some of the various white bands. That is, describe what happens in those regions.*

4. Get the Y Vectors:

YV *<Enter>*

and verify that y1 is set to 0 (the default value of y1).

5. The objective is to plot (RHO,C1)-values for which the trajectory through y1 has a positive Lyapunov exponent. For this example, our criterion is to plot the point (small box) in parameter space if the approximate Lyapunov exponent is positive after 200 iterates of y1. Fetch the Basin of attraction Menu:

BM <Enter>

Set the Maximum Check to be 200:

MC *<Enter>*   **200** *<Enter>*

Choosing MC to be 1000 as in Figure 2-24 is preferable, but the computation time is 5 times as long.

6. We choose to have a grid of High Resolution to be tested. Retrieve the Basin of attraction Menu:
<Enter>

Get the Basin Resolution Menu and set the resolution to be the High Resolution:

BRM <Enter>

RH *<Enter>*

7. Retrieve the Basin Resolution Menu:

<Enter>

hit

<Esc>

to get the parent menu, the Basin of attraction Menu. Now enter the BAS command to plot the chaotic parameter set:

BAS *<Enter>* *<Enter>* *<Enter>*

The resulting picture will be a mixture of Figure 2-24a and Figure 2-24b.

Example 2-25: Box-counting dimension of attractors and chaotic saddles

The **box-counting dimension** of a picture on your computer screen is a number. If the picture is a single point its box-counting dimension is 0. Any line segment has box-counting dimension 1. A square or a disk has box-counting dimension 2. See the DIMension menu DIM.

The Box Dimension routine BD examines a large box in the center of your screen to estimate the box-counting dimension of the picture on your screen. It checks to see that some of your picture is in this box. Next, it divides this box into 4 smaller boxes and counts how many of these contain points of your picture. Then it repeats the procedure several times, always dividing each box into 4 boxes and counting the number of the smaller boxes containing points of your picture. If the number of boxes counted increases by an average factor of 2^D when the box size is halved, then the box-counting dimension is equal to D. In the plane the box-counting dimension of a set will never exceed 2. For more details, see Chapter 12.

You can compute an estimate of the box-counting dimension of any chaotic attractor or chaotic saddle you have created. For example, the Henon attractor in Example 2-4, or the chaotic saddle in Example 2-19.

The purpose of this example is to compute the box-counting dimension of the chaotic attractor of the time-2π map of the forced damped Pendulum differential equation (P)

$$X''(t) + 0.1*X'(t) + \sin X(t) = 0.15 + 1.5*\cos(t))$$

Start the program by typing

dynamics *< Enter >*   **P** *< Enter >*

to get the Main Menu for the forced damped Pendulum. The parameters RHO, C1, C2, C3, and PHI are by default 2.5, 0.2, 1, 0, and 1, respectively.

1. Set the parameter RHO to be 1.5:

RHO *< Enter >*   **1.5** *< Enter >*

2. Set the parameter C1 and C3 to be 0.1:

C1 *< Enter >*   **0.1** *< Enter >*

C3 *< Enter >*   **0.1** *< Enter >*

4. Since we are interested in the box-counting dimension of an attractor, set the number of Pre-Iterates to 1000:

PI *< Enter >*   **1000** *< Enter >*

4. Plot the trajectory:

$$T < Enter > < Enter >$$

A resulting picture of the chaotic attractor is given in Figure 2-25.

5. Fetch the Numerical Explorations Menu and get the Menu for estimating DImension:

$$NEM < Enter >$$

$$DIM < Enter >$$

Compute the box-counting dimension of the attractor:

$$BD < Enter >$$

After some computing, the box-counting dimension estimate of the attractor appears on the screen.

The routine T continues to plot points. It is important to get a detailed picture as shown in Figure 2-25 in order to reduce the error in the dimension. You can continue to enter BD occasionally. The error will become smaller as the picture is plotted in greater detail.

The routine BD works with less resolution if the size of the core is 640 by 480 (see the Size of Core Menu SCM for values of HIGH and WIDE). You may wish to use command SQ or SQ5 before creating the picture.

Note. The DImension Menu has three different ways of estimating dimension. All definitions of dimension require various limiting procedures. None of these can be implemented on a computer so that we can only obtain estimates. For more details, see Chapter 5 and Chapter 12.

Note. A dimension often used is the "Correlation Dimension". This involves the frequency with which the trajectory enters various regions. You can obtain a Correlation Dimension estimate as follows by entering command CD. There is a slightly better approach. First, clear the screen and enter the command IN which plots the trajectories incrementally. The first time a screen point is plotted with IN invoked, it will plot with color number 1 (that is usually blue). Points on the screen (or in the core picture) are called "pixels" for picture element. The second time that a pixel is hit, it will be colored color number 2.

Command CT will display the screen colors and their color numbers. Then while a trajectory is plotting, enter CD again. If you want to see how many pixels have been assigned each color, use the CouNT command CNT. The command IN makes no difference for the box-counting dimension.

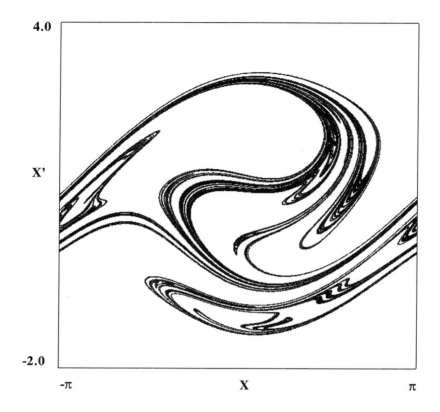

4.0

X'

-2.0

-π   X   π

Figure 2-25: Box-counting and Lyapunov dimension for an attractor
*This figure shows the chaotic attractor of the time-2π map of the forced
damped Pendulum differential equation (P)*

$$X''(t) + 0.1*X'(t) + \sin X(t) = 0.15 + 1.5*\cos(t))$$

*The attractor is chaotic, since one of the Lyapunov exponents of the
trajectory is positive. (Use command L to compute the Lyapunov exponents
while plotting a trajectory; see Example 2-2b.)*

*The program has 3 methods for computing dimension; see Chapters 5 and
12 for details. For an attractor, the Lyapunov dimension is determined from
the Lyapunov exponents, and is estimated to be 1.588 for this attractor
(the number of dots plotted is about 4×10^7). For this picture (SIZE = 16
1024 1024), the routine BD gives 1.626 ± 0.043 as an estimate of the box-
counting dimension and the routine CD gives 1.311 ± 0.046 as an estimate
for the correlation dimension. The box-counting dimension is usually
greater than the Lyapunov dimension, and similarly the Lyapunov dimension
is usually greater than the correlation dimension.*

Example 2-26: Zooming in on the attractor of the Tinkerbell map

There is a chaotic attractor for the default parameter values of the Tinkerbell map (see also Example 2-22). In this example, we want to present a simple procedure for zooming on toward some point of the attractor of the Tinkerbell map (T) with the default parameter values:

$$X \rightarrow X*X - Y*Y + 0.9*X - 0.6013*Y$$

$$Y \rightarrow 2*X*Y + 2.0*X + 0.5*Y$$

Type

dynamics *< Enter >*

T *< Enter >*

to get the Main Menu for the Tinkerbell map.

1. If you wish to verify that the parameters C1, C2, C3, and C4 have the default values, you can fetch the Parameter Menu:

PM < Enter >

to see their values.

2. Enter command T to plot a trajectory:

T *< Enter > < Enter >*

See figure 2-26 upper left window for a resulting picture.

3. The purpose is to zoom in at a point on the attractor shown in the upper left window of Figure 2-26. The routine for zooming in uses the vector y1 as its focus point. Therefore, y1 has to be set to that point on the attractor at which you want to zoom in. After a while, Pause the program:

P *< Enter >*

Fetch the Vector Menu and Replace Vector y1 by vector y0:

VM < Enter >

RV *< Enter >* 1 *< Enter >* 0 *< Enter >*

In this example, we have y1 ≈ (-0.386, -0.479). This has a shorthand command I1.

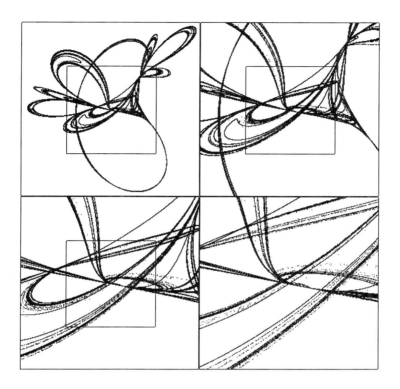

Figure 2-26: Zoom in toward a point on the chaotic attractor
This figure shows some consecutive pictures resulting from zooming in toward a point on the chaotic attractor of the Tinkerbell map (T)

$$(X, Y) \ \rightarrow \ (X*X - Y*Y + 0.9*X - 0.6013*Y, \ 2*X*Y + 2*X + 0.5*Y)$$

In the upper left window, the chaotic attractor and the box that would be used by ZI are shown. The upper right window displays the trajectory restricted to the small box of the upper left window together with the box that will be used when zooming in a second time. In the lower left window, the trajectory restricted to the small box of the upper right window together with the box that will be used when zooming in a third time are shown. The lower right window contains the trajectory restricted to the small box of the lower left window.

In all the four windows, the variable X is plotted horizontally and the variable Y is plotted vertically. In the upper left window, the X Scale is from -1.3 (left) to 0.5 (right) and the Y Scale is from 0.6 (bottom) to -1.6 (top).

4. Fetch the ZOOm Menu:

ZOOM < Enter >

Verify that the default of the Zoom Factor is 2. In this example, we use the default of the Zoom Factor ZF.
Draw the box that would be used by the Zoom In routine (see upper left window in Figure 2-26):

ZB *< Enter >*

5. Hit

< Enter >

to return to the ZOOm Menu and Zoom In towards the small cross (which is at y1) by a factor ZF:

ZI *< Enter >*

Hit

< space bar >

to continue plotting of the trajectory.

6. After a while, hit
< Enter >

to retrieve the ZOOm Menu and draw the box that would be used by the Zoom In routine (see upper right window in Figure 2-26):

ZB *< Enter >*

Hit

< Enter >

to return to the ZOOm Menu and Zoom In towards the small cross:

ZI *< Enter >*

7. Repeat Step 6 as many times as you wish.

Figure 2-26 shows some pictures resulting from zooming in three consecutive times.

Example 2-27: Period plot in the Mandelbrot set

The purpose of this example is to generate a "period plot" in the Mandelbrot set which was created in Example 2-14. A **period plot** is the set of parameter values (C5,C6) such that the trajectory through (0,0) is approaching a periodic orbit under iteration of the complex quadratic map.

Start the program and enter C to get the Cubic map:

$$\textbf{dynamics} < Enter > \quad \textbf{C} < Enter >$$

and the Main Menu for the Cubic map appears on the screen. With the default values RHO = 0, C2 = C3 = C4 = 0, this map reduces to the quadratic map

$$Z \;\rightarrow\; Z*Z \,+\, C5 + iC6$$

with Z = X + IY where X = y[0], Y = y[1]. A **period-m plot** for the complex quadratic map is the set of parameter values (C5,C6) such that the trajectory through (0,0) is approaching a periodic orbit of smallest period m under iteration of this map. The goal is to create a picture exhibiting period-m plots, where m = 1, 2, 3,..., 12. Since the Mandelbrot set is the set of complex numbers C5+iC6 for the trajectory through 0+i0 remains bounded, under iteration of the complex quadratic polynomial map, the period-m plots are contained in the Mandelbrot set. Verify that the initial point y1 has coordinates (0,0), so the initial point is as needed.

1. Change the horizontal coordinate to be the parameter C5:

$$\textbf{XCO} < Enter > \quad \textbf{C5} < Enter >$$

2. Change the vertical coordinate to be the parameter and C6:

$$\textbf{YCO} < Enter > \quad \textbf{C6} < Enter >$$

3. Fetch the Basin Menu, get the BAS Plotting scheme Menu and select the "BAS plots periodic attractor period (and chaotic attractors)" coloring option:

$$BM < Enter >$$

$$BASPM < Enter >$$

$$\textbf{BP} < Enter >$$

4. Enter command BAS:

$$\textbf{BAS} < Enter > \; < Enter >$$

and the union of period-m plots with m = 1, 2, 3, ..., 12 resembles the Mandelbrot set, or as good as an approximation as we get with a 100 by 100 grid. See figure 2-27 for the resulting picture with a high resolution and for the period-1, period-2 and period-3 plots.

For more details on period plots, see Chapter 7.

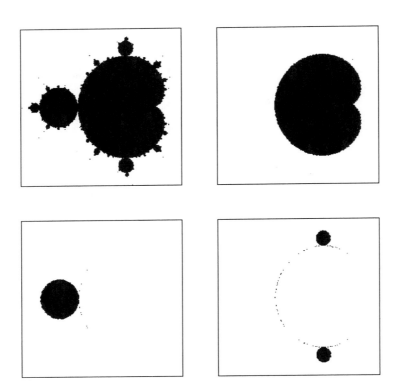

Figure 2-27: Period plots in the Mandelbrot set

The black area in the upper left window is the set of parameter values
(C5,C6) such that the trajectory through (0,0) approaches a periodic
attractor of smallest period m, where m = 1,2,3,..., 12 under iteration of
the complex quadratic map (the complex Cubic map (C) with RHO = 0)

$$Z \rightarrow Z*Z + C5+iC6$$

The white area is the set of parameter values (C5,C6) such that the
trajectory through (0,0) is unbounded under iteration of the complex
quadratic map. Compare the black area in this figure with the Mandelbrot
set in Figure 2-14. On a color screen, the black region is plotted in a
variety of colors, different colors for different periods.

The upper right window shows the period-1 plot, that is, all
parameters (C5,C6) for which the trajectory of (0,0) approaches a fixed
point attractor. The lower left window shows the period-2 plot, and the
lower right window shows the period-3 plot.

The period-m plot is the resulting picture of the picture in the upper
left window, by deleting all colors except color m. Before creating this
picture, the resolution command RH was invoked, and MC was set to be 1000.

APPENDIX

COMMAND FOR PLOTTING A GRAPH

G

The "plot Graph" command G plots the graph of 1-dimensional functions. The routine G calculates the function value for a fixed number of points (see WIDE in the Size of Core Menu to find this value).

If the Iterates Per Plot IPP is changed from 1 to 3, for example, then G will graph the third iterate of the function. You may wish to CONnect the consecutive dots by first entering the command CON. See command CON in Chapter 4. Enter CON again after creating the graph to turn that feature off. This command is **not in the menus**.

COMMANDS FROM THE NUMERICAL EXPLORATIONS MENU

T

The "Trajectory" command instructs the computer to plot the dots of a trajectory. The maximum number is specified by the command DOTS. Usually DOTS is set to such a large value that it will not be reached. Use X COordinate and Y COordinate commands XCO and YCO to specify which coordinates will be plotted against which horizontally and vertically. See XCO. Usually XCO and YCO are set to be the space coordinates of the process. For example, in Henon the default setting of XCO is 0, which means y[0] is plotted horizontally, and YCO is 1 which means y[1] is plotted vertically. However as shown below XCO and/or YCO can be set to be parameters of the map.

You can specify that one of the axes is a parameter of the system, for example the horizontal axis could be for the parameter RHO (if that is one of the parameters of the currently selected process), and then RHO would be varied over a range specified using the "X Scale" command XS. Of course, the trajectory would then lie on a single line corresponding to that parameter value. The sequence of commands

XCO *<Enter>*   **RHO** *<Enter>*

YCO *<Enter>*   **1** *<Enter>*

says that XCO is RHO, that is, RHO will be plotted horizontally, that is the X coordinate, while coordinate 1 of the state vector will be plotted vertically.

Another possibility is to plot one coordinate against itself, in which case the current value of the coordinate is plotted vertically while the horizontal coordinate is the **previous** value (the value from the previous dot).

After entering the Trajectory command T, an options menu appears. One of the options is to follow a number (TN) of trajectories simultaneously. Usually routine T plots a single trajectory, but if the Trajectory Number TN is greater than 1, the routine T plots TN trajectories simultaneously. If you set TN to be 100 for example, 100 initial points will be chosen. See commands TN, TNB, and TNE for more information.

DYN

The command DYN is used to quit and to restart *Dynamics*. This command allows selection of another differential equation or map; most parameters are reset to the default values.

OWN

The command OWN is used to quit the program *Dynamics* for creating your OWN process (map or differential equation).

P

The command P causes the program to Pause. Hit the < space bar > or enter the "Un-Pause" command UP to end pause mode. If you are going to start a process like T or UM and you want it to be paused initially, then this is the command to use. The pause initiated by < · > while plotting has a different effect: the program will automatically unpause from the < · > pause if you start a new process.

Q, QX

The command Q is for Quitting the program. Enter this command and a menu appears on the screen before exiting the program, to allow saving of the data.

The command QX is for Quitting and eXiting *Dynamics*. This command terminates the program; data that has not been saved will be lost.

UP

Command UP Undoes the Pause mode of the program. The command UP is **not listed in the menus**.

Plotting multiple trajectories simultaneously

After invoking the command T, the following "Actions, Hints and Options for Trajectories" appears on the screen:

ACTIONS, HINTS and OPTIONS for TRAJECTORIES
OK   – proceed with T
Cancel – do NOT plot trajectory
I   – Initialize trajectory at small cross
IN   – INcrement color number when plotting: OFF
P   – trajectory will initially be Paused: NOT paused
MENU for plotting trajectories simultaneously
TN   – Number of simultaneous Trajectories plotted by T: 1
TNB   – for TN > 1; now: OFF
If TNB=ON, then TN initial points are chosen in small Box;
if TNB=OFF, then they are chosen on line ya to yb
TNE   – for TN > 1; now: OFF
If TNE=ON, then each trajectory point is Erased before
the next iterate is plotted
Hit <Tab> while plotting for information update

Usually routine T plots a single trajectory, but if the Trajectory Number TN is greater than 1, the routine T plots TN trajectories simultaneously. If you set TN to be 100 for example, 100 initial points will be chosen. See commands TN, TNB, and TNE for more information.

TN

The "Trajectory Number" command TN specifies the number of trajectories that the routine T will follow simultaneously. If TN = 1, then the routine T plots a single trajectory starting from the point y[].

If TN > 1, then the routine T plots TN trajectories simultaneously; each time you enter command T, these TN initial points are chosen anew.

The initial conditions can be either on the line from ya to yb if TNB is "OFF", or if TNB is "ON", they are chosen in the "small box"; the "small box" is either the whole screen or the current window, or the small box that is selected using the mouse or the arrow keys and is drawn using the Box command B.

See also commands TNB (TN Box) and TNE (TN Erase).

TNB

The "TN Box" command TNB is a toggle; it is either "ON" or "OFF".

When TN > 1 and TNB is OFF, the trajectory command T chooses TN initial points equally spaced points are chosen on the segment from ya to yb. These points include ya and yb, and they are used as initial points for trajectories. If you set TN to be 100 for example, 100 initial points will be chosen on the line from ya to yb. (See Y Vectors command YV to see what these values are at any time. Initially they are in the lower left and upper right corners of the screen so the line is the diagonal connecting them.) Then T will follow 100 trajectories simultaneously. You can see the process more clearly if before entering the command T you first enter the Pause command P. The when T is entered, the program will be paused. Hit the period key $< \cdot >$ and all 100 will be advanced one iterate. Hit $< \cdot >$ again and all will advance one more step.

When TN > 1 and TNB is ON, the trajectory command T chooses TN points at random from a rectangle which is either the current window (or whole screen) or if the small box has been set, it chooses the initial points in that box. You can see the small box, if it has been set, using command B.

TNE

The "TN Erase" command TNE is a toggle; it is either "ON" or "OFF".

Command TNE may be useful or interesting for the case where TN > 1. When TN > 1 and TNE is ON, each of the trajectories has its current point erased just before its next trajectory point is plotted.

When using the command TNE, the refresh command has side effects and generally "Clear Screen" command CS is preferable. Recall that CS does not erase the core picture.

CHAPTER 3

SCREEN UTILITIES

3.1 BASIC SCREEN FEATURES (Screen Menu SM)

```
SM        – –      SCREEN MENU

C        –  Clear current window
CC       –  Complete Clear
CS       –  Clear Screen (but not core copy)
R        –  Refresh screen
ROT      –  ROTate square picture 90 degrees clockwise
SHIFT    –  SHIFT or flip picture
TL       –  Text Level (or commands T0,...,T4):  3
TS       –  Size of Type used with command TYPE:  1
TYPE     –  TYPE text on screen at small cross

            MENUS
AXM      –  AXes Menu
BXM      –  BoX Menu
CM       –  Color Menu
CTM      –  Color Table Menu
EM       –  Erase color Menu
KM       –  Kruis (=cross) Menu
SCM      –  Size of Core Menu
WM       –  Window Menu
ZOOM     – ·ZOOm in/out Menu
```

From Example 2-1 you know some commands to clear or refresh the screen. These and related commands are in the Screen Menu. Enter the command SM:

$$SM < Enter >$$

and the Screen Menu appears on the screen. The menus of the Screen Menu will be discussed in separate sections in this chapter.

Commands for clearing the screen

C

The "Clear" command C just clears the active window (and the core copy). That is, while plotting in a window (either one of the quadrants of the screen or the entire screen), command C clears just that quadrant. To clear the entire screen and the core copy, use command CC (Complete Clear).

● To eliminate a window, use command CC. See command CC.

CC

The "Complete Clear" command CC clears the entire screen and the core copy. If the active window is the entire screen, the commands C and CC are equivalent. See also command C.

To eliminate a window, use command CC. Note that this will clear the entire screen. All the windows will be gone but can be re-entered using the appropriate function key, < F1 >, ..., or < F4 > (or type *F1 < Enter >*, ..., or *F4 < Enter >*).

CS

The command CS Clears the Screen but not the core copy. The refresh command R will restore the picture. This command is useful when you want to see what is being plotted in a complicated picture.

Commands for controlling the screen

R

The program keeps a copy of the picture in core that is or can be of higher resolution than the copy on the screen. That copy is clean in that almost no text is written on it. The command R Refreshes the screen, eliminating extraneous text by retrieving it from the core memory copy. The "core" memory copy is the copy of the picture that is used for printing high resolution copies on a printer.

ROT

If you have a core picture that is square, (720 pixels by 720) for example, you can rotate your picture. This does not changes the scale and future plotting will be unaffected. This command can be useful for rotating a bifurcation diagram after you have created it so you can print it in a rotated format. Use the command SCM (Size of Core Menu) to find out or set the core size. This command is often used before printing bifurcation diagrams. See also commands FLIPH and FLIPV.

SHIFT

The command SHIFT allows you to move the picture. If your picture consisted of a single window with a picture in it, you could shift it anywhere else on the screen. The shift also affects the core memory. It does not affect the coordinates of XS and YS. The screen and core are treated like a torus, so any part of the picture that is shifted off the edge of the screen will reappear on the other side of the screen.

You must specify at the menu how many pixels to shift the picture horizontally and how many vertically. Choose non-negative numbers.

Enter SHIFT and the Actions, Hints and Options menu for SHIFT appears and it includes the following:

```
ACTIONS, HINTS and OPTIONS for SHIFT

OK        – proceed to SHIFT
Cancel    – do not shift

XSHIFT  – number of pixels to shift horizontally:  0
YSHIFT  – number of pixels to shift downward:  0

FLIPH   – FLIP picture Horizontally
FLIPV   – FLIP picture Vertically
```

FLIPH, FLIPV

The command FLIPH flips the picture horizontally switching the left and right sides. The command FLIPV flips the picture vertically switching the top and bottom. No parameters are changed.

If the picture is square, the commands FLIPH and FLIPV can sometimes be used with ROT which rotates a square picture 90 degrees.

XSHIFT, YSHIFT

The command XSHIFT allows you to specify the number of pixels that the core picture shifts horizontally to the right when SHIFT is executed. Points shifted off the screen on the right hand side, appear on the left side of the screen.

The command YSHIFT must be used to specify the number of pixels that the picture shifts downward when SHIFT is executed. Points shifted off the screen on the bottom of the screen, appear on the top of the screen.

If the core picture is 720 pixels wide by 720 pixels high and if XHSIFT = 720 and YSHIFT = 720, then the picture will be shifted onto itself with no apparent change.

Level of text output

TL and T0, T1, T2, T3, T4, T5, and T6
The "Text Level" command TL determines the amount of text that is being printed on the screen or sent to the printer while the program is plotting. Enter the command TL and the menu appears.

```
TL  –   TEXT LEVEL for SCREEN and PRINTER

T0  –  no text
T1  –  minimal text
T2  –  ideal text
T3  –  lots of text (default)
T4  –  send current y values to printer
T5  –  send continuous stream of y values to
            printer as they are computed
T6  –  send all screen text to the printer
```

The higher the value of the Text Level, the more data will be displayed. The Text Level starts with value 3 but can take on values 0, 1, 2, 3, 4, 5, or 6. The higher the number, the more text one gets. Enter command T2, and TL is changed to 2; or enter T1 and it is changed to 1, and so on.

The setting TL = 3 prints on the screen intermediate results that may not be needed. For example, while plotting a trajectory and computing its Lyapunov exponents, the exponents will automatically be printed frequently. The printing process often slows the computer significantly (sometimes by more than a factor of ten), so it is preferable to switch from TL = 3 to TL = 2. It takes the computer much longer to print the results for typical maps than to compute the data.

If the program is printing continuously (as is the case with the default setting of 3 when Lyapunov exponents or stable and unstable manifolds are being computed), you may want to type in

$$T2 < Enter >$$

(You still get lots of data, but not so much that computing is significantly slowed). Use Text Level 1 for just a little information.

Warning: do not enter commands T4, T5, or T6 before consulting the information on Text Level in the section entitled Printer Utilities (Section 4.6) because these send text to the printer (or to a file in the case of Unix).

Writing on pictures

TYPE, TYPE0   (DOS only)

Position the small cross at the point where you want to begin typing, for example, by using the left mouse button. Enter the command TYPE and type you message. One line only, please. If you want more lines, you must enter the command again. See command TS for setting the size of the type and command TC for setting the color of the type.

The command TYPE0 is like TYPE but it puts the text at the position of y instead of y1.

TC   (DOS only)

The "Type Color" command TC allows you to select a color (number) in which the text appears on the screen and/or picture when using the commands TYPE and TYPE0. The command TYPE allows you to write text on the screens. The color of the text is the current active color.If you want a different color, use the Type Color command TC. The command TC is **not listed in the menus**.

TS   (DOS only)

The command TS allows you to select the Type Size used in pictures (DOS only). Commands B1 and B2 draw boxes and in DOS write the values of the coordinates on the screen. The command TYPE allows you to write text on the screens. If you want a different size type, use the Type Size command TS. The larger TS, the larger the size of the type. Normal type corresponds to TS = 1. The initial text on the screen only approximates the position of the core copy, so Refresh the screen (command R) to see the actual position and size.

3.2 THE ARROW KEYS AND BOXES (BoX Menu BXM)

The small cross shows the position on the screen of the initialization vector y1. That is, the initialization vector y1 and the position vector of the small cross are the same vector. That vector, discussed in Example 2-1i of Chapter 2, is represented by the small cross. The small cross can be moved via the arrow keys or via the mouse by clicking the left mouse button. Hence, when moving the small cross, the initialization vector y1 changes its coordinates to correspond to those of the small cross. For example the < arrow upward > moves the initialization vector upwards and puts a cross at the new location. When the key is pressed repeatedly, or more simply, when it is held down, the cross moves at increasing speed. When an arrow key is hit, the small cross turns on. If there was a small cross already there, it is moved to the new position. To get y to take on this value, enter the "Initialization" command I as described in Chapter 2:

$$I < Enter >$$

These arrow keys can be used for drawing a box on the screen, or for changing the scale of the screen. Enter the "BoX Menu" command BXM:

$$BXM < Enter >$$

and the BoX Menu will appear on the screen

```
BXM  – –    BOX MENU

B      –  draw Box
B1     –  draw Box with tic marks
B2     –  draw Box with double tic marks
BB     –  set Box equal to entire window
BLL    –  set Box Lower-Left corner at y1
BUR    –  set Box Upper-Right corner at y1


          MENUS
AXM    –  AXes Menu
KM     –  cross (=Kruis in Dutch) Menu
WM     –  Window Menu
```

Arrow keys move small cross which is at y1.
Right mouse button is for positioning box:
move mouse arrow to desired position of one
corner, depress and hold right button, moving
to opposite diagonal corner and release.

Selecting and drawing a box on the screen

After a picture has been plotted on the screen (for example, a trajectory of the Ikeda map I with the default parameter values), select a rectangular region to be enlarged. See Figure 3-1. If there is a menu on the screen, hit <space bar> so that either the arrow keys or mouse can be used. One way to select a box is by using the right mouse button (provided there is no menu on the screen). Move the mouse arrow to one corner of the desired box. Depress the right button and hold it down while moving the mouse arrow to the opposite diagonal corner. Release the button.

If you wish to use the arrow keys, move the small cross using the arrow keys to the lower left hand corner of that region.

Enter the "Box Lower-Left corner" command BLL:

BLL <*Enter*>

to declare that this point is the lower left hand corner of the desired box.

Next move the small cross to the upper right corner of the new box and enter the "Box Upper-Right corner" command BUR:

BUR <*Enter*>

to declare that this point is the upper right corner of the box. Now enter the "Box" command B:

B <*Enter*>

to draw the box you have selected and be assured that the desired region has been boxed off. See Figure 3-1a for a result. To make a blow-up picture (as shown in Figure 3-1b), see the Section entitled Windows and rescaling (Section 3-5).

If the command BUR is not executed, then the default upper right corner of the box is the upper right corner of the screen. If BLL is not executed, the computer will use the lower left corner of the screen for lower left corner of the box.

See the command BB for making the box equal to the entire screen.

Note. In this chapter and Chapter 4 we present basic instructions useful in creating many kinds of pictures. Some of these commands are illustrated in the figures. Rather than creating separate examples, the instructions used in making the pictures have been placed in the captions of the figures.

Commands for drawing boxes

B

The "Box" command B causes a box to be drawn. The default position of the box is the entire screen (or window if you have rescaled to a quadrant). To create a box, use the arrow keys to move the small cross to the point where the lower left corner of the desired box should be. Then enter the "Box's Lower-Left corner" command BLL. Next move the small cross to the point where the upper right corner of the desired box should be. Then enter the "Box's Upper-Right corner" command BUR. Enter command B to draw the box that has been created. Hence, for drawing a box, first set the positions of the lower left and upper right corners of the box by moving the small cross and then enter "BLL" or "BUR" followed by command B. See also the commands OW (Open new Window), BLL and BUR.

If you enter B and no box is drawn, this usually suggests the box is outside the screen area, due to your change in coordinates. Use BB to reset the box to be the whole screen, and try B again.

B1, B10 (DOS only)

Command B1 draws a box and then draws tic marks, from 4 to 9 tic marks on a side, choosing the number so as to give natural divisions. For example, if a horizontal or vertical scale ran from 0.38 to 6.62, then the tic marks would be positioned at 1.0, 2.0, ..., 6.0.

Note. If the Text Level is 2 or higher, then this command will also display the numbers corresponding to the tic marks. See also the "Text Level" command TL. This feature works for MS DOS computers only.

Use command B10 (= command B1 with text level 0) to draw the box and tic marks, suppressing the printed coordinates.

● Use the Type Size command TS to set the Type Size of the numbers in the pictures.

B2, B20 (DOS only)

Command B2 draws a box and then draws a double set of tic marks. The first set is identical with the tic marks of command B1. The second set is a finer set of small tic marks, smaller in size so the two can be distinguished. The distance between the small tic marks is 1/5 the distance between the big tic marks.

Note. If the Text Level is 2 or higher, then this command will also display the numbers corresponding to the larger tic marks on the axis. See also the "Text Level" command TL.

Command B20 does the same as B2 but without the printed values.

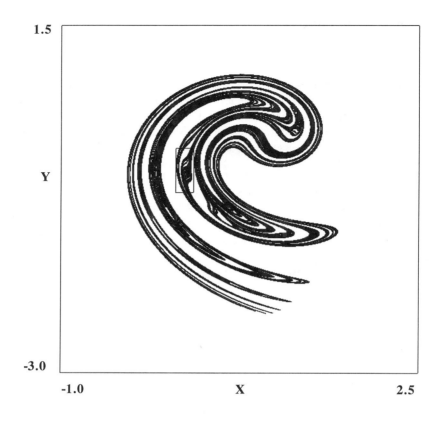

1.5

Y

-3.0

-1.0   X   2.5

Figure 3-1a: A trajectory
This figure shows a trajectory of the Ikeda map (I)
$$Z \rightarrow 1 + 0.9*Z*exp\{i[0.4 - 6/(1 + |Z*Z|)]\}$$
where $Z = X + iY$.

To create a small box like the one shown in this figure, move the mouse arrow to one corner of the desired box, depress the right button and hold it down while moving the mouse arrow to the opposite diagonal corner and release the button. A box will appear on the screen.

If you wish to use the arrow keys, move the small cross using the arrow keys to the point where the lower left corner of the desired box should be, and enter the "Box Lower-Left corner" command BLL. Next move the small cross to the point where the upper right corner of the desired box should be, and enter the "Box Upper-Right corner" command BUR. Now enter "Box" command B to draw the box that has been created.

Topic of discussion. *As RHO = 1 is increased slightly the attractor and its basin disappear. Describe how this destruction occurs. This event is called a "boundary crisis".*

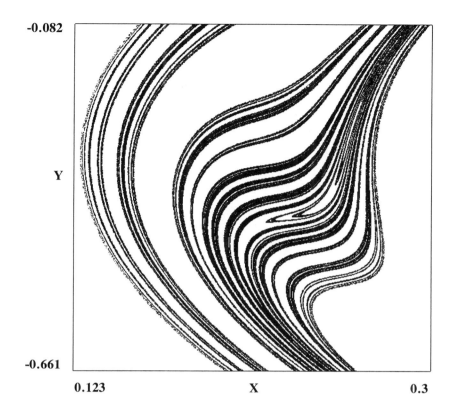

Figure 3-1b: Blow-up of small box in Figure 3-1a

To make this picture, create a small box as in described in Figure 3-1a.
Enter the "Open Window" commands OW and OW0, and the screen rescales, using
the coordinates of the box that was created for the coordinates of the
screen. The scale of the screen is now the scale of the small box.

Topic of discussion. *This box shows possibly the most irregular*
feature of Figure 3-1a. As RHO = 1 is decreased to 0.98, describe what
happens in this box.

BB

Command BB sets the area of the box to be all of the active window. When you enter command BB, no box is drawn, but if you then enter command B, a box will be drawn around the edges of the active window. This is useful when you have a couple of windows open and you want to see how they are related to each other. When you enter BB and then switch to another window with larger scale and then enter command B, a box will be drawn in the second window showing which part of the active window is represented by the previous window.

Commands for selecting rectangular regions

The following commands are for selecting a rectangular region, a box, on the screen. A simpler alternative is to use the right mouse button, if a mouse is available. See the beginning of Section 3.2 and Figure 3-1a.

BLL

Command BLL is used when creating a box in junction with command BUR. To create a box, use the arrow keys to move the small cross to the point that will become the lower left corner of the box. Then enter the "Box Lower-Left corner" command BLL to declare that this point is the Lower Left corner of the Box. Next move the small cross to the point that will become the upper right corner and enter "Box Upper-Right corner" command BUR. Enter command B to draw the box that has been created. If BLL is not executed, the computer will use the lower left corner of the screen for the box. If BUR is not executed, the computer will use the upper right corner of the screen for the box. See also command B.

BUR

For command BUR, see Box Lower Left command BLL.

3.3 INITIALIZING TRAJECTORIES, PLOTTING CROSSES, DRAWING CIRCLES AND THEIR ITERATES (cross Menu KM)

Recall from Section 3.2 that the storage vector y1 is used for initializing the system and that the arrow keys can move the small cross representing this initializing vector y1. From Example 2-1g you know that a cross can be plotted at the current position of a trajectory. These, and other related commands can be found in the Kruis (cross) menu KM. Type

$$KM < Enter >$$

and the Kruis (cross) Menu is called to the screen.

```
KM   - -      KRUIS (CROSS) MENU

I      -  Initialize y using y1
II     -  plot y1 & its MI = 1 Iterates
K      -  big cross along trajectory:  OFF
KK     -  permanent cross at y
KK1    -  permanent cross at y1
KKK    -  permanent cross at each point
              as a trajectory is plotted:  OFF
KKS    -  permanent cross Size:  6 by 6
O      -  draw circle at y1
OI     -  draw circle at y1 & its MI Iterates
W      -  show coordinates of y1

Arrow keys move small cross which is at y1
or click left mouse button at desired spot.
```

Commands for initializing trajectories

I

The "Initializing" command I reinitializes the trajectory vector y to be the current position of y1; the latter's position is shown by the small cross. (It may be necessary to hit the < space bar > key while plotting to make the small cross appear.) The new horizontal and vertical coordinates of the trajectory are printed on the screen. The dot number is set to 0 and Lyapunov exponents are reinitialized if they are being computed.

The vector y1 is special in that it also contains values for the parameters of the system. For example, if you are using RHO for one of the axes, say the horizontal axis, then as the left or right arrow keys move the small cross horizontally, the position of the small cross designates a changing value of RHO. The program keeps track of that value of RHO. Command I sets the trajectory to begin at the small cross, changing the programs value of RHO to the value represented by the small cross. This feature is useful in combination with the "Follow Orbit" command FO.

II

The "Initialize and Iterate" command II is designed to be used together with the arrow keys. This command puts the program in a mode where it plots the position of the small cross and the first MI iterates of the small cross, where MI is the Maximum number of Iterates to be plotted. Then the program pauses, waiting for the user to move the small cross with the arrow keys. MI has default value 1 and may be changed using command MI. When any arrow key is hit or when the left mouse button is clicked, the small cross moves and the first MI iterates of the new position of the small cross are automatically plotted. This command can be used to find periodic points of period MI. See also command RP for finding periodic points on the screen.

Commands for drawing crosses

K

The "Kruis" command K (Kruis means cross in Dutch) results in a cross being plotted as each dot is plotted; the cross is erased when a new dot is plotted. If the new dot is out of the screen, so that it cannot be plotted, the old cross is left on, and stays on until a new dot is plotted within the window.

K is a toggle, so if you enter K again this feature turns off.

KK, KK1

When command KK is entered, a permanent cross turns on at the current position of the trajectory, y. "Permanent" means it is also drawn on the core copy of the picture. The command R refreshes the screen and replots this cross, so this cross remains until the window is erased. The cross stays on as y moves away. To get a series of a few permanent crosses while plotting a trajectory, you can pause the program (Enter P or hit < · >), and use command KK followed by < · > repeatedly. Note that <Return> or <Enter> follows command KK but not the interrupt < · >. Hence, while the program is plotting a trajectory, the keystrokes

$$< · > KK <Enter> \quad < · > KK <Enter> \quad < · > KK <Enter>$$

result in three crosses at three consecutive dots along a trajectory. For another method of getting crosses, see command KKK below.

The command KK1 draws the permanent cross at y1 instead of y.

KKK

The command KKK draws a permanent cross at each point of the trajectory.

KKK is a toggle, so if you enter KKK again this feature turns off.

KKS

The command KKS sets the Size of the permanent crosses created using commands KK and KKK. KKS will give you the current size.

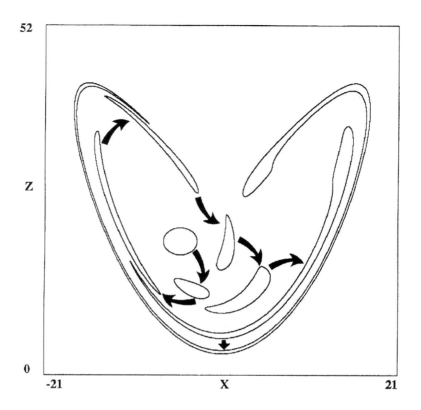

52

Z

0

-21   X   21

Figure 3-2a: Circle and its iterates
This figure shows a circle and 7 of its iterates for the Lorenz system of differential equations (L)
$$X' = 10*(Y-X), \ Y' = 28*X - Y - X*Z, \ Z' = X*Y - 2.666667*Z$$
*To make the picture, enter the 2-dimensional mode (command D2) and move the small cross (using arrow keys) to where the circle should be centered (here y1[0] = -5, y1[1] = 1, and y1[2] = 20). Use the "RAdius" command RA to set the radius (here RA = 2.0). Set the Maximum number of Iterates MI = 7, and the command OI draws a circle (which is a broad ellipse in these coordinates) and its 7 iterates. The arrows in the picture were added manually. The iterates are **not** consecutive time steps. They are 20 Runge-Kutta steps apart. This is achieved by setting the Iterates Per Plot to IPP = 20.*

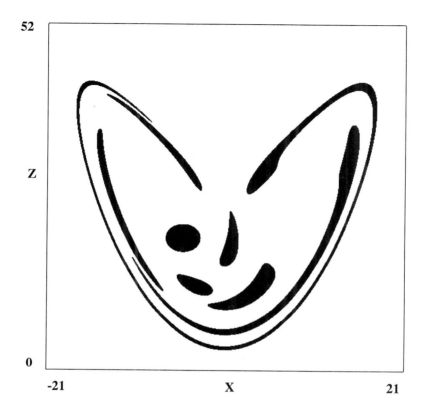

Figure 3-2b: Solid disk and its iterates
This figure shows a solid disk and 7 of its iterates for the Lorenz system of differential equations (L). The values of the parameters (default), the scales and the settings for the vector y1, RAdius RA, Maximum number of Iterates MI and Iterates Per Plot IPP are the same as in Figure 3-2a.

To make the picture, carry out the steps of Figure 3-2a. Before entering command OI, the toggle SO should be turned ON. The routine OI now plots a disk centered at y1 (filled-in circle) and its MI = 7 iterates.

Topic of discussion. *Notice that the second and third iterate of the disk are quite stretched but the fourth is not, but the sixth and seventh are. Is it possible that later iterates will again shrink to a small disk? Notice that the origin (0,0,0) is a saddle.*

Commands for plotting circles (and their iterates)

O, OYK, and Ok, where k = 1,...,9, a,...,e

The command O instructs the program to draw a circle with radius RA, centered at y1 (the small cross). The name of the command was chosen because the letter 'O' looks like a circle. This routine can be terminated by hitting <Esc> once. See command OI.

The command OYK in the Actions, Hints and Options for O, draws a circle centered at a storage vector that you select.

The command Ok instructs the program to draw a circle centered at yk, where k = 1,···,9, a, ···, e.

OI, OIYK, and OIk, where k = 1,...,9, a,...,e

The "circle and its MI Iterates" command OI tells the program to draw a circle with radius RA (centered at y1) and to plot the MI iterates of that circle. If say, MI = 2, it plots points on the circle and the first and second iterates of those points on the circle. Terminate this routine by hitting <Esc> once. Set RA and MI using the commands RA and MI. OI is useful in basin routines for determining if the circle with a specified center and radius is actually in the basin of attraction, though of course y1 must be moved to the center.

If you have a color screen and the Maximum number of Iterates MI is smaller than the number of colors (usually 16) minus one, then the consecutive iterates will be drawn in different colors. Command CT will display the color table. Color 0 is on the left, and it is the background color. Next is color 1, etc. usually up to color 15 on the right. If MI is 10, then the first iterate will be plotted using color number 1 (that is, color number 1 in the Color Table CT), the second iterate will be plotted using color 2, etc. Color 0 is not used because that is the background color. If the Maximum number of Iterates MI is larger than the number of screen colors available, then each color may correspond to more than one consecutive iterates. If for example, MI = 150, and the number of colors is 16, then colors 1 to 15 are allocated as follows: the first 10 iterates are plotted using color 1, the next 10 iterates using color 2, etc.

If the point to be plotted has already been plotted, then it will only plot the point if the new color number is lower than the number of the plotted point. Hence, if the routine arrives at the same pixel from two different points on the circle, the color seen indicates the earliest arrival time. Hence, the color of a plotted point tells how fast trajectories can get to different points when starting from points on the circle. If you want to see the image of the circle after say 20 iterates and do not want intermediate iterates of points plotted, then set MI = 1 and the Iterates Per Plot IPP to 20. Then call OI. Do not forget to reset IPP to 1 afterwards.

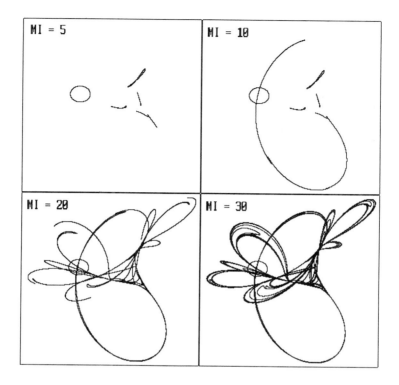

Figure 3-2c: Circle and its MI iterates, where MI = 5, 10, 20, 30
This figure shows a circle and its MI iterates for the Tinkerbell map (T)

$$X \rightarrow X*X - Y*Y + 0.9*X - 0.6013*Y$$
$$Y \rightarrow 2*X*Y + 2.0*X + 0.5*Y$$

In all four windows, the variable X is plotted horizontally, and the variable Y is plotted vertically. The X Scale is from -1.8 (left) to 1.2 (right) and the Y Scale is from 1 (bottom) to -2 (top).

In the figure, y1 = (-0.72,-0.64) and RA = 0.1 (default values). For the upper left window (window W1) MI = 5. For the three remaining windows W2, W3, and W4, the Maximum number of Iterates was set to be 10, 20, and 30, respectively. To create the picture, enter commands OW and OW1 to open window W1, set MI = 5, then command OI draws a circle and its MI = 5 iterates. Next, open window W2 (enter commands OW and OW2), set MI = 10, and command OI draws a circle and its MI = 10 iterates. Similarly, for the windows W3 and W4, set MI = 20 and MI = 30, respectively.

The initial points on the circle are plotted using whatever the plotting color was before O was invoked.

If you have a VGA screen and you enter "COLor" command COL, then you will have 256 colors available. The command COL would wipe out the current picture. In either case, whether 16 or 256 colors, commands such as BRY (establishing a Blue-Red-Yellow color table) are useful for ordering the colors in a visually meaningful scheme.

Warning: When using command COL, if your computer does not have a super VGA capability, the text is printed in a large font and is irregularly strewn across the screen, and so this mode is unpleasant to work in, but the resulting pictures are often rewarding. See command COL.

The command OIYK in the Actions, Hints and Options for O, draws a circle centered at a storage vector that you select and its MI iterates.

The routine OIk draws a circle centered at yk and its MI iterates, where $k = 1, \cdots, 9, a, \cdots, e$.

SO

The "SOlid" command SO is a toggle. If SO is ON, the routines O, OI, OY, OIY plot solid balls instead of circles. (The commands OY and OIY are described in Chapter 7). If SO is OFF, these routines draw circles. See the commands O, OI, OY and OIY. The command SO is **not listed in the menus**.

SF

The "Screen Fraction" command sets the minimum distance (as a fraction of the screen) that the small cross moves when an arrow key is hit. Sometimes it is important to move the small cross a tiny distance, and this cannot be achieved with a mouse. This command is **not listed in the menus**.

W

The "Where" command W causes the position of the small cross to be printed on the upper right of the screen; the values are updated when the small cross is moved. This command can be used to help locate the small cross. It can also be used for finding the coordinates of an interesting feature on the screen. Enter W and then move the small cross to the feature and its coordinates will be displayed. The command W is a toggle; enter it again and this feature is turned off.

This command also prints the position of y (marked by the big cross) on the upper left of the screen once, when W is turned on. Those coordinates are not updated after the key is hit. To find just the coordinates of y, then enter W twice (W < Enter > W < Enter >), turning it on and off. Then the small cross position will not be continuously printed out as it moves. (Recall < Tab > will give you the scale of the current window.)

3.4 DRAWING AXES (AXes Menu AXM)

From Example 2-1h you know that axes can be drawn. Axes can also be drawn with either a single set of tic marks or a double set of tic marks. These, and other related commands can be found in the AXes Menu AXM. Type

AXM < *Enter* >

and the menu appears on the screen.

```
AXM    – –      DRAW AXES MENU (2 dimensional)

XAX    –  X AXis
XAX1   –  X Axis with tic marks
XAX2   –  X Axis with double tic marks
XAXI   –  X Axis Intercept (height) : 0

YAX    –  Y AXis
YAX1   –  Y Axis with tic marks
YAX2   –  Y Axis with double tic marks
YAXI   –  Y AXis Intercept (X value): 0

          MENU
AX3M   –  rotate AXes Menu  – –  3 dimensional
```

Commands for drawing axes (2-dimensional)

XAX
The "X AXis" command XAX causes the X AXis to be drawn.

XAX1
The command XAX1 draws the X AXis and then draws tic marks, from 4 to 9 tic marks on a side, choosing the number so as to give natural divisions. For example, if the horizontal scale runs from -0.62 to 5.62, then 6 tic marks would be positioned at 0.0, 1.0, 2.0, 3.0, 4.0, 5.0.

Note. If the Text Level is 2 or higher, then this command will also display the numbers corresponding to the tic marks on the axis. See also the "Text Level" command TL. This feature works for MS DOS computers only.

XAX2

The command XAX2 draws the X AXis and then draws a double set of tic marks. The first set is identical with the tic marks of command XAX1 and the second set is a finer set of small tic marks, smaller in size so the two can be distinguished. The second set is spaced 5 times closer together than the large tic marks.

Note. If the Text Level is 2 or higher, then this command will also display the numbers corresponding to the larger tic marks on the axis. See also the "Text Level" command TL. This feature works for MS DOS computers only.

XAXI

The "X AXis Intercept" command XAXI is for setting the height of the horizontal axis. The default value for XAXI is 0.

YAX, YAX1, YAX2, YAXI

The "Y AXis" commands YAX, YAX1, YAX2 and YAXI are analogous to the X AXis commands XAX, XAX1, XAX2 and XAXI.

THE ROTATE AXES MENU AX3M (3-dimensional)

Enter the "rotate AXes Menu" command

AX3M *< Enter >*

and the menu appears on the screen.

```
AX3M    - -  ROTATE AXES MENU (3 dimensional)

ANGLE  -  ANGLE of rotation:  5 degrees
AD     -  Axes Default orientation
AX     -  make the X axis the rotation Axis
AY     -  make the Y axis the rotation Axis
AZ     -  make the Z axis the rotation Axis

D2     -  2 Dimensional mode
D3     -  3 Dimensional mode
ZCO    -  Z COordinate, currently NOT SET
ZS     -  Z Scale

3-DIM INTERRUPTS:
+/- :  ROTATE  5  degrees
X Y Z (upper case only):  have the same
         effect as commands AX, AY, and AZ
```

Commands for rotating axes (3-dimensional) See also Figure 2-16.

ANGLE

While plotting in 3-dimensional mode, keys < + > and < - > will rotate the coordinates about an axis by ANGLE degrees. The default axis of rotation is the horizontal X axis. The default value of ANGLE is 5 degrees.

AD

The "Axis Default orientation" command AD is to set the 3-dimensional rotation Axes in the default position.

AX, AY, AZ

The command AX is to set the 3-dimensional rotation Axis to be the horizontal X axis. The command AY is to set the 3-dimensional rotation Axis to be the vertical Y axis.

The command AZ is to set the 3-dimensional rotation Axis to be the perpendicular-to-the-screen Z axis.

D2

The command D2 is for switching back to plotting in 2 dimensions.

D3

The command D3 says trajectories should be plotted in 3 dimensions when using command T (Trajectory). Specify the third variable to be plotted with the "Z COordinate" command ZCO. When plotting in 3 dimensions, the coordinates can be rotated about either the horizontal axis, the vertical axis, or the axis that is perpendicular to the screen. These are called the X, Y, and Z axes. The "active axis" is the axis that will be used when $<+>$ or $<->$ is hit. Initially the active axis is the X axis. Hitting $<+>$ or $<->$ while plotting will rotate the figure by 5 (default value of ANGLE) degrees about the axis, and then the coordinate box is drawn.

Change the active axis of rotation either by using commands AX, AY, or AZ, or by hitting the **capital** letters $<X>$, $<Y>$, or $<Z>$ while plotting. Do not hit $<Enter>$ when using the $<X>$, $<Y>$, or $<Z>$ option. This implies that to use a command beginning with x, such as XX, it is necessary to use lower case letters if you are actively plotting in 3 dimensions.

The scales XS, YS and ZS refer to the horizontal, vertical, and perpendicular to the screen scales, and these do not rotate. Suppose the Z scale is set to -30 30. The lower end of the Z range (Z = -30) corresponds to the surface of the screen, while the upper range (Z = 30) is behind the screen. Points in the screen area are plotted in white (color 15) if they are outside the 3-dimensional coordinate frame.

See D2 to change to two dimensional plotting, and commands AX, AY, and AZ for commands equivalent to interrupts $<X>$, $<Y>$, and $<Z>$.

ZCO

The value of the ZCO is the name of the perpendicular-to-the-screen COordinate variable. This command is used for setting the variable or parameter that is to be plotted perpendicular to the screen, and is otherwise similar to the commands XCO and YCO described in Chapter 4.

ZS

The "Z Scale" command ZS allows you to change the perpendicular-to-the-screen scale. The first number is the coordinate of the surface of the screen and the second one is behind the screen. You must enter both with a space but no comma between them. The first number can be larger than the second if you want a decreasing scale. See also commands XS and YS in Chapter 4.

3.5 WINDOWS AND RESCALING (Window Menu WM)

You can split the screen into quadrants if desired. Each quadrant can be a window for plotting. The different windows may have different scales or even different coordinates for plotting, and (if desired) a trajectory can be plotted simultaneously in all windows. Enter the "Window Menu" command:

WM < *Enter* >

and the menu will appear on the screen.

```
WM      - -  WINDOW MENU

C       -   Clear current window
CC      -   Complete Clear
OW      -   Open new Window


            MENUS
AXM     -   AXes Menu
BXM     -   BoX Menu
KM      -   Kruis (cross) Menu
RDM     -   Read Disk Menu

OPENING A WINDOW:
Having selected a box (see BoX menu BXM), use command OW.

SWITCHING WINDOWS:
Use <F1>, ..., <F4>, <F10> to switch between open windows.
Do NOT use command OW unless you want to create a window.
Use <F10> to switch back to the whole screen.

Each window is a quadrant of the screen, situated as follows:
                <F1>              <F2>
                <F3>              <F4>
            <F10> represents the whole screen
```

Assume a box has been created as described above by using the commands BLL and BUR as described in Section 3.2. Enter the "Open new Window" command OW:

OW < *Enter* >

followed by

$$\text{OW0} < Enter > < Enter >$$

This rescales the screen, using the coordinates of the box that was created for the coordinates of the screen. The screen will be cleared since the old picture is obsolete. The scale of the screen is now the scale of the old box. The screen is erased, and the picture can be recomputed to reveal greater detail.

The command OW can be used to create a window in any of the four quadrants of the screen. If you enter OW, the program will ask in which quadrant of the screen you want the newly scaled picture to appear. The simplest answer is to enter the command OW0 as just described. If instead you reply

$$\text{OW1} < Enter > < Enter > \quad \text{or} \quad \text{OW2} < Enter > < Enter >$$

or

$$\text{OW3} < Enter > < Enter > \quad \text{or} \quad \text{OW4} < Enter > < Enter >$$

the new window will be a quadrant of the screen, and you can consecutively try them all. The pattern of the four windows above shows which quadrant you get; for instance, command OW2 creates a window in the upper right quadrant of the screen. The function keys $< F1 >$, $< F2 >$, $< F3 >$, $< F4 >$, and $< F10 >$ are for switching between the windows. Once the windows have been opened, hitting these function keys (without hitting $< Enter >$) shifts the program into that quadrant again (making it the current active window) and continues plotting. You can change coordinates in the current active window or even change which variable or parameter is being plotted for each axis of the window.

When you open a new window, the small cross is placed at the center of that window. You can use BLL (set Box's Lower-Left corner) and/or BUR (set Box's Upper-Right corner) to rescale any active window or use the right mouse button to select the box.

When losing track of where the small cross is, enter the CENTer command CENT:

$$\text{CENT} < Enter >$$

and the small cross will show up in the center of the active window.

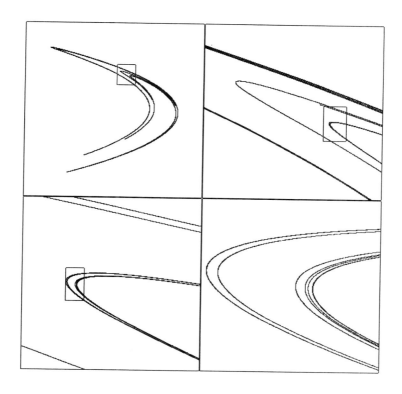

Figure 3-3a: Attractor and consecutive blow-ups

The upper left window shows the attractor of the Henon map (H) with the same parameter values and scale as shown in Figure 2-4. See Example 2-4.

    To make this picture, enter the commands OW and OW1 to open window W1, set PI = 100, and plot the trajectory. Enter the box command B1 to write the scale. Next, create a small box (using arrow keys), and enter the commands BLL, BUR, and B. Then enter the commands OW and OW2 to open window W2, and plot the trajectory. After a while, create a small box, open window W3, and plot the trajectory. Similarly, plot the trajectory in window W4.

    Topic of discussion. *If we had instead selected a series of blow-ups centered at the fixed point on the attractor, how would the eigenvalues of the Jacobian matrix at this fixed point be related to what is seen in the series?*

Figure 3-3b: Basin and consecutive blow-ups

The black area is the basin of attraction of a stable period 11 orbit of the complex quadratic map (C) using the same parameters as shown in Figure 2-15. See also Figure 7-3.

To make the picture, enter OW and OW1 to open window W1, and plot the basin of the stable period 11 orbit using the routine BAS (see Chapter 7). Next, create a small box, and enter OW and OW2 to open window W2, and plot the basin. Similarly, create a small box in window W2, open window W3 and plot the basin of the period 11 orbit. Finally, create a small box in window W3, open window W4, and plot the basin. To draw the small box that is window W4 in all the windows, hit <F4> to activate window W4 and enter command BB, then hit first <F3> and enter the "Box" command B, Next, hit <F2> and enter B. Finally, hit <F1> and enter B.

Topic of discussion. *The black set is the union of connected black regions. Each of these connected regions maps onto another. What is the pattern? Which regions map to which?*

Detailed view on the structure of an attractor

The Henon attractor has a nice banded structure that can be viewed in detail using windows as shown in Figure 3-3a. Remember when using the arrow keys to define the new coordinates, you are also changing the initialization vector y1.

As an example of the windowing procedure, start the program plotting a Henon trajectory:

dynamics <*Enter*>

H <*Enter*>

T <*Enter*> <*Enter*>

Choose a small rectangular region of the screen containing interesting features of the trajectory. Either use the right mouse button or using the arrow keys, move the small cross to the lower left-hand corner of that rectangle. Command BLL (Box's Lower Left corner) can be used to specify that the point where the small cross is will be the lower left corner of the box:

BLL <*Enter*>

Now move the cross to the upper right-hand corner of the rectangle and enter BUR (Box's Upper Right corner):

BUR <*Enter*>

telling the program where the upper right corner is. Now enter the "Box" command B:

B <*Enter*>

and a box will be drawn around the rectangle.

Now enter the command OW and then enter command OW3

OW <*Enter*>

OW3 <*Enter*> <*Enter*>

A window will be drawn and cleared in the lower left quadrant of the screen. The coordinates of this window are the coordinates of the rectangle you constructed. Though all points must be re-computed, the program will plot in this window which is a blowup of the above rectangle. To return the plot to the full screen, hit the function key <F10> and then to go back to the window hit <F3>. This time do not use the command OW. Command OW is used only when a window is initially opened, that is, when it is created.

It is not used to transfer to a window that is already open.

To retain the trajectory's position (see Section 4.2), store it first with

$$RV < Enter >$$

$$A < Enter > \quad 1 < Enter >$$

(replacing storage vector ya's position by the position of storage vector y1, thereby storing the coordinates of y1 in ya). Then after rescaling the screen, you will be able to reinitialize the trajectory with that value using the "Replace Vector" command RV and "Initialize" command I. See command RV. You alternatively could have stored it with the command

$$RVA1 < Enter >$$

Commands for opening windows

OW and the function keys < **F1** >, < **F2** >, < **F3** >, < **F4** >, or < **F10** >

Command OW stands for Open a new Window. First move the small cross and use the commands BLL (set Box's Lower-Left corner) and BUR (set Box's Upper-Right corner) to designate the two corners of the desired box. Then enter OW.

Finally, enter one of the window commands OW0, OW1, OW2, OW3, or OW4. If you enter OW0 then the entire screen rescales to the scale of the box. Entering OW1, OW2, OW3, or OW4 will instead rescale it to occupy one quadrant of the screen using the coordinates of the box: upper left, upper right, etc. The window in question will be cleared. You can go back and forth between the sections by hitting the appropriate function keys < F10 >, < F1 >, < F2 >, < F3 >, or < F4 >, or use the commands F10, F1, F2, F3, or F4.

F1, F2, F3, F4, F10

The result of command Fk is equivalent to the function key < Fk >, where k = 1, 2, 3, 4, 10.

3.6 ZOOMING IN or ZOOMING OUT (ZOOm Menu ZOOM)

Zooming in toward a specified point, for example on a fractal basin boundary, may reveil interesting properties. Enter the command ZOOM

$$ZOOM < Enter >$$

and the ZOOm in/out Menu ZOOM appears on the screen.

```
ZOOM   – –     ZOOM MENU

ZF      –  Zoom Factor (usually  >  1):  2
ZB      –  draw the Box that would be used by ZI
ZI      –  Zoom In towards the small cross by a factor ZF
ZO      –  Zoom Out away from the small cross by a factor ZF
DW      –  restore Default coordinates and scales in current Window
```

Commands for zooming in or out

ZB
    The "Zoom Box" command ZB draws the Box used by ZI.

ZF
    The "Zoom Factor" command ZF is the enlargement factor used by ZI. The default value is 2.

ZI
    The "Zoom In" command ZI enlarges a region by a factor ZF and zooms in on the position of the small cross. The simplest operation procedes as follows. Position the small cross at some point in the screen. Enter the Zoom In command ZI. The program rescales the screen by a factor of 2 (or more generally by a factor of ZF) so that each corner moves twice as close to the small cross. In the new coordinates, the new cross will appear to be where it was. The screen is erased and you must recompute what you want to see. If you are in doubt as to what the picture will look like, command ZB draws a Box that would be extent of the screen if ZI is called.

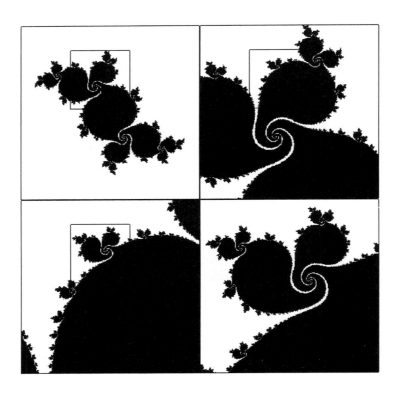

Figure 3-4a: Zooming in toward a basin boundary point
This figure shows some consecutive pictures resulting from zooming in toward a point on the basin boundary of the complex quadratic map (C)

$$Z \;\rightarrow\; Z*Z - 0.11 + 0.6557i$$

where $Z = X + iY$. The black area is the basin of a stable period 3 orbit; the white are is the basin of infinity. See also Figure 7-5.

First, some points of a Basin boundary Straddle Trajectory on the common boundary of the two basins are computed (see Example 2-21 and Chapter 7). The point P used here is $P = y \approx (-0.2597, 0.8712)$ and zoomed in toward to $y1 = P$ (replace $y1$ by y using command RV10). In window W1, the two basins and the box that would be used by ZI are shown; here $ZF = 3$. Window W2 displays the basins restricted to the small box of window W1 together with the box that will be used when zooming in a second time. In window W3, the basins restricted to the small box of window W2 together with the box that will be used when zooming in a third time are shown. Window W4 contains the basins restricted to the small box of window W3. In all the windows, X is plotted horizontally and Y is plotted vertically. In the upper left window, both the X and Y Scales are from -1.5 to 1.5.

Figure 3-4b: Zooming in toward a point on an attractor

This figure shows some consecutive pictures resulting from zooming in toward a point on the attractor of the Van der Pol's diff. eq. (VP)

$$X''(t) - 0.1*X'(t)*(1-X(t)*X(t)) - 2.5*X(t) + 2.5*X(t)*X(t)*X(t) = 5*sin(t)$$

In the upper left window, the chaotic attractor and the box that would be used by ZI are shown; here ZF = 2. The point P on the attractor used here is P = y ≈ (-0.609, 0.772) and zoomed in toward to y1 = P (replace y1 by y using command RV10). The upper right window W2 displays the trajectory restricted to the small box of window W1 together with the box that will be used when zooming in a second time. In the lower left window W3, the trajectory restricted to the small box of window W2 together with the box that will be used when zooming in a third time is shown. The lower right window W4 contains the trajectory restricted to the small box of window W3.

In all the four windows, the variable X is plotted horizontally and the variable X' is plotted vertically. In the upper left window, the X Scale is from -2.0 to 2.0 and the Y Scale is from -2.5 to 4.5.

DW

The Default Window command DW restores the scale to the default scale. It also resets XCO, YCO, and SD to the default values. This is particularly useful after several zooms and moves of the small cross. This gets you back to the original coordinate system.

The command DW should also reset SD and reinitialize the 3D axes.

ZO

The "Zoom Out" command ZO shrinks the screen by a factor ZF, with the position of the small cross being the point zoomed out on. Zoom Out reverses ZI; see command ZI.

Detailed view on the structure of an attractor using the Zoom In feature

The attractor for the Van der Pol differential equation has a complicated structure that can be viewed in detail using the Zoom In feature as shown in Figure 3-4b.

As an example of the Zoom In procedure, start the program plotting a Van der Pol trajectory:

dynamics *< Enter >*

P *< Enter >*

T *< Enter > < Enter >*

After a while, single step through the trajectory by hitting < · >. repeatedly until the big cross is at a position at which you want to zoom in. Terminate the routine T by hitting *<Esc>* (or Pause the program).

Next, replace y1 by y:

RV10 *< Enter >*

Then choose a value for the Zoom Factor ZF, say 4, and continue as follows:

ZF *< Enter >*   **4** *< Enter >*

ZI *< Enter >*

T *< Enter > < Enter >*

You can zoom in as many times as you wish.

3.7 SETTING COLORS (Color Menu CM and Color Table Menu CTM)

The high resolution "core" copy in memory from which the program refreshes the screen, uses 16 colors in a 720 by 720 grid of **pixels**, that is, picture elements or dots. (These numbers can be changed, for example, by using the command SIZE.) In other words, there is a 720 by 720 array of numbers that can have values between 0 and 15. The "Color Table" command CT will display the current colors. They are numbered from left to right starting with 0 on the left. You should imagine here an artist's palette with 16 positions numbered 0 to 15, and the computer (which is the artist) uses a paint by numbers scheme, ignoring the actual paint colors that are deposited on those positions. The default for position 0 is blue or black. When we refer to the **color number** we actually mean the **CT position number**; that is, the number of the position in the color table. Hence, **color k** is **CT position k**, where k assumes the values 0 through 15. You can change the colors on the color table with EGA, VGA, and Unix/X11 graphics. On PC's, changes in the color table immediately result in changes in the colors seen on the screen. On Unix/X11 it is necessary to Refresh the screen (command R) to see the result. See the "Set Color" command SC and the "Set Color Table" command SCT to change the colors that are on the color table.

In summary, there are two very different kinds of color changes available to you:

● Change the number of the color being used, switching for example from color 4 to color 5. See interrupt < F7 > through < F9 > in the Color Menu below.

● Change the color table, using commands in the Color Table Menu.

In the default color scheme the difference between the left half of the color table (color 0 to color 7) and the right half of the color table (color 8 to color 15) is mainly brightness. Many color screens have two brightness and/or intensity knobs. Often one of those knobs greatly affects the difference between those two sets. Color 0 and color 8 for example may appear the same if that knob is turned off.

Color screens

The program automatically plots in color when using a color screen with an EGA or VGA board or Unix/X11. The screen has the following default resolution:

	pixels wide	pixels high	#colors
VGA monochrome or color	640	480	2 or 16
EGA monochrome or color	640	350	2 or 16

When invoking the command COL (depending on the capabilities of your computer) the resolution becomes

		pixels wide	pixels high	#colors
VGA color	either	640	480	256
	or	320	200	256

After invoking COL, the mouse cannot be used.

Core copy of the picture

The core copy of the picture is used for refreshing the screen and other tasks. For MS-DOS computers *Dynamics* keeps a copy of the picture with a resolution of 720 pixels (or dots) wide and 720 pixels high allowing 16 colors if the computer has sufficient memory available. These values can be changed by using the command SIZE. If you do not intend to print the picture, then there is no advantage to the higher resolution and you can use the command SAME which decreases the core resolution to be the same as the screen's resolution (640 by 480 for VGA screens). This smaller core will speed refresh slightly and will result in slightly smaller files when you save the pictures using the "To Disk" command TD.

Using many colors on a low resolution screen

If you have a VGA color screen without super VGA capabilities, then depending on your computer, the 256 color mode may be of very low screen resolution, only 320 pixels by 200. Then the low resolution of the screen means the text is large and only limited amounts of text will fit on the screen. As a result, the point-and-shoot version of the menus is not available. You must type in the commands. We recommend that you set the parameters of the process before you enter COL. Exit this mode by using the "Screen DEFault" command SDEF. Both COL and SDEF destroy the current picture.

Color planes

The number of color planes determines the number of colors the core picture permits. Four color planes are necessary for the core copy of the picture to record 16 colors and eight for 256 colors (see command COL). If you have only 3 color planes, the core has 8 colors. The program *Dynamics* currently needs slightly more than 1 Megabyte of memory to run with 4 color planes with 720 by 720 resolution. Use the DOS command MEM to see how much memory your computer has available. *Dynamics* uses either expanded or extended memory.

Smalldyn is a version of the program, taylored for computers with one megabyte of memory. It has virtually all the commands of the main version, but is restricted in the size of the picture it keeps in memory. It also runs slower for computationally intensive projects. *Smalldyn* will run with at least four color planes when 522KB is available. *Smalldyn* may reduce the size of the screen's picture if it is short of memory. If your computer does not have enough memory available for running *Smalldyn* in 16 colors, you may wish to find out which items have been added to the *config.sys* and possibly *autoexec.bat* files, that consume lots of memory, and you may decide to delete them. For example, with many computers, you can start the computer from a floppy disk which has *autoexec.bat* and *config.sys* files which make fewer demands on the computer memory. Then run *Smalldyn*.

Color slides

It is our experience from a variety of color terminals with a variety of computers that one way to get a high quality permanent record of a color picture on the screen is to photograph it. We recommend color slides. Color printers often do not reproduce the brilliant colors of the screen. The slide gives you an actual visible record of the colors while color prints give a color reversed image. Color prints may be disappointing since your photo shop will not know what the image should look like. Commercial photo equipment aims at good flesh tones, which are irrelevant here. Experienced astronomers recommend using photo stores with one-hour processing (which may require print film), at times when they are not busy, since such machines allow some control over the final product. Be prepared to pay for the ones you do not like.

If you want to create a high quality color picture with the objective of photographing the screen, you should create the picture, enter P if necessary to pause, and switch to the minimal text level T0. Then refresh the screen using command R, and snap the picture. The disk copies of pictures includes the color number for each pixel, but **not** the particular color table used if you use more than 16 colors. It only stores the color table of the first 16 colors.

Note. For printing color pictures, see Section 4.6.

THE COLOR MENU

Enter the Color Menu command CM:

CM < *Enter* >

and the Color Menu will appear on the screen.

```
CM    - -      COLOR MENU

CTM   -  Color Table Menu  - -  set/count/view colors
EM    -  Erase color Menu

CI    -  get Color Information
CNT   -  CouNT pixels of each color in picture
COL   -  256 colors on the screen
IN    -  INcrement color number when plotting:  OFF
SDEF  -  Screen DEFault mode

         COLOR FUNCTION KEYS
F7    -  decrease color number by 1
F8    -  increase color number by 1
F9    -  choose color number, now:  14
```

Commands for setting or changing colors

CMAX

The "MAXimum number of Colors" command CMAX is used for setting the maximum number of colors in the color table. If you enter the command COL, thereby providing 256 colors, you may wish to decrease that number somewhat, say to 64. Just set CMAX to 64. A second similar use occurs when you want fewer than 16 colors. When your core picture has 8 colors (color 0 through 7) and your screen has 16, all features that do not necessarily deal with the core allow 16 colors. For example the "Color Table" command CT will display 16 colors, and the "Red-Green-Blue" command RGB will distribute those colors across the 16 available spots on the color palette. On the other hand, IN (INcremental) plotting is restricted to 8, since it deals with the core. In this case you may wish to set CMAX to 8. The command CMAX instructs the program to work with CMAX colors when setting the color table using commands RGB, RRR, etc. to a fixed color table and when displaying the color table. CT will now display only CMAX colors. You should set CMAX to a value greater than 0 and less than the number of colors. The command CMAX is **not listed in the menus**.

CNT

This command reports how many pixels there are of each color number. It lists how many have been colored using color 0, color 1, color 2, ... , and it reports what fraction of the pixels that represents. Nothing is printed out for color numbers that have not been used. Reduce the text level below 2 to reduce the text printed.

COL

This command selects a color table with more than 16 colors. For many computers, the standard screen resolution is unchanged, but for some the screen resolution will only be 320 X 200. Use the "Screen DEFault" command SDEF to undo COL. Commands COL and SDEF will destroy the current picture.

The command COL can be used with command IN (see below) in combination with a trajectory command like T and a command like RGB which sets the color table. On PC's, the mouse will not work in 256 colors.

Warning: When using command COL, the text may be printed in a large font and is somewhat unpleasant to work in.

Note. When using more than 16 colors on the screen, the mouse may not work.

IN

This command changes the color of the point being plotted INcrementally. When a dot is to be plotted under the IN regime, the program examines the current color number (CT position number) of the pixel where the plot is supposed to occur. Suppose for example, initially the screen is black (color 0). The CT position number (color number) runs from 0 to 15 in the standard mode, or 0 to 255 if command COL is executed. Each time a pixel is plotted, the color number is increased by one, so if a trajectory is being plotted, the first time a specific pixel is plotted, the pixel is turned to color 1. The next time that pixel is hit, it is changed to color 2, continuing until it reaches color 15 (in the 16 color case) and from then on that pixel will stay at color 15, despite successive hits. Actually, the pixels of the core copy of the picture, not the screen copy, determine this behavior. Hence, the computer does not have to read the screen. If you have used command SAME, then the core copy has the same resolution as the screen.

IN is a toggle, allowing you enter command IN again to turn it off.

MONO

This command is for monochrome performance. The number of colors becomes 2; undo with SDEF. The command MONO is **not listed in the menus**.

SDEF

This command returns the Screen to the DEFault mode for the program, restoring the original color table, the original values of HIGH and WIDE (that is, the size of the core), and the original number of colors. If the command COL succeeds in getting you 256 colors, then you can undo the COL command with this command. This command returns the core copy of the picture to the default resolution and undoes the commands COL and MONO.

Warning: SDEF destroys the current picture.

F7 and **F8** and **F9**

Interrupt <F7> (or command F7) decreases the color number by 1 until the color number gets to 0 and interrupt <F8> (or command F8) increases the color number by 1 until the color number reaches the maximum. Interrupt <F9> (or command F9) allows you to specify the color number. <F9> must be followed by the desired color number. It sets the color number. By <F7> we mean hit the function key and do not hit <Enter>. By command F7 we mean the letter F and the digit 7 followed by <Enter>.

COLOR TABLE MENU

Enter the command CTM

$$\textbf{CTM} < Enter >$$

and the Color Table Menu appears on the screen.

```
CTM  – –     COLOR TABLE MENU

CT   –  display Color Table
CI   –  get Color Information
SC   –  Set Color
SCT  –  Set Color Table

BGY  –  Blue to Green to Yellow
BRI  –  BRIght colors
BRY  –  Blue to Red to Yellow to white
DEF  –  DEFault color table
RGB  –  Red to Green to Blue to white
RNB  –  RaiNBow: Red Orange Yellow --
             Green Blue Indigo Violet
RRR  –  dim Red to Red to bright Red
RYB  –  Red to Yellow to Blue
RYW  –  Red to Yellow to White
WYR  –  White to Yellow to Red

         MENUS
BCM  –  Background Color Menu
CM   –  Color Menu
```

Commands for setting colors or the color table

CI

  Dynamics uses a slight modification of standard VGA colors. In particular, the background color (color #0) is dark blue rather than black. Furthermore, you can reset colors. Use command CI to list the 16 colors being used, and the color of the vectors y and y1.

CT

This "Color Table" command displays the current color table at the bottom of the screen (or of the active window).

SC

The command SC sets one color in the color table. Recall, the color number refers to the position in the color table, where color 0 is the leftmost color in the table, color 1 is next, and so on. The color numbers usually run from 0 to 15 on EGA, VGA or Unix color screens. The actual color represented say by color 4 is by default a shade of red, but this or any other color number can be reset to be another combination of red, green and blue. Color monitors are "RGB" monitors, which means that the primary colors or phosphors of the screen are red, green, and blue.

If you have a mouse available and have 16 colors, you can use the left mouse button to make your selection.

Four horizontal bars will appear. Just click the left mouse button in the top bar (not on the numbers) to select a color to be set and then use the other three bars to select intensity. Set as many colors as you want.

Here we describe the procedure in which no mouse is available. If a mouse is available but the number of colors exceeds 16, then you also must use this procedure. The computer prompts:

```
The colors in the Color Table are numbered as follows:
Color 0 (the background color) is the leftmost color;
Color 15 is the rightmost.

Enter color number to be changed, between 0 and 15 :  __
```

Then enter the color number, for example color 4:

$$4 <Enter>$$

The computer prompts:

```
Now setting color 4
Choose 3 color strengths (each between 0 and 63),
        for RED, for GREEN, and for BLUE:  __ __ __
Sample setting for bright yellow -- 63  63  0   <Enter>
```

Type 3 integers for the amounts of red, green, and blue for color 4, and finally hit <Enter>. The amounts of red, green, and blue are all numbers between 0 and 63, with 0 meaning no color and 63 being the maximum allowed.

On EGA screens, fewer combinations of red, green, and blue are available, so you may find slightly different specifications result in the same screen color. For example,

$$63 \ 0 \ 0 < Enter>$$

means "color 4 is set to maximum red, no green and no blue". The zeros are nonetheless necessary.

The colors in *Dynamics* have the following default settings; see also command CI.

color number	red	green	blue
0	0	0	0
1	0	0	42
2	0	42	0
3	0	42	42
4	42	0	0
5	42	0	42
6	42	15	0
7	52	52	42
8	21	21	21
9	21	21	63
10	21	63	21
11	21	63	63
12	63	21	21
13	63	21	63
14	63	63	21
15	63	63	63

Notice the pattern: colors 1, 2, and 4 are the screen's fundamental colors blue, green, and red, while color 8 is dim white. Other colors are appropriate sums of these. For example color 5 is color 1 plus color 4.

SCT

The command SCT sets one or more colors in the color table. Master the command SC before using this command. See the command SC for a description of how to set the color of a single position in the color table. Before you enter command SCT you are advised to refresh the screen by invoking the command R. Here we describe the procedure in which no mouse is available. Upon entering command SCT, the computer responds just as with command SC:

> The colors in the Color Table are numbered as follows:
> Color 0 (the background color) is the leftmost color;
> Color 15 is the rightmost.
>
> Enter color number to be changed, between 0 and 15 : __
> Hit <Enter> for No Change

Then enter the color number, for example color 4:

$$4 < Enter >$$

The computer prompts now as if you were setting a color with command SC:

> Now setting color 4
> Choose 3 color strengths (each between 0 and 63),
>   for RED, for GREEN, and for BLUE: __ __ __
> Sample setting for bright yellow -- 63   63   0   <Enter>

Then enter the red, green and blue levels for the first color number to be changed, that is, color 4 in this example. For example,

$$63 \ 0 \ 0 < Enter >$$

means "color 4 is set to maximum red, no green and no blue".
The computer makes the change in the color table and then prompts:

> Enter the new color number to be changed, between 4 and 15 : __
> OR   Hit <Enter> to TERMINATE color specification
>
> ALERT -- if you now set a new color number > 4+1, the program
>   will also reset the intermediate colors by interpolation.

If you do not want to change others, hit <Enter> again. The next color number entered must at least 4. You may enter the same color number twice, in case you don't like the color you get the first time. If you enter '6' and thereafter

$$0 \ 63 \ 0 < Enter >$$

the program will not only set color 6 to be maximum green as specified, but also set all color numbers between color 6 and the previous color number

---

specified (and that is just color 5). These intermediate positions are set by interpolation, so that in this case color 5 will be set to be 31 31 0.

If you enter a second color number after being prompted, the color number must be larger than the first. If the color numbers were, say, 4 and 12, then the program would also assign shades to 5 through 11, linearly interpolating the shades. If you then entered another color number and a set of 3 integers, such as for color 15, it would assign 15 as requested and would assign 13 and 14 by linearly interpolating.

Commands for selecting a color table

BGY

The "Blue to Green to Yellow" command sets the color table to run from blue to green to yellow.

BRI

This command sets a BRIght shade of the colors, especially color 1 through color 7. Color 4, for example, is the maximum level of red. The background (color 0) is black. This selection is sometimes useful when photographing the screen.

BRY

The "Blue to Red to Yellow" command sets color bands. This command sets the entire palette, whether 16 colors or 256. Color 0 is the unchanged background color; then the color table runs from blue to magenta to red to greenish red to yellow to white.

DEF

The "DEFault color table" command changes the color table back to the default setting on VGA screens and on EGA to a color table that is similar to the default.

RGB

The "Red to Green to Blue" command sets the color table to run from red to green to blue. Color 0 is the background color and is not changed.

RNB

The "RaiNBow" command sets the color table to run from red to orange to yellow to green to blue to indigo to violet, the so-called ROY G BIV pattern.

RRR

The "dim Red to Red to bright Red" command sets the color table to run from dim red to red to bright red.

RYB

The "Red to Yellow to Blue" command sets color bands. This command sets the entire palette, whether 16 palette positions or 256. Position 0 is the background color; then the color table runs from red to yellow to blue.

RYW

The "Red to Yellow to White" command sets the color table to run from red to yellow to white. Color 0 is the background color and is not changed.

WYR

The "White to Yellow to Red" command sets the color table to run from white to yellow to red. Color 0 is the background color and is not changed.

THE BACKGROUND COLOR MENU BCM

The background color (color #0) can be set using command SC. Alternatively, you can use the Background Color Menu BCM which has four different preset background colors. Enter the command BCM

$$BCM < Enter >$$

and the Background Color Menu appears on the screen.

```
╔═══════════════════════════════════════╗
║ BCM   – –   Background Color Menu       ║
║                                         ║
║ BLA   –   BLAck background color        ║
║ BLU   –   BLUe background color         ║
║ GRE   –   GREy background color         ║
║ WHI   –   WHIte background color        ║
╚═══════════════════════════════════════╝
```

Commands for changing the background color
The following commands are for changing the variables that are listed in the Background Color Menu above. Color 0 is the background color, so it can be changed to any desired color using "Set Color" command SC.

BLA

The "BLAck" command BLA resets the background color to black.

BLU

The "BLUe" command BLU resets the background color to blue.

GRE

The "GREy" command GRE resets the background color to grey.

WHI

The "WHIte" command WHI resets the background color to white.

THE ERASE COLOR MENU EM

Enter the command EM

EM *< Enter >*

and the Erase color Menu appears on the screen.

```
EM    – –       ERASE COLOR MENU

EK    –   Erase color K where K  <  16
EOD   –   Erase all ODd numbered colors
EEV   –   Erase all EVen numbered colors
PAT   –   erase using shaded PATtern 0 (=0 means total)
```

Commands for erasing colors

EK

The command EK can be used to erase color K, where K < 16.

EOD

The command EOD can be used to Erase all ODd numbered colors.

EEV

The command EEV can be used to Erase all EVen numbered colors.

PAT

Many printers allow only black and white without shades of grey. All colors are printed as black. You can nonetheless create pictures that print various greys. Before using this and EK, we recommend that you first store the picture with command TD (To Disk) in case you wish to change the pattern.

The PATtern command PAT can be used in junction with the command EK to create shading by erasing color K using a prescribed pattern, where K < 16.

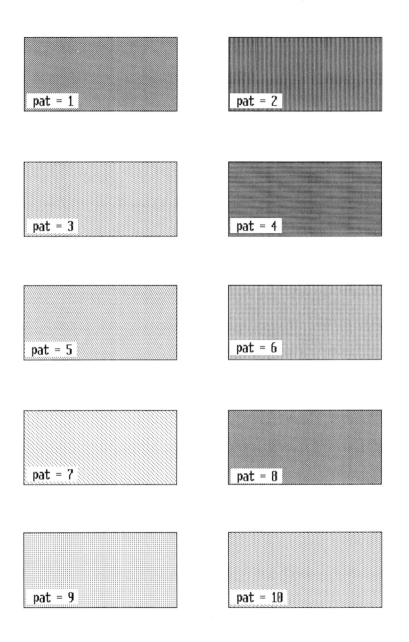

Figure 3-5: Erasing colors using PAT
This figure shows the result of erasing colors using PATtern PAT = k, where
k = 1, 2, ..., 10.

PAT = 0 means total erasure

PAT = 1 means 1/3 darkened pattern in core (2/3 erasure):
                    O O O O O O O O O O
                     O O O O O O O O O O
                    O O O O O O O O O O

PAT = 2 means vertical stripes in core (50% erasure):
                    O O O O O O O O O O
                    O O O O O O O O O O
                    O O O O O O O O O O

PAT = 4 means horizontal stripes in core (50% erasure):
                OOOOOOOOOOOOOOOOOOOOOOO

                OOOOOOOOOOOOOOOOOOOOOOO

PAT = 8 same as 1 but slanting the opposite way (2/3 erasure)

PAT "adds" so PAT = 2+4 means intersect the 2 patterns. Use PAT = 0, 1, 2, 3 = 1+2, 4, 5 = 1+4, 6 = 2+4, 7 = 1+2+4, 8, 9 = 1+8, 10 = 2+8, 11 = 1+2+8, 12 = 4+8, 13 = 1+4+8, 14 = 2+4+8, 15 = 1+2+4+8.

All of the above erase 50% or more of color K (when EK is called). Using negative numbers results in complementary patterns, representing 50% or less erasure. The default value of PAT is 0.

CHAPTER 4

UTILITIES

4.1 SETTING PARAMETERS (Parameter Menu PM)

The Parameter Menu lists the values of the parameters of any process. The parameters are chosen to have reasonable default values. For the process of the Henon map, enter the "Parameter Menu" command PM,

$$PM < Enter >$$

and both the map and the menu appear on the screen.

```
PM       – –      PARAMETER MENU

OK       –  parameters are fine as set
XCO      –  X COordinate to be plotted:  y[0]
XS       –  X Scale: from  -2.5  to  2.5
YCO      –  Y COordinate to be plotted:  y[1]
YS       –  Y Scale: from  -2.5  to  2.5
C1       –  C1 = -0.30000000
RHO      –  rho = 2.12000000
SD       –  Screen Diameters:  1.00


             MENUS
VM       –  Vector Menu for initializations, etc.
WWM      –  When and What to plot Menu
```

The process parameters are set using C1, C2, ... , RHO, PHI, etc. Only those process parameters that are used in the equations appear in the Parameter Menu. In the same sense, command STEP for changing the Runge-Kutta step size, appears in the Differential Equations Menu only when it is appropriate.

Commands for setting parameters of the process

The following commands are for changing the process parameters. The most commonly used parameters are C1, C2, ..., and RHO. Enter the parameter's name (like RHO) and the program will prompt for a new value.

Ck, where k = 1, ..., 9

This command is for changing the value of Ck, where k = 1, ... , 9.

PHI

This command is for changing the value of phi. This parameter occurs only in the periodically forced ordinary differential equations, like the Pendulum equation P and the Duffing equation D.

RHO

This command is for changing the value of rho.

Commands for setting parameters not being process parameters

The following commands are for changing the variables listed in the Parameter Menu above that are not process parameters. Enter the variable's name (like SD for Screen Diameters) and the program will prompt for a new value or values. As shown in Figure 3-3, the screen can be split into windows. Each window has its own values of XS, YS, XCO, YCO, and SD. If you change these values, you change it only for the active window.

XS

The "X Scale" command XS allows you to change the horizontal scale of the screen for plotting dots; the first number is the left hand side and the second is the right one. The two numbers must be entered with a space but no comma between them. (The first number can be larger than the second one if you want a decreasing scale.) For example:

XS *<Enter>* **-3.5 5.0** *<Enter>*

YS

The "Y Scale" command YS allows you to change the vertical scale of the screen. The first number is the coordinate of the bottom of the screen and the second one is the top. You must enter both with a space but no comma between them. The first number can be larger than the second if you want a decreasing scale.

XCO

Most variables take on numerical values. The value of the variable XCO is a name, the name of the horizontal COordinate variable. This command is for specifying the variable or parameter that is to be plotted horizontally. For the Henon map, the default value of XCO is "y[0]". With this setting the numerical value of y[0] is plotted horizontally. The value of XCO can either be a coordinate of the process (like y[0]) or a process parameter (like RHO or C1). To specify a coordinate like y[1] for the horizontal axis, type XCO<Enter> and then enter just the number of the coordinate. In the case of y[1], type

XCO<*Enter*> **1**<*Enter*>

To plot a parameter of the process, either RHO, PHI, or C1, C2, ... C9, type the name of the process parameter and <Enter>. For example,

XCO<*Enter*> **C1**<*Enter*>

If you choose a parameter and plot a trajectory, then the trajectory will appear on a vertical line, the line corresponding to that parameter value. Choosing a parameter can be appropriate in a variety of cases, including plotting basins (especially the basin of infinity, that is, points going to infinity as the parameter varies), or with commands FO and FOB for following periodic orbits, or for plotting trajectories, where the arrow commands are used for reinitializing to different parameter values.

As listed at the top of the Main Menu and the Parameter Menu, the Lorenz differential equations have four variables in this implementation and they are numbered as follows:

$$y[0] = X; \quad y[1] = Y; \quad y[2] = Z; \quad y[3] = \text{time}.$$

In the initial (default) setting the program plots variables y[0] (horizontally) against y[2] (vertically) when plotting 2 dimensionally. See command PT for plotting time as the horizontal coordinate of a map.

YCO

The value of YCO is a name, the name of the vertical COordinate variable. This command is for setting the variable or parameter that is to be plotted vertically, and is otherwise similar to XCO. You can assign either a variable of the process or a parameter of the process. You can even assign one parameter horizontally (using XCO) and another vertically in order to plot analogues of Mandelbrot sets (using command BAS or other related commands in the Basin of attraction Menu, see BM).

There is a trick that is available with XCO and YCO. Both can be set equal to the same variable. One would expect to get points plotted on the diagonal, but that would not be useful. Suppose for example, both coordinates are set to be variable 0. Then the program plots vertically the specified coordinate of the trajectory, but horizontally it plots that coordinate from the previous time. That is, the vertical coordinate is delayed. To make the delay several time steps instead of 1, use IPP to change the number of iterates per plot.

See PT for plotting Lyapunov exponents as a horizontal variable. See PRM for the vertical coordinate in bifurcation diagrams.

SD

A trajectory that goes "far" from the screen will be assumed to be "diverging", and is likely to be on route to infinity. So how far is "far"? The "Screen Diameters" parameter SD usually has a default value 1.0. Whenever it is time to plot a point, the program checks to see where the point is. If, for example, the point is more than SD widths of the screen left or right of the screen, or is above or below the screen by SD times the height of the screen, the program says the trajectory diverges. When such an event occurs, the program is paused, and can be restarted by hitting <space bar>. The trajectory must be reinitialized first, for example, by using the arrow keys and then command I. You can turn this feature off by setting SD to a negative number.

This parameter SD is also used by the basin boundary and saddle straddle trajectory routines to determine if the trajectory is diverging. If you want to specify that the trajectory has diverged as soon as it leaves the screen, then set SD equal to 0.

When the screen is rescaled while plotting using command OW, the program automatically rescales diameters, increasing it accordingly if a small section of the screen is blown up. Each window has its own value of SD. Command SD changes the value of SD for the currently active window (which is most often the whole screen).

THE WHEN AND WHAT TO PLOT MENU WWM

The Parameter Menu has one submenu, namely the When and What to plot Menu. Enter the "When and What to plot Menu" command

WWM < *Enter* >

and the When and What to plot Menu appears on the screen.

```
WWM  - -      WHEN  AND  WHAT  TO  PLOT  MENU

ALL    –  plot dots simultaneously in ALL small open windows: OFF
CON    –  CONnect consecutive dots:  OFF
DOTS   –  DOTS to be plotted:  100000000;  dot number now:  0
IPP    –  Iterates Per Plot:  1
MI     –  Maximum Iterates for ABI, II, OI:  1
PI     –  PreIterates before plotting:  0
PRM    –  PaRaMeter to be varied by  +/- :  "rho"
             Allowed settings for PRM:  C1   RHO
PS     –  Parameter Step that +/- change "rho": 1.0000000000
PT     –  T Plots Time horizontally if ON; now   OFF
V      –  inVert process:  AVAILABLE
```

Commands for changing "When and What to plot" variables

ALL

ALL is for plotting in ALL active small windows. If more than one small window has been opened for plotting, then command "all" causes points to be plotted in all the small windows simultaneously. This capability is turned off by entering the command again or by hitting one of the function keys that tells the computer to go to a specific window.

ALL is a toggle: if ALL is ON, enter it again and it turns OFF.

CON

The "CONnect consecutively plotted dots" command CON causes successive dots to be connected by straight lines. It can also be used for maps, though the results are strange-looking graphs. This feature turns itself off momentarily in a number of situations, such as if one dot is outside the screen area, since then it may not make sense to connect dots before and after the outlier. Also, if a coordinate of y is changed using the modulo calculations, for example, if the angle variable in the pendulum goes beyond π and is decreased using a mod calculation, then the resulting dot is not connected to the previous dot. While plotting the time-2π return map, one may find that between two consecutive points, the program had to alter one of the phase space coordinates using a modulo calculation, so those points will not be connected.

CON is a toggle: if CON is ON, enter it again and it turns OFF.

CONT

The "CONTinue picture" command CONT is used when a command has been stopped and should be restarted. It indicates that when a trajectory is plotted, the dot number should CONTinue from the current value, assuming it is less than DOTS. This command is particularly useful in continuing the computation of a basin, since if this command is not invoked, the "BASin" command BAS will result in the computation beginning anew from the upper left of the picture. This command is **not listed in the menus**.

DOTS

DOTS is for setting the total number of dots. This variable has a very large default value so it does not need to be set unless you want the program to stop at a specific dot. This is the total number of dots that will be used in making a picture, including attempts to plot dots that are off the screen. The loop variable 'DOT' that runs up to 'DOTS' is reset to zero **while plotting** whenever the trajectory vector is initialized using command I or is replaced using command RV. In addition, the command

$$RV < Enter > \ 0 < Enter > \ 0 < Enter >$$

(that is, replace vector 0 with itself) is a dummy Replace Vector command that resets the DOT counter without an actual replacement of a vector.

A basic unit in the program is the amount of computation required for calculating one point for plotting. Hence you specify the total number of dots to be plotted. If some turn out to be outside the area of the screen, they still count in the total. The basic "do-loop" uses DOT as a counter. Command PI lets you specify the number of preiterates, that is, the number of dots computed before the plotting begins. The do-loop begins at minus the number of PreIterates and actually only begins plotting when the counter reaches 0. This is useful for getting pictures of chaotic attractors in cases where one wants to give the trajectory time to reach the attractor. Command DOTS gives the upper end of the loop.

IPP

IPP is to set the Iterates Per Plot; typically the value is 1. The process is iterated IPP times before each plot. This command is particularly useful for obtaining the time-2π map from a differential equation that is forced periodically. If SPC (Steps Per Cycle) in the Differential Equation Menu is set to say 25, then setting IPP to 25 will mean that one point is plotted every cycle.

This command can also be useful for studying maps even if they do not come from differential equations. To examine the behavior of a period 5 point of the Henon map, you may find it convenient to look only at every fifth iterate.

MI

The "Maximum number of Iterates" command MI is useful when carrying out routines using the commands II, OI, and ABI (see Chapter 9 for the command ABI). For example, if MI = 10 then invoking the command OI results in drawing a circle (centered at y1) with radius RA and its first 10 iterates.

PI

PI is for setting the number of PreIterates, that is, the number of iterates computed before the first point is plotted. When iterating a process, it may be iterated PI times before plotting begins. The number of PreIterates PI should be at least 0. The usual default is 0.

PRM

The command PRM is for specifying the PaRaMeter, for example, the parameter to be varied in bifurcation diagrams. As with the commands XCO and YCO, the value of the variable PRM is a name. The value of PRM is the name of the parameter to be varied. PRM specifies which parameter is varied in making bifurcation diagrams, using commands BIF and BIFS. It also specifies the parameter used in the "Follow Orbits" commands FO and FOB. Initially it is RHO, but can be changed to any of the process parameters, namely RHO, PHI, C1, C2, ...

The value of variable PRM is also the name of the parameter that is changed while the computer is plotting if you hit the $<+>$ and $<->$ keys. Typing $<+>$ or $<->$ while plotting increases or decreases the numerical value of the parameter PRM by the Parameter Step amount PS.

PS

Hitting $<+>$ or $<->$ while plotting changes the numerical value of the parameter PRM. (See the PaRaMeter command PRM.) The Parameter Step variable PS is the size of the change that $<+>$ and $<->$ make. PS is the amount that

RHO (or whatever process parameter is named by PRM) is changed by when typing $<+>$ or $<->$ while the program is running.

The simplest use of command PS is for changing RHO using the $<+>$ and $<->$ keys while the computer is plotting. While plotting, this step size can be changed without returning to the menu by using either the $<$Home$>$ key on the numeric key pad (to cut in half the amount that RHO changes by) or the $<$PgUp$>$ key to double the step size.

PT

The "Plot Time" command PT is a toggle. When PT is ON, the Trajectory command T plots the time horizontally. Vertically it plots either the Lyapunov exponents or y[1] depending on whether $L > 0$ or $L = 0$.

When the Lyapunov exponents are not being computed ($L = 0$) and PT is ON, then T plots y[1] vertically versus the time horizontally.

When the Lyapunov exponents are being computed ($L > 0$) and PT is ON, then T plots the Lyapunov exponents vertically versus the time horizontally.

The consecutive dots can be connected by invoking the "CONnect consecutive dots" command CON.

V

Command V is for making certain routines (like the trajectory T) run backward in time. We say the command V "inVerts" the process. It is available for all differential equations and most maps. For maps the inverse needs to have been programmed into the program. It is a toggle, so entering command V again returns the process to normal. This command can be used for examining features on attractors, and seeing where the point comes from. Pause the trajectory before entering V. The command V is often used with the "Trajectory" command T and the "Initialize and Iterate" command II.

Example. While plotting a trajectory of the Henon map, to find where some point came from, that is, what the inverse of the Henon map is for that point, hit $< \cdot >$ to pause the program, and move the small cross to the point in question. Enter I to Initialize, that is, to move the large cross to the position of the small cross. Invert the system by entering command V. Now hit $< \cdot >$, which makes the inverted map take one step, (sometimes it may be necessary to hit it twice) and the new point will be shown by the position of the large cross. Hitting $< \cdot >$ again shows where that point came from. Do not forget to invert the system again to return to the normal mode of operation. Hit $<$*space bar*$>$ and the program resumes plotting.

4.2 SETTING AND REPLACING A VECTOR (Vector Menu VM)

Recall from Chapter 2 that the storage vector y1 is used for initializing the system and that the mouse and the arrow keys can move the small cross representing this initializing vector y1.

The vectors y0, y1, y2, y3, ..., y9 and ya, yb, ..., ye are storage vectors. Each of these vectors can store some data, and every vector can be replaced by any other vector, that is, every vector can be set equal to any other vector. To do this, enter the "Replace Vector" command RV by typing

$$\text{RV} <Enter>$$

followed by

$$\text{k} <Enter>$$

where k = 0, 1, 2, ..., 9, a, b, ..., e. Then type in

$$\text{m} <Enter>$$

where m = 0, 1, 2,..., 9, a, b,..., e. This means that the vector yk will be set equal to the vector ym, or in other words, the data in vector yk is replaced by the data in vector ym. When the program is plotting a trajectory, vector y (or y0) contains the current coordinates of the trajectory. To store the current position in one of the storage vectors, type

$$\text{RV} <Enter> \quad 2 <Enter> \quad 0 <Enter>$$

which results in the data in storage vector y2 being replaced by the data in storage vector y0. You are advised to enter the command YV (Y Vectors) to convince yourself of the desired storage of data in vector y2.

Note. There is an alternative short hand version, namely command RVkm, where k and m use the above values. For example,

$$\text{RV} <Enter> \quad 2 <Enter> \quad 0 <Enter>$$

can be replaced by

$$\text{RV20} <Enter> \ .$$

To see what this command does, we assume the program is plotting a trajectory of the Henon map. Hit the period key $<.>$ to pause the program. Hit $<.>$ to single step the trajectory until, for example, it is near a particular feature of the attractor. Enter the command RV, that is, type RV$<$Enter$>$, hit 0 or 1 or 2 or ... 9 or a or b ... or e followed by $<$Enter$>$ to "Replace this Vector" by the vector corresponding to another location (that is, store in this vector a different set of coordinates). Then hit 0 or 1 or 2 or ... 9 or a or b ... or e followed by $<$Enter$>$. For example, assuming you have paused the program by hitting $<.>$, to store the current position in storage vector y2, type RV$<$Enter$>$ followed by 2$<$Enter$>$ and 0$<$Enter$>$. The data in storage vector y2 is replaced by the data in storage vector y0. The vector y0 is unaffected by this change.

In particular, typing RV < Enter > 1 < Enter > 0 < Enter > will reset the coordinates of y1 to the current state of y (or y0). The current state of the trajectory is stored and the coordinate values of y remain unaffected by this change. As discussed in Chapter 2, when you enter the initialization command I, the vector y is replaced by the vector y1 so the trajectory will start from the initialization vector that was just stored.

The Vector Menu VM

The Vector Menu displays the coordinates of the storage vectors. After entering the command **VM** < *Enter* > the Vector Menu will appear on the screen.

VM   – –   VECTOR MENU
CENT – CENTer small cross (y1)
DIAG – DIAGonal: set ya = lower left screen corner and yb = upper right screen corner
I – Initialize: change y to equal y1
I1 – Initialize y1: change y1 to equal y
RV – Replace one storage Vector by another vector
SV – Set values of a storage Vector
YK – list coordinates of yK for K = 0,1,...,9 or a,b,c,d,e
YV – list coordinates of several storage vectors
Initial conditions for trajectory vector y reside in vector y1; To change coordinates of y1, use command SV and select SV1, or while plotting, use the left mouse button. To initialize y, use command I.

After entering the command YV, the Y Vectors appear on the screen.

YV – – Y VECTORS		
Vectors -- y0 = y = current state, y1 = cursor position		
state vec y	storage vec y1	storage vec y2
y[0] = 0;	y1[0] = 0;	y2[0] = NOT SET
y[1] = 2;	y1[1] = 2;	y2[1] = NOT SET
storage vec ya	storage vec yb	storage vec ye
ya[0] = -2.5;	yb[0] = 2.5;	ye[0] = NOT SET
ya[1] = -2.5;	yb[1] = 2.5;	ye[1] = NOT SET

Y Vectors

The Y Vectors list displays some of the storage vectors. The vectors of the program are denoted y (alias y0), y1, y2, ... , y9, and ya, ... , ye. Each of these vectors can have up to 21 coordinates, depending on the process and on whether Newton's method is being used or Lyapunov exponents being calculated.

"Own" and the coordinates of y[ ]

When using "Own" to define your own map or differential equation, the primary variables allowed are r, s, t, x, y, and z. Those used in "Own" are assigned to y[0], y[1], ... in alphabetical order. If you use s, x and y, then they become y[0], y[1] and y[2], respectively. For more details, see Chapter 13.

Vector Coordinates

The coordinates are numbered starting with 0 (as is forced upon users of the language C). The coordinates are indicated with square brackets [ ] so coordinate 2 of the initialization vector y1 is written y1[2].

What is in the Y Vectors list?

Vectors y, y1, y2, ya, yb and ye are always listed. The vectors y3, y4, y5 are listed only when at least one of these three vectors is set. The vectors y6, y7, y8, y9, yb, and yd are never listed.

Some of the (storage) vectors have special purposes:

● The vector y1 (marked by the small cross) is the initialization vector and y is set equal to y1 when entering the command I. It is the point that the Newton method takes as its initial point. This is the only vector that can store parameter information. When one of the screen's coordinates is a parameter, you can still move the small cross along that coordinate using the arrow keys thereby changing that parameter value, associated with y1. Then y and y1 will differ in that coordinate. Another way the parameter values stored in y1 can be used is with the "Follow Orbit" commands FO and FOB (Follow Orbit Backwards), which change a parameter. See command PRM which names the parameter being varied. While the parameter value of the parameter named by PRM changes, the initial value is stored in y1 and the parameter value can be reset using command I.

● The vectors y2, y3, y4 are used in the basin routines (see BM) to designate points of an attractor whose basin points will be plotted.

● The vectors y5, y6, y7 are used in the basin routines to designate points of an attractor whose basin points will **not** be plotted.

● The vectors ya and yb denote two ends of a line segment to be used in the straddle trajectory calculations; see commands BST (Basin boundary Straddle Trajectory) and SST (Saddle Straddle Trajectory). See also command DIAG (DIAGonal) which sets ya to be the lower left corner of the active window and sets yb to be the upper right corner of this window.

Any of the vectors can be used to save temporary data.

Commands for setting and replacing vectors

CENT

The command CENT is to CENTer small cross (y1). When losing track of where the small cross is, enter the command CENT and the small cross will show up in the center of the active window.

DIAG

The storage vectors ya and yb play an important role in the straddle trajectory calculations (ya and yb denote two ends of line to be used in these calculations). When you enter the command DIAG (DIAGonal) the vector ya is set to be the lower left corner of the active window and yb is set to be the upper right corner.

I1

The command I1 is a combination of the commands "Initialize" and "Replace Vector". The command I1 replaces the initialization vector y1 by the state vector y, and it initializes dot and Lyapunov exponents.

RV

The "Replace Vector" command RV is for replacing a storage vector by another one; that is, it sets the coordinates of a vector equal to the coordinates of the vector by which it will replaced. When the command RV is entered, the program first asks for the vector to be replaced, that is, the vector whose coordinate values will be changed; the second vector will replace the first vector, that is, the coordinates of the second vector will be the new coordinates of the first vector. The trajectory vector is y0 (or y), the small cross is at y1. Replacing y resets the dot number to 0 and reinitializes the computation of the Lyapunov exponents. See also the discussion at the beginning of this section.

SV

The "Set Vector" command SV is for setting the coordinates of a storage vector to desired values.

YK where K = 0, 1, ..., 9, or a, ..., e.

This command is for listing the coordinate values of the storage vectors y0, ..., y9, ya, ..., ye.

YV

The "Y Vectors" command YV displays some of the storage vectors.

Setting vectors with the mouse

When there are no menus on the screen, move the mouse arrow to the desired location. Depress the left button (without moving the mouse) until you hear a beep (about one second) and release it. A menu of options MVM will appear.

```
┌─────────────────────────────────────────────────────────────┐
│                  MOUSE SET VECTOR MENU                        │
│                                                               │
│  The following use the position of the mouse at the beep.     │
│  MV     –  set y and y1 to current mouse arrow position       │
│  cancel –                                                     │
│  MV0    –  Set Vector y          MV1   –  Set Vector y1       │
│  MV2    –  Set Vector y2         MV3   –  Set Vector y3       │
│  MV4    –  Set Vector y4         MV5   –  Set Vector y5       │
│  MV6    –  Set Vector y6         MV7   –  Set Vector y7       │
│  MV8    –  Set Vector y8         MV9   –  Set Vector y9       │
│                                                               │
│  MVA    –  Set Vector ya         MVB   –  Set Vector yb       │
│  MVC    –  Set Vector yc         MVD   –  Set Vector yd       │
│  MVE    –  Set Vector ye                                      │
│                                                               │
│  MQ9    –  apply Quasi-Newton 9 times                         │
│  MWW    –  initialize and WW:                                 │
│              check When and Where the trajectory goes.        │
└─────────────────────────────────────────────────────────────┘
```

4.3 SETTING STEP SIZE (Differential Equation Menu DEM)

If you enter the command DEM (that is, if you type **DEM** <*Enter*>), while in the Main Menu for the Henon map, the Differential Equation Menu appears on the screen.

```
DEM      – –      DIFFERENTIAL EQUATION MENU

OK       –  The current process is not a differential equation
EULER    –  Euler solver, fixed step size
RK4      –  4th order Runge-Kutta solver, fixed step size
RK5      –  5th order Runge-Kutta solver, variable step size
RK6      –  6th order Runge-Kutta solver, fixed step size

E        –  compute single step Error every  5  steps:  OFF
LE       –  Local Error for RK5:  1e-007
RET      –  Poincare RETurn map (for process HAM) and for
            periodically forced diff. equations (using phi)
DF       –  plot the Direction Field
```

The above menu also declares: "The current process is not a differential equation". This is due to the fact that the menu is presented for the Henon map, which is not a differential equation. For the Lorenz and Pendulum differential equations, there are some additions to the above.

If you select the DiffEqM or if you enter the command DEM, while in the Main Menu for the Lorenz equations, the following menu appears on the screen.

```
DEM      – –      DIFFERENTIAL EQUATION MENU

OK       –  The current solver has order 4
EULER    –  Euler solver, fixed step size
RK4      –  4th order Runge-Kutta solver, fixed step size
RK5      –  5th order Runge-Kutta solver, variable step size
RK6      –  6th order Runge-Kutta solver, fixed step size

E        –  compute single step Error every  5  steps:  OFF
LE       –  Local Error for RK5:  1e-007
RET      –  Poincare RETurn map (for process HAM) and for
            periodically forced diff. equations (using phi)
DF       –  plot the Direction Field
STEP     –  STEP size for differential equation:  0.01
```

The Lorenz differential equations are the following.

$X' =$ sigma*(Y-X), $Y' =$ rho*X - Y - X*Z, $Z' =$ X*Y - beta*Z.

Numbering of variables: y[0]=X, y[1]=Y, y[2]=Z, y[3]=time.
For the Lorenz equations, the step size can now be selected if the solver used is either Euler, RK4, or RK6.

The forced damped Pendulum equation is the second order differential equation
$$X'' + C1*X' + C2*\sin X = \text{rho}*(C3+\cos(\text{phi}*t))$$

The numbering of variables is y[0]= time t (mod 2π/phi), y[1]=X, y[2]=X', y[3]=t.

If you select the DiffEqM or if you enter the command DEM while in the Main Menu for the forced damped Pendulum, the menu appears on the screen in approximately the following format. Notice that the last command SPC has changed.

```
DEM    – –    DIFFERENTIAL EQUATION MENU

OK      –  The current solver has order 4
EULER   –  Euler solver, fixed step size
RK4     –  4th order Runge-Kutta solver, fixed step size
RK5     –  5th order Runge-Kutta solver, variable step size
RK6     –  6th order Runge-Kutta solver, fixed step size

E       –  compute single step Error every  5  steps:  OFF
LE      –  Local Error for RK5:  1e-007
RET     –  Poincare RETurn map (for process HAM) and for
              periodically forced diff. equations (using phi)
DF      –  plot the Direction Field
SPC     –  Steps_Per_Cycle:  30.000000
```

Commands for differential equations

DF, FG

The "Direction Field" command DF is for plotting the Direction Field of a differential equation.

The "Field Grid" command FG is for specifying the grid used by DF (Direction Field) and VF (Vector Field).

E

The "Errors" command E is for differential equations when using fixed step size differential equation solvers; it computes estimates of the local error, that is, it reports how big the error is in taking a single step. E does this repeatedly, about every 5 Runge-Kutta steps. Command E is a toggle: enter E again and it turns off.

To carry out E, the program compares the last differential equation step with the result of two steps half as big that start from the same point. For fourth order Runge-Kutta, we can expect the two small steps to be about 32 times as accurate as the big step. The reported error is simply the distance between the two results. This is just the single step (local) error, not the cumulative error which gets arbitrarily large exponentially fast in a chaotic system. The ratio of that error to the actual size (Euclidean norm) of the step is also given. The maximum of these individual errors is also stored. One may note missing points in the plotted trajectory, since such evaluated points are not printed. In this way, one can tell which step was tested. Use command EE to retrieve results of E.

EE

The "Express Error" command EE results in printing results of command E including the maximum local error encountered thus far. The command EE is **not listed in the menus**.

EULER

EULER is for choosing Euler differential equation solver. This command makes the differential equation solver a single step Euler solver.

LE

LE is for setting the Local Error for RK5. This command sets the target error to be LE for each individual time step, and the size of the time step is varied accordingly. The error estimate is obtained by comparing the fifth order solver with the sixth order solver.

RET

For the Hamiltonian differential equation HAM, this "RETurn map" command causes command T to produce Poincare return map iterates. It works only for the process HAM and for the periodically forced differential equations (using phi). To use the Poincare return map of the Lorenz system, choose the process LPR in the PROcess Menu.

RK4

This command makes the differential equation solver a fourth order fixed-step-size Runge-Kutta solver. This is the usual default solver, the exception being HAM which has RK5 as the default solver.

RK5

This command makes the differential equation solver a fifth order variable step size Runge-Kutta solver. This should not be used with differential equations that use the parameter PHI, since the map that corresponds to one time-cycle expects a fixed number of time steps, and so a variable step solver cannot be used. The Runge-Kutta-Verner method used here estimates the error for a single step of the differential equation using the RK6 solver. See Verner (1978).

RK6

This command makes the differential equation solver a sixth order fixed step size Runge-Kutta solver. See references for RK5.

SPC

The "Steps Per Cycle" SPC command is used instead of STEP when the differential equation has periodic forcing; if the period of the forcing is 2π/phi (as is true in several cases) you must tell the program the number of Steps Per Cycle period rather than the exact step size.

STEP

STEP is the STEP size of the differential equation solvers RK4 and EULER.

VF, FG

The "Vector Field" command VF is for plotting the Vector Field of a differential equation.

The "Field Grid" command FG is for specifying the grid used by DF (Direction Field) and VF (Vector Field).

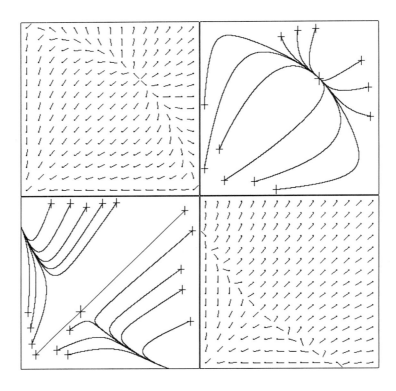

Figure 4-1a: Direction field and trajectories of ODE

In all the pictures of Figure 4-1 the direction field and some trajectories are shown for the Lotka/Volterra differential equations

$$X' = (C1 + C3*X + C5*Y)*X, \quad Y' = (C2 + C4*Y + C6*X)*Y$$

where X is plotted horizontally and Y is plotted vertically. The initial value of each trajectory is indicated with a small cross; a larger cross indicates an equilibrium. Set CON to be ON or set STEP = 0.001.

In the upper windows, C1 = 0.5, C2 = 0.5, C3 = -0.0005, C4 = -0.0005, C5 = -0.00025, and C6 = -0.00025. The X Scale runs from 0 (left) to 1000 (right), and the Y Scale runs from 0 (bottom) to 1000 (top).

In the lower windows, C1 = 0.5, C2 = 0.5, C3 = -0.0005, C4 = -0.0005, C5 = -0.00075, and C6 = -0.00075. The X Scale runs from 0 (left) to 1200 (right), and the Y Scale runs from 0 (bottom) to 1200 (top).

Topic of discussion. *The eigenvalues and eigenvectors of the derivative of the vector field at an equilibrium determine the behavior near the equilibrium. What do the pictures in all of Figure 4-1 reveal about the eigenvalues and eigenvectors at the equilibria?*

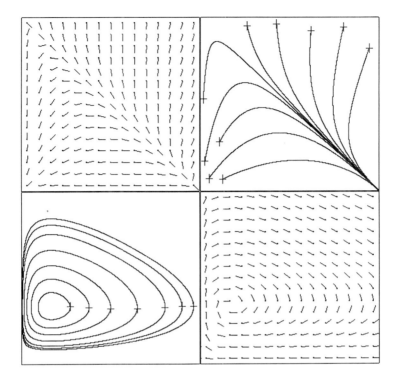

Figure 4-1b: Direction field and trajectories of Lotka/Volterra equations
    In the upper windows, C1 = 0.5, C2 = 0.5, C3 = -0.0005, C4 = -0.0005, C5 = -0.00025, and C6 = -0.00075. The X Scale runs from 0 (left) to 1000 (right), and the Y Scale runs from 0 (bottom) to 1000 (top).
    In the lower windows, C1 = 0.5, C2 = -0.1, C3 = 0, C4 = 0, C5 = -0.01, and C6 = 0.0002. The X Scale runs from 0 (left) to 3000 (right), and the Y Scale runs from 0 (bottom) to 150 (top).

    Topic of discussion. *What happens to the plots of the closed orbits shown in the lower left window if a larger step size STEP is used by the Runge-Kutta differential equation solver? What if the Euler method is used to solve the equation? See the Differential Equations Menu DEM.*

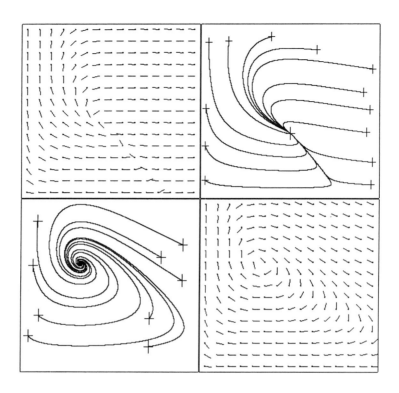

Figure 4-1c: Direction field and trajectories of Lotka/Volterra equations
In the upper windows, C1 = 0.5, C2 = -0.1, C3 = -0.000625, C4 = 0, C5 = -0.01, and C6 = 0.0002. The X Scale runs from 0 (left) to 1000 (right), and the Y Scale runs from 0 (bottom) to 50 (top).

In the lower windows, C1 = 0.5, C2 = -0.1, C3 = -0.00025, C4 = 0, C5 = -0.01, and C6 = 0.0002. The X Scale runs from 0 (left) to 1500 (right), and the Y Scale runs from 0 (bottom) to 60 (top).

4.4 SAVING PICTURES AND DATA (Disk Menu DM)

The Disk Menu displays an overview of using the disk. Enter the command DM, (that is, type **DM** <*Enter*) and the Disk Menu will appear on the screen.

DM	– –	DISK MENU
OK	–	Current root name for disk files: h
RN	–	Root Name (to change root h)
DD	–	Dump Data to disk file h.dd
PCX	–	copy picture to PCX file h.pcx readable by MS Windows
		The following 5 commands use the file h.prn
PPP	–	if on, Printer output goes to file h.prn: OFF
PSC0	–	copy picture to PostScript Color file with no text header
PSC2	–	copy picture to PostScript Color file with text header
PSP0	–	copy picture to PostScript file with no text header
PSP2	–	copy picture to PostScript file with text header
REC	–	RECord all keystrokes in file h.rec: OFF
TD	–	copy picture and parameters To Disk file h.pic
TDD	–	call TD every 0 Dots (no call if TDD = 0)
TDM	–	call TD every 0.0 Minutes (no call if TDM = 0)
		MENU
RDM	–	Read Disk files Menu

Commands DD and TD save current data/picture in h.dd or h.pic.
  Data can be later restored by starting the program by
  DYNAMICS h.dd < Enter >   or   DYNAMICS h.pic < Enter >

Creating a batch file of commands

A batch file of commands can be created using an editor or word processor, or it can be done automatically using the command DD (Dump Data) listed above. As an example we have used an editor to create *lorenz.dd* which is for the Lorenz differential equations. It causes one of the variables, $X(t)$, to be plotted vertically against time horizontally (instead of the default for the Lorenz map: $X(t)$ horizontally against $Z(t)$ vertically). This requires resetting a number of parameters. You can see all these settings by viewing the file with an editor before running the program, that is, when one has the DOS prompt. Command DD creates similar but longer files. You never have to personally read a file created by the DD command since the program will do it. So in fact, one needs not worry about how to interpret it.

The comments below should **not** be in the actual file.

L	*Code for the Lorenz system*
XCO	
3	*Coordinate 3 is time t (horizontal axis)*
YCO	
0	*Coordinate 0 is X(t) (vertical axis)*
STEP	*Set differential equation time step size at:*
0.005	*0.005*
XS	*Set horizontal coordinate to run from*
0 60	*0 to 60.*
YS	*Set vertical coordinate to run from*
-22 22	*-22 to +22*
D2	*Two dimensional plotting*
FileEnd	*Take input from keyboard.*

When the program is plotting, the interactive keys such as $<.>$, $<$Tab$>$, $<$Esc$>$, $<$space bar$>$, and the function keys $<$F1$>$ - $<$F10$>$ are operational, even if FileEnd has not yet been encountered.

Commands for saving pictures and data

The following commands are for changing the variables that are listed in the Disk Menu above. The description of PPP can be found in section 4.6.

RN

Pictures can be stored on disk with all current parameter values. The root name must be permissible by the operating system. MS-DOS allows a root name up to 8 letters, then a period followed by up to 3 letters of an extension provided by *Dynamics*.

If you store a picture on the disk or to retrieve it from the disk, you have to specify the root name or use the default name. The "Root Name" command RN is for changing the disk file's Root Name. You provide the root name and *Dynamics* provides the tailing "." and the extension. You can precede the name by the disk designation, so you might use a name like "A:h" and the program will use A:h.pic or A:h.dd or A:h.prn depending on what is in the file. The picture that is stored using the "To Disk" command TD is the high resolution copy, not a copy of the screen. Current default resolution is 720 dots horizontally by 720 vertically for PC's if your computer has sufficient memory. Substitute for "A:" above the appropriate disk drive. If the picture is in the current disk drive directory, no drive designation is necessary, so you would have "h.pic" where ".pic" is added by *Dynamics* to denote the file contains a picture. If you store just the data and parameters in a file, then the file name is the same except that after the period the program automatically inserts "dd".

If you are going to create disk files to save data or pictures that you have spent time creating, then make sure the disk has enough space to save it. Before saving data with the commands DD (Dump Data) or TD (To Disk), be sure to have space available. The files h.pic and h.dd are not actually created unless you use the commands TD and DD respectively.

If you store data using command PPP or if you are on a Unix computer and send a picture or data to the printer, then the data will be stored in a corresponding file with the tail ".prn". On Unix computers you cannot send files directly to the printer so it is sent to a printer file. Using the "Print Picture" command PP then results in a file such as h1.prn. It uses the root name you have provided or the default. Using RN to create a new root name causes existing files to be closed, so they can be accessed later.

Note. If the file is stored on one disk (for example using "A:" as part of the name), it becomes difficult to read it from a different disk drive because the name says it is on the disk in the "A-drive".

TD

The "To Disk" command TD copies the current picture To a Disk file. See command RN (Root Name) for the name of the file. The picture also remains in the program. The disk file includes almost all current values of the parameters (and your description of the map or differential equation if made with "Own"). The values are stored in 15 digit format and so the values are not exactly the same as the computer binary representation. The disk space needed varies depending on the number of parameters and vectors to be saved, but you can estimate 70 KBytes.

● On Unix computers, be sure you have permission to create files.

● If a file with that name already exists, the program will write over the old file, destroying the old one.

The file is mixed binary and text, with all current values of the parameters in text.

Now if you have created a file named h.pic, the only way to get *Dynamics* to access the parameter settings in this file is to use the file to start the program from DOS using the DOS line

dynamics h.pic < *Enter* >

This resets the parameters and vectors to their previous value according to the values in the file. See also the commands RN (Root Name), DD (Dump Data) and FD (From Disk). You can use commands FD and AFD (Add From Disk) to retrieve the picture while the program is running.

DD

The "Dump Data" command DD creates a batch file that contains the process name and the values of most constants and vectors. If you used "Own" to create a map or differential equation, it is also included. The file is just like the one created using command TD (To Disk) except the picture is not saved. To use such a file, you must not be using *Dynamics*. This dump data file is a batch file for starting *Dynamics*. Ordinarily you would start the program with the command (the Unix operating system uses case sensitive file names and there the program name is all lower case letters):

dynamics < *Enter* >

Now if you have created a data dump file named lorenz.dd the program can be started from DOS using the DOS line

dynamics lorenz.dd < *Enter* >

This resets the parameters and vectors to their previous value according to the values in the file. The picture is **not** contained in this file. See also the commands RN (Root Name) and FD (From Disk).

PCX

The command PCX instructs the program to create a file called "root.pcx", a copy of the current picture, in PCX format. This format is used by Microsoft Windows and Microsoft Paintbrush. Windows can then be used for printing the picture. The picture can be read into Windows/Paintbrush by the DOS command

win pbrush root.pcx

Substitute the actual root name for **root**. See also "Printing color pictures" in Section 4.6.

PSC0, PSC2

Commands PSC0 and PSC2 are shorthand commands that combine what would otherwise take a series of steps:

PSC0 = T0 + PSC + PPP + PP and
PSC2 = T2 + PSC + PPP + PP

then PPP is turned off; see also PPP (section 4.6). They directly send the picture to your disk in PostScript Color format. PSC0 creates the file with no header. Command PSC2 includes the header that describes the process and parameter values.

PSP0, PSP2

Commands PSP0 and PSP2 are shorthand commands that combine what would otherwise take a series of steps:

PSP0 = T0 + PSP + PPP + PP and

PSP2 = T2 + PSP + PPP + PP

then PPP is turned off; see also PPP (section 4.6). They directly send the picture to your disk in PostScript format. PSP0 creates the file with no header. Command PSP2 includes the header that describes the process and parameter values.

REC

The "RECord" command REC is a toggle, and can be invoked before or after a process is selected. When REC is turned ON, it instructs the program to store all the keystrokes in a file. The file name ends in ".rec" and is printed in the Disk Menu DM.

TDD

The "To Disk Dots" command TDD causes the program to execute the "To Disk" command TD at regular intervals while plotting. It saves the current picture and data in the file whose name is specified in the Disk Menu DM. The saving occurs every TDD dots while plotting. Since the same file is used repeatedly, old results are wiped out. This is designed for routines that take a long time to run, to make sure results are not lost by a power failure. For example, TDD can be chosen to be 100000 (dots).

When the window and core copy are cleared (command CC), TDD is automatically set to 0 to avoid accidentally destroying a valuable file that you have created.

TDM

The "To Disk Minutes" command TDM causes the program to execute the "To Disk" command TD every TDM minutes while plotting. See command TDD. Since the same file is used repeatedly, old results are wiped out. For example, TDM can be chosen to be 100 (minutes).

When the window and core copy are cleared (command CC), TDM is automatically set to 0 to avoid accidentally destroying a valuable file that you have created.

Figure 4-2: Trajectories of the Gumowski/Mira map

The figure shows four trajectories of the initial condition (x_0, y_0) of the Gumowski/Mira map (GM) for C1 = 1, C2 = 0 and RHO = -0.1

$$(X, Y) \rightarrow (Y + F(X), -X + F(Xnew)$$

where $F(u) = -0.1u + 2.2u^2/(1+u^2)$, X is plotted horizontally and Y is plotted vertically. The X Scale and Y Scale both run from -30 to 30.

Upper left window: $(x_0, y_0) = (1,1)$; upper right window: $(x_0, y_0) = (11,9)$; lower left window: $(x_0, y_0) = (18,18)$; lower right window: $(x_0, y_0) = (15,15)$

After the pictures have been created and stored To Disk, they have been Added From Disk by setting AFDR = 0.9 and using the commands AFD1, AFD2, AFD3 and AFD4.

Topic of discussion: *What is the nature of the plotted trajectories? Assuming that the trajectory in the lower left window is dense on a closed curve C. Is is true that any trajectory that starts at a point in the inside region of C will stay in that inside region forever?*

THE READ DISK FILES MENU RDM

Enter "Read Disk files Menu" command

RDM *< Enter >*

and the Read Disk files Menu appears on the screen.

```
RDM    - -    READ DISK FILES MENU

AFD    -  Add picture From Disk file h.pic into core picture
AFD1   -  Add picture into window 1 (reducing size of pic)
AFD2   -  Add picture into window 2
AFD3   -  Add picture into window 3
AFD4   -  Add picture into window 4
AFD5   -  Add picture From Disk in format and position you specify
AFDR   -  AFD Ratio changes size of window for AFD1,...,AFD4:   1
FD     -  read picture From the Disk file h.pic
```

Commands for reading disk files

AFD

The "Add From Disk" command AFD adds the picture from Disk file into core picture, so that the two pictures are overlaid. You will be prompted for the name of the file.

The disk file copy remains on the disk unaltered. This command enables you to create a number of files with different parts of a picture, and then add them together.

AFD1

The "Add From Disk to W1" command AFD1 is like AFD, but they differ in that the picture will now appear in the upper left quarter of the screen. This command does not open window W1.

This command together with the commands AFD2, AFD3, and AFD4 enables you to create four different pictures, and then add them together as 4 windows in one picture. Storing such 4 window pictures and repeating this process permits 16 pictures in one figure.

Figure 4-3a: Computer art: Fractals and Dynamics

You may create the picture in this figure as follows. First, start with a triangle inside a triangle using the command AB (setting the vectors ya, yb, yc and yd = ya). If you use the Tinkerbell map, for example, call this picture t0.pic and store it To Disk. Next, set XPS to be 180 540 and set YPS to be 0 360 and carry out the following steps.

Step 1: AFD5<Enter> <Enter> <Enter>
*   AFD3<Enter> <Enter>   AFD4<Enter> <Enter>*
*   RN<Enter>   T1<Enter>   TD<Enter> <Enter>*

Step 2: AFD5<Enter> <Enter> <Enter>
*   AFD3<Enter> <Enter>   AFD4<Enter> <Enter>*
*   RN<Enter>   T2<Enter>   TD<Enter> <Enter>*

Step n: AFD5<Enter> <Enter> <Enter>
*   AFD3<Enter> <Enter>   AFD4<Enter> <Enter>*
*   RN<Enter>   Tn<Enter>   TD<Enter> <Enter>*

After a number of steps, the resulting picture resembles a fractal known as Sierpinski's triangle. In this figure, n = 8.

Figure 4-3b: Computer art: Fractals and Dynamics

You may create the picture in this figure as follows. First, start with a triangle inside a triangle using the command AB as in Figure 4-3a. If you use the Tinkerbell map, for example, call this picture t0.pic and store it To Disk. Clear screen and change XS and YS such that the Tinkerbell attractor fits is surrounded by the inner triangle (plot Trajectory). After a while, add the triangles (command AFD) and store the new picture into file ta0.pic.

Next, set XPS to be 180 540 and set YPS to be 0 360 . Proceed as in Figure 4-3a and have, for example,

Step n: AFD5 <Enter> <Enter> <Enter>
AFD3 <Enter> <Enter> AFD4 <Enter> <Enter>
RN <Enter> TAn <Enter> TD <Enter> <Enter>

AFD2, AFD3, AFD4

The "Add From Disk to W2" command AFD2 is like AFD, but they differ in that the picture will now appear in the upper right quarter of the screen. This command does not open window W2. See also command AFD1.

The "Add From Disk to W3" command AFD3 is like AFD, but they differ in that the picture will now appear in the lower left quarter of the screen. This command does not open window W3. See also command AFD1.

The "Add From Disk to W4" command AFD4 is like AFD, but they differ in that the picture will now appear in the lower right quarter of the screen. This command does not open window W4. See also command AFD1.

AFD5

The "Add From Disk to flexible window" command AFD5 is like AFD and AFDk, where k = 1, 2, 3, 4, but it differs in that the picture will now appear in a rectangular region of the screen which you can set by using the commands XPS and YPS. This command enables you to compress a picture in the format and position you want. See also commands XPS and YPS.

AFDR

The command "AFD Ratio" reduces the size of the picture for AFD1,..., AFD4. When the AFD Ratio is less than 1, the commands AFD1, $\cdots$, AFD4 yield smaller pictures. See also command AFD1.

FD

The "From Disk" command FD makes the program read the picture from the disk into core memory. You will be prompted for the name of the file. If the file name is set to be A:h.pic for example, then the program expects to find the file on the disk in the "A-drive", and it will not look elsewhere.

Invoking FD does not change any parameters of the current process.

XPS, YPS

You can set the pixels into which a picture should be compressed. The command "X Position Scale" XPS sets the horizontal pixel numbers and the "Y Position Scale" command YPS sets the vertical pixel numbers. For example, if you want to compress a 720 pixels by 720 pixels picture into a 560×560 flexible window whose center is the center of the screen region, enter

$$\text{XPS} <Enter> \quad \textbf{80} \ \ \textbf{640} <Enter>$$
$$\text{YPS} <Enter> \quad \textbf{80} \ \ \textbf{640} <Enter>$$

Now enter AFD5 to actually add the picture into the format you have specified.

4.5 SETTING THE SIZE OF THE CORE (Size of Core Menu SCM)

If you enter the command SCM (that is, type **SCM** <*Enter*>) , the Size of Core Menu appears on the screen.

```
SCM    - -  SIZE OF CORE MENU

Changing Size of Core destroys current picture
    (Core copy of a picture is what gets printed)

OK       -  OK as is
SAME     -  set core picture size equal to screen
SIZE     -  set core picture, width, height, no.colors
SQ5      -  core picture 512 pixels wide by 512 high
SQ7      -  core picture 720 pixels wide by 720 high
SQ9      -  core picture 960 pixels wide by 960 high
HIGH     -  core picture is 720 dots HIGH
WIDE     -  core picture is 720 dots WIDE
COLP     -  set number of COLor Planes:  4

Core now saves      16 colors
The core is         720 pixels wide by 720 high
The screen is       640 pixels wide by 480 high
```

Commands for changing the core size

COLP
The command COLP sets the number of color planes available for core pictures. If COLP is $n \geq 1$, then the core picture uses 2^n colors. The screen will still only show 16 colors unless you call command COL. The maximum number of COLP is 16. The default value is 4. The number of color planes available, depends on your computer. If your computer permits you to use 16 color planes, then this means that 16,536 core colors are available. When plotting a trajectory incrementally with IN and T, you may have large numbers of colors. The distribution of colors can be obtained using the count command CNT.

HIGH
The command HIGH sets the height (in pixels) of the internal core picture. The default value is 720.
Changing the core size destroys the current picture.

SAME

The command SAME sets the core picture size equal to that of the screen. It has the effect of speeding refresh (command R).

Changing the core size destroys the current picture.

SSAME

The "Screen SAME" command SSAME decreases the screen picture to match the core picture, if the core picture is narrower or shorter than the screen picture. If the screen is smaller than the core, then SSAME does nothing. If the screen width is greater than the core width, the effective screen width is decreased, and if the core height is less than the screen height the effective height of the screen is decreased. The picture then plots in the size specified, cramming the picture into the upper left.

For Unix workstations the picture read in will sometimes not correspond to the window size; if the window is too large, this may handle the problem. It does not clear the core picture.

Changing the core size destroys the current picture. This command is **not listed in the menus**.

WIDE

This command sets the number of columns of dots in the core picture. That is, it sets the width (in pixels) of the internal core picture. The default value is 720. Hewlett-Packard printers will not accept a value greater than 960.

Changing the core size destroys the current picture.

SQ5, SQ7, SQ9

Command SQ5 makes the core picture 512 pixels high by 512 pixels wide. See also Chapter 5. Changing the core size destroys the current picture.

Command SQ7 makes the core picture 720 dots high and 720 dots wide. Changing the core size destroys the current picture.

Command SQ9 makes the core picture 960 dots high and 960 dots wide. Changing the core size destroys the current picture. This command is for Unix computers.

SIZE

The SIZE command allows you to set HIGH, WIDE, and the number of color planes COLP with one command. See also command COLP.

4.6 PRINTING PICTURES (PriNter Menu PNM)

If you enter the command PNM (that is, type **PNM** <*Enter*>), the PriNter Menu appears on the screen.

PNM   – –   PRINTER MENU
OK   – Current selection: NONE EP   – EPson or compatible dot matrix printer HP   – HP printer PSC   – PostScript Color printer PSP   – PostScript Printer MENUS DM   – Disk output and printing Menu PNOM – PriNter Options Menu SCM   – Size of Core Menu

To get info on PCX (for all other printers) enter "*PCX".
Option 1: To print grey scales, (i) use command PAT
        (e.g., set PAT = 2); then partially erase color K
        via command EK; E2 erases color 2 using pattern
        set under command PAT, or (ii) send a PSC
        output file to PostScript monochrome printer.
Option 2: For color output, either (i) use command PCX or
        (ii) use PSC and send to PostScript Color printer.

Commands for specifying printer

EP

The "Epson Printer" command EP declares that the user has an Epson MX 80 or compatible matrix printer.

HP

The "Hewlett-Packard printer" command HP declares that the user has a Hewlett-Packard LaserJet II or compatible printer. Many HP printers may be compatible. This is not a PostScript printer.

When printing to an HP printer from a Unix computer using HP formatting, you must instruct the computer you are sending binary output. You may need the system manager to use its superuser status to give you permission to print binary.

PSC

The "PostScript Color printer" command PSC declares that the user has a PostScript Color compatible printer. Creating pictures in PostScript form requires that the computer has the file *yps.txt* on the *Dynamics* disk.

See also commands PSC0 and PSC2 in Section 4.4.

PSP

The "PostScript Printer" command PSP declares that the user has a PostScript compatible printer. Most PostScript printers should be compatible. Creating pictures in PostScript form requires that the computer has the file *yps.txt* on the *Dynamics* disk.

See also commands PSP0 and PSP2 in Section 4.4.

Encapsulated PostScript by **Tim Sauer**

To create a PostScript picture to be encorporated in a document, you must create an Encapsulated PostScript file. You must first create a PostScript file with no header.

To do that, create the picture on the screen and enter

PSP0

This creates a PostScript file of your picture (with no header) and sends it to your disk. The PostScript file you create will end in ".prn". To change this to Encapsulated PostScript, find the third line of your file:

%%BoundingBox: (atend)

and change it to

%%BoundingBox: 100 300 500 700

This file can then be inserted into a Latex document using the epsf.sty style file.

PRINTER OPTIONS MENU PNOM

Enter "PriNter Options Menu" command

PNOM < *Enter* >

and the PriNter Options Menu appears on the screen.

```
PNOM  - -      PRINTER OPTIONS MENU

FF      -  Form Feed, eject printer sheet
PP      -  Print Picture (core copy of picture)
PP0     -  T0 + PP + restore old text level
PP2     -  T2 + PP + restore old text level
PPP     -  Printer output goes to file h.prn :   OFF
TL      -  Text Level (or commands T0,...,T3) :   3
```

Commands for printer options

FF

The "Form Feed" command FF instructs the printer to advance the paper to the next sheet.

HPW

The command HPW instructs the HP printer that you want the picture printed Wide, that is 100 dots per inch instead of 150. This will only work if the picture you try to print has a width with WIDE of at most 640. You can make the height HIGH 800 or less if you print the picture with command PP0 (meaning Print Picture with no header). Remember you must set HIGH and WIDE before you make the picture, since these numbers tell the program how to make the picture. This command is **not listed in the menus**.

PH

The command PH instructs the HP printer to insert a number of blank lines on top of the (header of the) picture when it gets printed. This number of blank lines is determined by the value of PH. The default value is PH = 0. The command PH is **not listed in the menus**.

PP

The "Print Picture" command PP sends to the printer the core copy of the picture. Select first the printer you have. For example, use command HP if you have a Hewlett-Packard printer. Printing can be terminated by hitting <Esc> once. The printing is terminated and the program continues. On Unix workstations, the picture is instead sent to a file with a name ending in ".prn"; (see Section 4.4 for a discussion of file names).

See also commands PP0 and PP2.

PP0 and PP2

The commands PP0 and PP2 commands are for setting the text level so that the printer heading is as desired.

The "T0 + PP + restore old text level" command PP0 sets text level to 0, Prints the Picture without text, and then restores the old text level.

The "T2 + PP + restore old text level" command PP2 sets text level to 2, Prints the Picture with text, and then restores the old text level.

PPP

The command PPP is a toggle. If you turn it ON, then the printer output from T4, T5, T6, or PP is directed to the file ending in ".prn". See the Disk Menu DM for the root name. See also the command TL.

If you will be using an HP printer and eventually want to send the file to a printer, you must enter command HP before picture or data is stored.

The PC DOS command

copy root.prn /B PRN

can be used to send a picture file "root.prn" to a printer. Substitute the actual file name for *root.prn*.

Warning. If you turn on the PPP feature, it will destroy any old copy of file root.prn so you should first change the name of the root.

SPACING

This is a rather specialized command for EPSON compatible printers that have a slightly different spacing between lines. If the printer works fine except for some spacing between printer lines, this command may help. The default value is 24. If that leaves small gaps between every eighth line in the picture, then 23 may be preferable. To create a gap with a standard Epson, use SPACING = 28, for example. The command SPACING is **not listed in the menus**.

Text Level output

TL and T0, T1, T2, T3, T4, T5, and T6

The "Text Level" command TL determines the amount of text that is being printed on the screen or sent to the printer while the program is plotting. Enter the command TL and the following menu appears.

```
TL   -   TEXT LEVEL for SCREEN and PRINTER

OK   -   currently: T3
T0   -   no text
T1   -   minimal text
T2   -   ideal text
T3   -   lots of text (default)
T4   -   send current y values to printer
T5   -   send continuous stream of y values to
             printer as they are computed
T6   -   send all screen text to the printer
```

The Text Level determines the amount of text that will be sent to the printer. The higher the value, the more data you get. The Text Level has a default value of 3 but can take on values 0, 1, 2, 3, 4, 5, or 6. Enter command T2 and TL is changed to 2; or enter T1 and it is changed to 1, and so on up to 6. See also the Text Level command in Section 3.1.

The setting TL=0 results in no header for a printed picture. The setting TL=1 results in a small header for the picture, and the settings TL = 2 or TL = 3 result in a substantial output to printer when printing a picture. The text will not appear on the picture itself. The settings TL = 4, TL = 5, and TL = 6 send data to the printer immediately.

Warning. Be careful when entering commands T5 or T6 if you have entered command PPP or if you have a Unix computer. For example, with TL = 5 you can rapidly send megabytes to your disk and you can fill up your entire disk. Therefore you must limit output by soon switching to T2.

The Text Level settings TL = 4 and TL = 5 cause some data to be sent to the printer. On Unix workstations, the data is instead sent to a file whose name ends in ".prn"; see the Disk Menu DM for the root of the name.

Setting TL to 5 causes position vector coordinates to be printed on the printer **every time** the program plots a point, or attempts to plot a point, which may be off screen. Make sure the printer is on. Type

$$\text{T5} < Enter >$$

This results in the trajectory's position (that is, the numerical value of all the components of y) being printed on the printer, dot by dot, following the header mentioned above. Try it. Then quickly type

T2 < *Enter* >

to get back to normal. If the printer has a buffer, then it may continue printing for a while.

Text to printer. If the Text Level is 6, then all the text that appears on the screen will be sent to the printer as it appears on the screen plus all the data that level T5 would send. Text already on the screen is ignored. If your printer is off or is not set for printing, then set it for printing or the computer will act dead until the printer is ready.

Example. You want to print the number of pixels of each color. The "CouNT" command CNT prints the values on the screen only. Use T6 to send this results to the printer. Type

T6 < *Enter* >

CNT < *Enter* >

and after the text has been recorded, type

T2 < *Enter* >

PRINTING COLOR PICTURES

The **core copy** of the picture is used for printing the picture and other tasks. For MS-DOS computers *Dynamics* keeps a copy of the picture with a resolution of 720 pixels (or dots) wide and 720 pixels high allowing 16 colors. These values can be changed by using the command SIZE. Procedures will differ slightly depending on whether you wish to create a file (which is always the case with Unix) or if you wish to send the picture directly to the printer.

The number of **color planes** determines the number of colors the core picture permits. Four color planes are necessary for the core copy of the picture to record 16 colors and eight for 256 colors (see command COL). If you have only 3 color planes, the core has 8 colors. The program *Dynamics* currently needs slightly more than 1 Megabyte of memory to run with 4 color planes with 720 by 720 resolution. Use the DOS command MEM to see how much memory your computer has available. *Dynamics* uses either expanded or extended memory.

It is our experience that a good and easy way to get a high quality permanent record of a color picture on the screen is to print it on a **color printer**. If you want to create a high quality color picture with the objective of sending it to a color printer, you should create the picture and save it to disk. Before sending the picture to the disk, you may wish to enter command **T0** in order that no text appears on top of the picture. The disk copies of pictures includes the color number for each pixel, but **not** the particular color table used if you use more than 16 colors. It only stores the color table of the first 16 colors.

For setting the colors in the picture to be printed, you may wish to use the commands CI (to get the Color Information on the colors used), CT (to view the Color Table) and SC (to set the colors); see Section 3.7 for details. The choice of colors is a matter of taste. An easy way to get started is to use the command BRI that sets a BRIght shade of the colors, especially color 1 through color 7. Since the background (color #0) is black with the command BRI, change the background to a white (command WHI) or to a grey background (command GREy). Sometimes we like a light blue background in which color #0 is set to, for example, 58 63 63. Consult Section 3.7 for more details on setting colors and color tables such as BRI and RNB (RainBow).

It will depend on which printer you have how to proceed.

Printing color pictures with PostScript color printer from *Dynamics*

If you have a PostScript color printer, just enter command **PP** (Print Picture), select the printer (command PSC) and the program sends to the printer the core copy of the picture.

If your printer is not connected to your computer, you may wish to create a PostScript file; the commands **PSC2** and **PSC0** create such files. with and without headers, respectively. If the name of such a file is root.prn, use the PC DOS command

copy root.prn /B PRN

to send the file to a PostScript Color printer.

Printing (color) pictures with any printer supported by MS Windows

The command PCX instructs the program to create a file called "root.pcx", a copy of the current picture, in PCX format. This format is used by Microsoft Windows and Microsoft Windows Paintbrush. Windows can then be used for printing the picture.

Many computers have printers that are not compatible with the ones in the program's menu. Computers with Microsoft Windows (3.0, 3.1 or 95) can still print pictures with whatever printer has been installed. If the printer is a color printer, the pictures will be in color! If the printer has only black print, it will convert the colors to shades of grey. Warning: A black background will come out black.

If you do intend to print the picture using PBRUSH (see discussion below), then you may wish to decreases the core resolution to be 640 by 640, since PBRUSH will print such pictures on a single page of paper. We have used PBRUSH to print color pictures that are 3000 by 3000. See note about Big Pictures below.

Here's how to proceed. Run *Dynamics* and create the picture on the screen. You can do this either in DOS or in a DOS window of Windows 3.0 or 3.1 or Windows 95. Enter command WHI (which creates a white background; this can be undone with command BLU which makes the background BLUe).

Command **PCX** now creates a file root.pcx (where root might be h if you are using the Henon map and have not changed the root name.) Before sending the picture to the disk using PCX, you may wish to enter command **T0** in order that no text appears on top of the picture.

Now you run the Microsoft Windows Paintbrush program PBRUSH. This program comes with windows and is in the \windows directory. As described below, you call up the new file into PBRUSH and print it.

Printing root.pcx using PBRUSH and MS Windows 3.1

If you use MS Windows 3.1, run PBRUSH as follows.

(1) Enter the Windows Program manager.
(2) Click your mouse on FILE and then click on "RUN..."; it prompts you for a Command Line name.
(3) Type PBRUSH, and then click on OK.
(4) Once in PBRUSH, click on File and click subsequently on "Open..."; then give it the name of the file you want to access, including the directory, for example,

C:\DYN\H.PCX

Part of your picture will appear on the screen in the window of PBRUSH.
(5) Click on File, click on Print... and then click on OK. (You may wish to select a smaller fraction than 100%, like 90% since a picture made with a resolution 720 by 720 will require more than one page to print.) Furthermore if you have a black and white printer, the Paintbrush program PBRUSH (a color based program) will turn your picture into shades. The blackest color will be one that appears black on your screen.
(6) The picture should now print.

Printing root.pcx using PBRUSH and MS Windows 95

If you use MS Windows 95, run PBRUSH as follows.

In the Windows Manager,

(1) Click the mouse on "Start", and then click on "Run".
(2) Type PBRUSH, and then click on "OK".
(3) Click on "File" and then click on "Open".
(4) Type the File name including directory (e.g., \dyn\h.pcx), and click on "Open".
(5) Click on "File" and then click on "Print", and then click on "OK".
(6) The picture should now print.

Note. Once the picture is on the screen when running PBRUSH, you can change the size of the picture by clicking on "Image" and then on "Stretch and Skew".

Big Pictures

Using PBRUSH, we can create pictures on one page if they are no more than 660 pixels wide by 900 high. If they are more, the picture is printed on several pages which can then be assembled into one high resolution large picture. In *Dynamics* we have set WIDE and HIGH to be 3000 and created a picture on the screen and saved it with PCX and printed it using PBRUSH as described above. Thanks to the magic of PBRUSH, it was printed on 20 sheets of paper automatically. We assembled it into a giant, beautiful picture. It took some experimenting to discover which color settings in *Dynamics* produced the best picture when printed. Blue in *Dynamics* always came out blue on the printer for example but the shades were quite different.

Troubleshooting for printing pictures from *Dynamics* using MS Windows 95

If you run *Dynamics* from a DOS window in MS Windows 95, you may encounter a problem in printing pictures from Dynamics. If you have a printer that is in the PriNter Menu of *Dynamics* (for example, a Hewlett Packard LaserJet printer) and if you have this printer **not** installed under Windows 95, then everything should work just fine. If you have installed a printer under Window 95 that is in the PriNter Menu of *Dynamics*, and if a picture does not get printed after you entered the command PP, the following options may be of some help.

1. Restart computer in MS DOS
(1) Click on "Start", and then click on "Shut Down".
(2) Click on "Restart the computer in MS-DOS mode" and then click on "Yes".
(3) After the MS-DOS reply is on the screen, run *Dynamics* and printing of the pictures should work fine. If you want to print a picture that you have saved on the disk, for example, h.pic, you may wish to enter dynamics h.pic, and print the picture.

Option 1 may be annoying when you want to switch from running *Dynamics* to another program in Windows 95 regularly. The following solution may be more appropriate and only has to be done once.

2. Change printer settings by **Brian R. Hunt**
(1) Click on "Start", select "Settings", and then click on "Printers".
(2) Double-click on the icon for your type of printer.
(3) Click on "Printer" (if you have selected the correct printer in Step 2, there should be a check mark next to "Set As Default"), and then click on "Properties.
(4) Click on "Details", and then click on "Port Settings...".
(5) Click on the check mark next to "Spool MS-DOS print jobs" to make it disappear.
(6) Click on "OK" in the windows for the previous two steps.

4.7 PLOTTING POINTS FROM A FILE

PLOT

The routine PLOT accepts pairs of double precision numbers and plots them on the screen if the pair is in the screen region. It is primarily for batch jobs. If you have a file of pairs of numbers, representing two dimensional points, such as a file named "points" consisting of the points:

$$0 \quad 0$$
$$1.2 \quad 2.4$$
$$5 \quad 5.0$$
$$0.0 \quad 0.0$$
$$-1.3 \quad -2.4$$

then you can plot the points in the file using the command PLOT. The file should contain no extraneous information. (Such a file can be created with the command TTD as described below.)

We begin with an example of how PLOT is used. Start the program; select the Henon map H; and enter the command PLOT; that is, type

> **dynamics** < *Enter* >
> **H** < *Enter* >
> **PLOT** < *Enter* >

At this point you will be prompted for the name of the file:

```
Enter the entire file name:_____
example -- h.dd
```

Now enter the name of the file, which we said is "points". The program prompts you with a menu

```
How many points do you wish to plot?
file: points

OK        - proceed and plot the points
Cancel    -
PLOTN     - the maximum Number of points to be PLOTted: 100000

PLOTN counts only points that are inside the screen area.
```

Hit "OK" since your file only has 5 points in it. The points are then plotted and a box appears on the screen:

```
4 points have been plotted.
< < HIT  <SPACEBAR>  TO CONTINUE  > >
```

The Henon map has default values for both XScale and YScale; they are from -2.5 to 2.5 and the points of the file are plotted as if they were a trajectory; that is, all but one of the four are in the screen area, while the point 5.0 5.0 is outside and so is not plotted and is not counted. The point 0 0 is in the file twice and is counted twice.

PLOTN

The "PLOT Number" command PLOTN is for setting the maximum number of points that will be plotted.

Creating a file of plotted points

TTD

A file of points like the above file "points" can be created using the command TTD followed by a trajectory command like T or the straddle trajectory commands BST, SST, ASST, ABST. (It cannot be used with Basin commands like BA or manifold commands of bifurcation commands, etc.) TTD does not save points by itself; you must then run a trajectory command.

After you have started Dynamics and have selected a process like the pendulum (P), entering the "Trajectory To Disk" command TTD will lead to a prompt in which you are told the file will be named "p.tdd", that is the root followed by ".tdd". You can then change the root if you wish. Only the points that are plotted inside the screen area will be saved in the file and only the two screen coordinates are saved.

TTD is a toggle so entering TTD again will turn it off.

ACTIONS, HINTS and OPTIONS for TTD
OK   – plotting a trajectory will store points Cancel – terminate and turn OFF the toggle TTD RN   – change Root of file Name, now p   Data will be stored in p.ttd TTDD – maximum number of dots that can be saved using TTD: 10000
Only points in the screen area will be saved.

TTDD

The Trajectory To Disk Dots command TTDD is for setting how many dots will be saved. The default value is 10,000 which will result in a file of about 500,000 bytes. You must be cautious so as not to create files so big that they fill the disk of your computer.

CHAPTER 5

LYAPUNOV EXPONENTS

5.1 INTRODUCTION and the METHODS

An important concept for (chaotic) dynamical systems is the notion of
Lyapunov exponents, introduced by Oseledec (1968). Lyapunov exponents are
numbers which describe the average behavior of the derivative of a map
along a trajectory. Let F be a differentiable map from the n-dimensional
phase space to itself. For each point x in the phase space, the **trajectory**
(or **orbit**) of x is the sequence x, F(x), F(F(x)), F(F(F(x))), ... For
each point x in the phase space, we consider the Jacobian matrix $DF^k(x)$ of
partial derivatives of the kth iterate of the map F at x, where k is any
positive integer. For the discrete time system $x_{k+1} = F(x_k)$ starting at x_0,
the Jacobian matrix $J_k = DF^k(x_0)$ is the product $DF(x_{k-1})DF(x_{k-2}) \cdots DF(x_0)$.
The matrices J_k can be used to estimate the exponential rate at which
nearby orbits are separated. For the discrete time system $x_{k+1} = F(x_k)$, the
separation of two initial points x_0 and y_0 after time k is $x_k\text{-}y_k$. If these
two initial conditions are close to each other, then the separation of
these two points under forward iteration of the map F is approximately the
matrix J_k times the difference vector $x_0\text{-}y_0$, $x_k\text{-}y_k \approx J_k(x_0\text{-}y_0)$.

Lyapunov exponents depend on a trajectory x_0, x_1, x_2, ..., x_k, ...,
where $x_{k+1} = F(x_k)$. Write B_0 for the unit ball in n-dimensional phase space
(that is, x is an n-dimensional vector), and denote the successive iterates
by $B_{k+1} = DF(x_k) B_k$. Notice that each B_k is an ellipsoid. Let $1 \leq j \leq n$.
Let $\beta_{j,k}$ = the length of the jth largest axis of B_k. Define the jth
Lyapunov number L_j of F at x_0 by $L_j = \lim_{k\to\infty}(\beta_{j,k})^{1/k}$; we assume here that
the limit exists. The trajectory would be extremely unusual if the limit
did not exist. Notice that the quantity L_j is the average factor by which

the jth largest axis grows per unit time. Here k is time. The **Lyapunov exponents** of the trajectory are the natural logarithms $\lambda_j = \log L_j$.

The trajectory of a point x is called a **chaotic trajectory** if (1) the trajectory of x is bounded and is not asymptotic to either a fixed point or a periodic orbit, and (2) F has at least one positive Lyapunov exponent at x. We call an attractor a **chaotic attractor** if it contains a chaotic trajectory; see Chapter 6 for the definition of attractor. Although the computer allows only a finite number of iterates, we will often speak of seeing chaotic trajectories on the computer screen. As in virtually all of our calculations, some imprecision remains. We never take a limit as $k \to \infty$, and we cannot guarantee the behavior of the approximate Lyapunov numbers or Lyapunov exponents would not change abruptly after, say, 1,000,000 iterates. If in fact we had to wait for an infinite time to make a tentative judgement of whether the Lyapunov exponents were positive, the concept of Lyapunov exponent would have little meaning in applications. Nonetheless, we speak loosely of Lyapunov numbers and Lyapunov exponents as if they can be computed with precision.

Computation of Lyapunov exponents: the numerical method

An efficient algorithm for the computation of Lyapunov exponents, which is based on Oseledec's theorem, was first given by Benettin, Galgani, and Strelcyn (1976). For improved methods, see for example Benettin, Galgani, Giorgilli, and Strelcyn (1980) and Wolf (1986).

The computation of Lyapunov exponents involves a fundamental step at a point x of a trajectory. Start with an orthogonal collection of unit vectors $u_1, \ldots, u_n$ (where u_i is an approximation of the direction of the ith axis of B_k for $x = x_k$). Define $v_i = DF(x)u_i$ where $i = 1, \ldots, n$. Use the Gram-Schmidt Orthogonalization procedure to compute the orthogonal vectors w_i ($i = 1, \ldots, n$) as follows.

The vector w_1 equals v_1. For $i > 1$, assume $w_1, \ldots, w_{i-1}$ have been defined. Then, denoting inner product by $< \cdot , \cdot >$ and Euclidean norm by $\| \cdot \|$,

$$w_i = v_i - \sum_{j=1}^{i-1} <v_i, u_j^*> u_j^*$$

where u_j^* is the unit vector $w_j / \|w_j\|$. If some w_j was 0, the computation would fail and would have to stop. Then u_j^* is a new orthonormal set of vectors, and the length $r_i = \|w_i\|$ is the growth in the ith direction. The u_j^* are approximations of the directions of the ith axis of B_{k+1}, and r_i is approximately the ratio of the length of the ith axis of B_{k+1} to that of B_k. This process is carried out on each iterate of F. We may write $r_{i,j}$ for the value r_i obtained on the jth iterate of F. Then

$$\lambda_i(k) = \frac{1}{k} \sum_{i=1}^{k} \log r_{i,j}$$

is an approximate value for λ_i which improves as k increases. Typically

$|\lambda_i - \lambda_i(k)|$ is proportional to $\sqrt{k}$, so convergence is slow.

For the computation of Lyapunov exponents or Lyapunov numbers only two commands (L and LL) are involved. To compute two Lyapunov exponents, enter the "Lyapunov exponent" command L and then enter 2. If you want to know what the results are for the Lyapunov exponents so far, enter the "List Lyapunov exponents" command LL.

Lyapunov dimension

While the phase space of the Henon map is 2-dimensional, the attractor (for the default parameter values) can be described as having a dimension between one and two. It has a Lyapunov dimension of (described below) approximately 1.245. In Figure 5-6a, the chaotic set seems to have nonzero area. When this area is indeed positive, the Lyapunov dimension is 2. The Lorenz attractor, which lies in a 3-dimensional phase space, has a Lyapunov dimension 2.06... . We now describe the definition of this dimension. When the Lyapunov exponents are printed on the screen, they are accompanied by a third number, which is the so-called Lyapunov dimension of the attractor. Assume the map F has an attractor A, and $\lambda_1 \geq \cdots \geq \lambda_m$ are Lyapunov exponents of F with respect to a certain measure; the **Lyapunov dimension** of attractor A (with respect to this measure) is defined to be
$$k + (\lambda_1 + \lambda_2 + \cdots + \lambda_k)/\lambda_{k+1},$$
where k is the maximum value of i for which $\lambda_1 + \ldots + \lambda_i > 0$. It is often used to estimate the dimension of the attractor since it is by far the easiest dimension estimate to evaluate. It is conjectured that the Lyapunov dimension of typical attractors is the information dimension of the attractor (Kaplan-Yorke conjecture). Young (1982) has shown that this conjecture is true for certain two dimensional systems, and Ledrappier and Young (1988) obtained results for higher dimensional random maps. It is widely believed that almost every attractor has its Lyapunov dimension equal to its information dimension. The Lyapunov dimension is bounded from above by n when the phase space is n-dimensional.

When the trajectory being plotted is a straddle trajectory, the trajectory is not on an attractor. Since Lyapunov dimension is for attractors only, it is not appropriate for straddle trajectories. See Chapter 8 for the notion of straddle trajectories.

Summary for computation of Lyapunov dimension and exponents
- Set the number of Lyapunov exponents to be computed (command L)
- Plot trajectory (command T, or SST, or BST etc.).

optional:
- Set Text Level to be 2 to speed up computation (command T2).
- Use "List Lyapunov exponents and numbers" command LL to get current estimates of the Lyapunov exponents and Lyapunov numbers.

Plotting of Lyapunov exponents versus time or parameter

The (approximate) Lyapunov exponents can be plotted as a function of time using "Plot Time" command PT. To plot approximate Lyapunov exponents as a function of time, set the horizontal axis scale (XS) to be the range of the time, and the vertical scale (YS) to be the expected range of the Lyapunov exponents, and turn the toggle PT on. Then enter a plotting command like the "Trajectory" command T, and the approximate Lyapunov exponents are plotted on the screen. See Section 5.2 for more details.

Using the features of plotting bifurcation diagrams, the approximate Lyapunov exponents can be plotted as functions of the parameter. The resulting Lyapunov exponent diagram shows in fact the long term values of Lyapunov exponents of the system, as some parameter is varied; see Chapter 6 for details.

Summary for plotting of Lyapunov exponents versus time

- Set the horizontal axis for the time scale (command XS).
- Set the vertical axis for the range of Lyapunov exponents (command YS).
- Set the number of exponents to be computed (command L).
- Turn the toggle PT on (command PT).
- Start a process like T or SST.

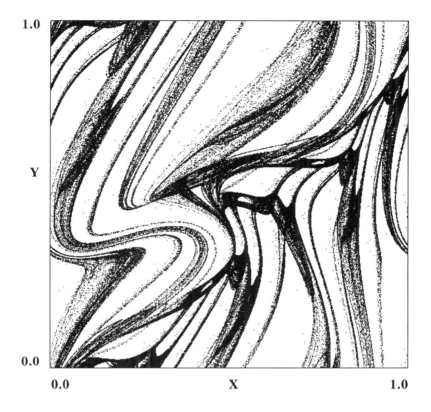

Figure 5-1a: Chaotic trajectory

Figure 5-1 shows plots of trajectories for the Quasiperiodicity map (Q)

$$(X, Y) \rightarrow (X + C1 + RHO*P1, \quad Y + C2 + RHO*P2)$$

where X and Y are computed mod 1 and where P1 and P2 are each periodic functions of X and Y with period 1 in both X and Y.

Except when specified otherwise, the pictures in Figure 5-1 were made as follows. The variable X is plotted horizontally and the variable Y is plotted vertically. The X Scale is from 0 (left) to 1 (right) and the Y Scale is from 0 (bottom) to 1 (top). The number of Pre-Iterates PI before plotting is 1000, the number of plotted DOTS of the trajectory is 1000000.

In the above figure C1 = 0.4, C2 = 0.3, and RHO = -1. The approximate Lyapunov exponents (at time 1000000) are 0.108 and -0.114. The estimate for the Lyapunov dimension of the attractor is 1.94.

The map Q is noninvertible for RHO = -1.

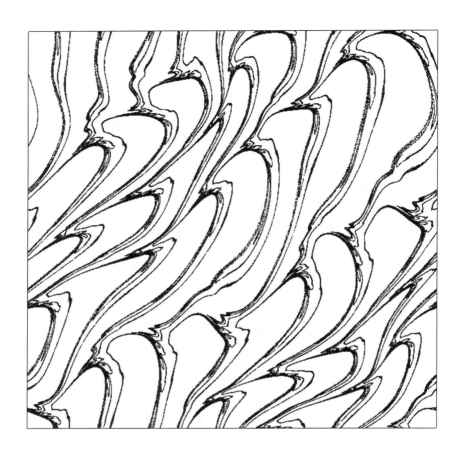

Figure 5-1b: Chaotic trajectory
C1 = 0.42, C2 = 0.3, and RHO = 0.6.

The approximate Lyapunov exponents (at time 1000000) are 0.008 and -0.014. The estimate for the Lyapunov dimension of the attractor is 1.56.

The larger than usual size (960 pixels wide by 960 pixels high) of Figures 5-1b, c, d, e, and g was made possible by computing the pictures on a Unix workstation. These pictures were printed on a HP laser printer.

The map Q is invertible for RHO = 0.6.

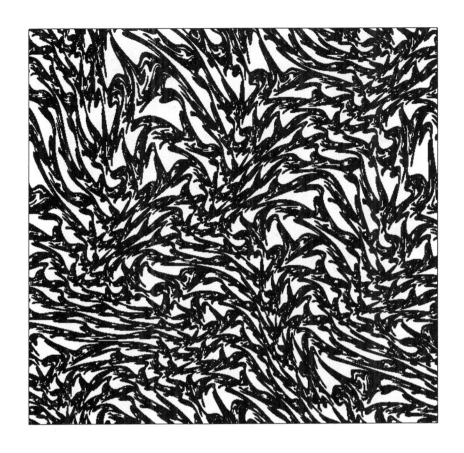

Figure 5-1c: Chaotic trajectory

C1 = 0.415, C2 = 0.735, and RHO = 0.605.

    The number of plotted DOTS of the trajectory is 50,000,000. The approximate Lyapunov exponents (at time 54,000,000) are 0.0013 and -0.0016. After 300,000 iterates the Lyapunov dimension of the attractor was estimated at 1.82, and after 50,000,000 it was 1.83.

    The map Q is invertible for RHO = 0.605.

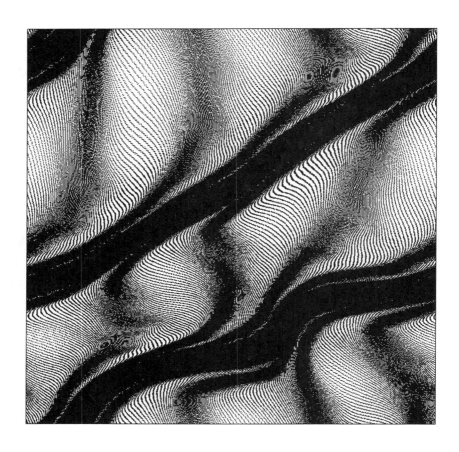

Figure 5-1d: (Nonchaotic) trajectory
C1 = 0.43, C2 = 0.73, and RHO = 0.57.

The approximate Lyapunov exponents (at time 800,000) are 0.000002 and -0.000002. The estimate for the Lyapunov dimension of the attractor is 2.00.

The map Q is invertible for RHO = 0.57. The actual Lyapunov exponents may be 0 giving a Lyapunov dimension of 2. The occurrence of quasi-periodicity versus chaos for the map Q has been discussed extensively in Grebogi, Ott and Yorke (1985).

Figure 5-1e: Chaotic trajectory
C1 = 0.55, C2 = 0.35, and RHO = 0.66.

The approximate Lyapunov exponents (at time 1,000,000) are 0.04 and -0.049. The estimate for the Lyapunov dimension of the attractor is 1.83. The map Q is invertible for RHO = 0.66.

Topic of discussion. *Why do the attractors in Figure 5-1 match at the tops and bottoms of the figures (Y = 0.0 and Y = 1.0), and at the sides (X = 0.0 and X = 1.0)?*

Figure 5-1f: Chaotic trajectories and a nonchaotic trajectory

The number of plotted DOTS of the trajectory is 500,000.

Upper left window: C1 = 0.51, C2 = 0.32, and RHO = 0.55. The approximate Lyapunov exponents are 0.017 and -0.025.

Upper right window: C1 = 0.44, C2 = 0.33, RHO = 0.55, and the approximate Lyapunov exponents are 0.000 and -0.001.

Lower left window: C1 = 0.55, C2 = 0.33, RHO = 0.66, and the approximate Lyapunov exponents are 0.025 and -0.072.

Lower right window: C1 = 0.952, C2 = 0.08, RHO = 0.5, and the approximate Lyapunov exponents are 0.018 and -0.027.

The map Q is invertible for all pictures.

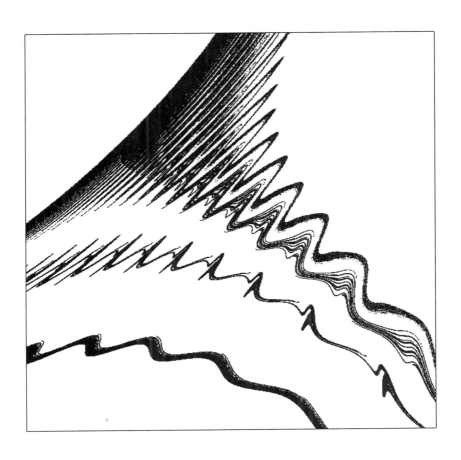

Figure 5-1g: Blow-up of small box in Figure 5-1f

The X Scale is from 0.231141 (left) to 0.335902 (right) and the Y Scale is from 0.0426566 (bottom) to 0.146142 (top). The number of plotted DOTS of the trajectory is 5,000,000. The approximate Lyapunov exponents (at time 5,000,000) are 0.018 and -0.027. After 100,000 iterates the Lyapunov dimension of the attractor was estimated at 1.66, and after 5,000,000 it was 1.67.

Topic of discussion. *There is a fixed point in this picture. How is it and its stable and unstable manifolds related to what is shown above?*

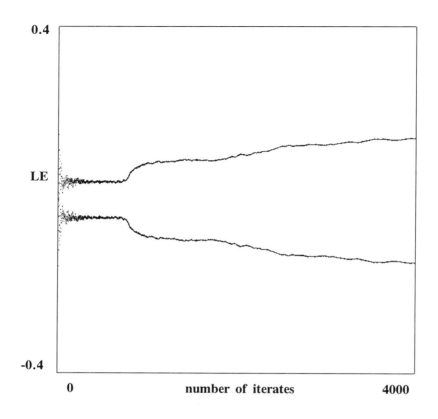

Figure 5-2: Approximate Lyapunov exponents

The two Lyapunov exponents are plotted for the pulsed Rotor map (R)

$$(X, Y) \;\rightarrow\; ((X + Y) \bmod(2\pi), \; Y + 1.08 * \sin(X + Y))$$

where the variable time (number of iterates) is plotted horizontally. The approximate Lyapunov exponents (LE) are plotted vertically. Since C1 = 1, the map is also called the **standard map** *with rho = 1.08.*

    This map has a large chaotic set through which trajectories wander (see the area in Figure 5-6a). The picture depends on the initial condition used but almost every initial condition in that chaotic set should yield the same limit value, as the number of iterates tends to infinity: Many initial conditions for this map yield trajectories that are not chaotic and both of the Lyapunov exponents are zero. The number of Pre-Iterates before plotting is 1000 to assure that the trajectory starts on the attractor.

5.2 LYAPUNOV MENU LM

Enter the "Lyapunov Menu" command:

LM < *Enter* >

and the menu appears on the screen in approximately the following format.

```
LM    - -    LYAPUNOV MENU

LL   –  List Lyapunov exponents (results of L thus far)
L    –  number of Lyapunov exponents:  0

If Lyapunov exponents are computed (L > 0), then:
PT   –  T Plots Time horizontally and Lyapunov exponents vertically
          if ON; now  OFF
BAS  –  plot chaotic basin or create Chaos plot
BIF  –  plot Lyapunov exponents vs. parameter on printer
BIFS –  plot Lyapunov exponents vs. parameter on the screen

        Related commands:
SD   –  Screen Diameters:  1.00
XS   –  X Scale: now from  -2.5  to  2.5
YS   –  Y Scale: now from  -2.5  to  2.5

Use with XCO and YCO to plot chaotic parameter values
```

Note. For the description of the commands for plotting Lyapunov exponents as functions of the parameter (which can be realized by using the features of plotting bifurcation diagrams), see Chapter 6. For the description of the commands for plotting pictures by using the features of plotting basins, see Chapter 7.

Commands for Lyapunov exponents
The following commands are listed in the Lyapunov Menu.

L
L is for specifying the number of Lyapunov exponents (and Lyapunov numbers) to be computed. Set the number of Lyapunov exponents and enter T or a straddle trajectory command like SST. Lyapunov exponents can be computed for most of the processes including the Lorenz system (L) and the Henon map (H). Such additional computations are time consuming and the computer slows substantially. Computing all three exponents for the Lorenz

---

system entails solving a 12-dimensional system, (not counting time, which is an extra variable) so these computations have been made optional. Type

$$L < Enter >$$

and a prompt asks for the desired number of Lyapunov exponents to be computed. This number of Lyapunov exponents to be computed is initially zero. The more exponents being computed, the slower the program runs. To compute two Lyapunov exponents, type

$$L < Enter > \quad 2 < Enter >$$

When a trajectory is plotted, the numbers are printed on the screen (Text Level 3). To speed up computation, set the text level to a lower value (for example, enter "Text level 2" command T2) and the numbers for the Lyapunov exponents do not appear anymore on the screen.

The trajectory commands with which the Lyapunov exponents can be computed are T, SST, BST, ABST, ASST, BIF, and BIFS.

LL

The "List Lyapunov exponents" command LL prints the current values of the Lyapunov exponents and Lyapunov numbers on the screen.

PT

The "Plot Time" command PT is for plotting time horizontally (see Section 4-1) and can be used for plotting the approximate Lyapunov exponents versus the time (the number of iterates). See Figures 5-3 through 5-6 and Example 5-1. To plot the Lyapunov exponents versus time, first do the following:

(1) Set the number of exponents to be computed (command L).

(2) Set the horizontal axis to be the range of the time scale (command XS). For example, 0 to 100,000 for Henon (H), or 0 to 1000 for the Pendulum (P), which is a slow process.

(3) Set the vertical scale to be the range of the Lyapunov exponents (command YS). Usually -10 to 10 is a safe first guess.

(4) Turn the toggle PT on (command PT).

(5) Initialize (command I), since if you have previously plotted a trajectory, the number of dots might be beyond the range of the screen.

Now enter (for example) the "Trajectory" command T, and the approximate Lyapunov exponents are plotted on the screen.

Note. Afterwards, the toggle PT must be turned off if you do not want to plot Lyapunov exponents vs. time in your next picture. The default value for PT is OFF. The menu LM tells whether PT is ON or OFF.

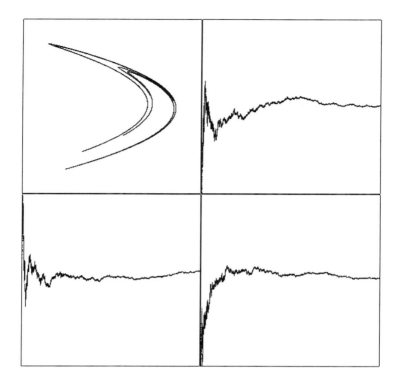

Figure 5-3: Chaotic trajectory and positive Lyapunov exponents
A trajectory and the approximate Lyapunov exponents are plotted for the Henon map (H)

$$(X, Y) \;\rightarrow\; (1.4 - X*X + 0.3*Y, \; X)$$

In the upper left window, the variable X is plotted horizontally, and the variable Y is plotted vertically, and the X Scale is from -2.5 (left) to 2.5 (right) and the Y Scale is from -2.5 (bottom) to 2.5 (top). In the three remaining windows, the positive approximate Lyapunov exponent is plotted versus the time for 3 different initial points. In all of these three windows, the X Scale is from 0 to 100,000 and the Y Scale is from 0.4 to 0.44.

The number of Pre-Iterates before plotting is 1000. Different initial points will yield different plots, but almost every initial point in the basin of the attractor should yield the same asymptotic values for the Lyapunov exponents.

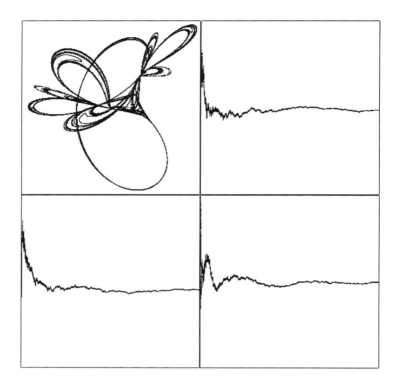

Figure 5-4: A chaotic trajectory and a Lyapunov exponent

A trajectory and the approximate Lyapunov exponents are plotted for the Tinkerbell map (T)

$$X \;\rightarrow\; X{*}X - Y{*}Y + 0.9{*}X - 0.6013Y$$
$$Y \;\rightarrow\; 2{*}X{*}Y + 2{*}X + 0.5{*}Y$$

In the upper left window, the variable X is plotted horizontally, and the variable Y is plotted vertically, and the X Scale is from -1.3 (left) to 0.5 (right) and the Y Scale is from 0.6 (bottom) to -1.6 (top). In the three remaining windows, the positive approximate Lyapunov exponent is plotted versus the time for 3 different initial points. In all of these three windows, the X Scale is from 0 to 100,000 and the Y Scale is from 0.18 to 0.22.

The number of Pre-Iterates before plotting is 1000. Different initial points will yield different plots, but almost every initial point in the basin of the attractor should yield the same asymptotic values for the Lyapunov exponents.

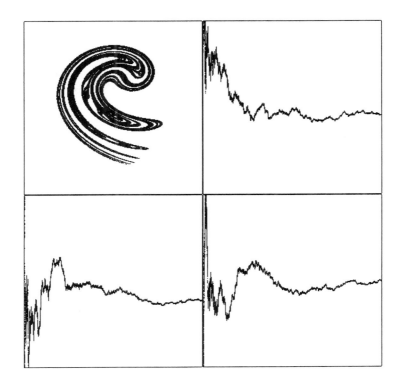

Figure 5-5: A chaotic trajectory and a Lyapunov exponent

A trajectory and the approximate Lyapunov exponents are plotted for the Ikeda map (I)

$$Z \rightarrow 1 + 0.9*Z*exp\{i[0.4 - 6/(1 + |Z*Z|)]\}$$

where $Z = X + iY$. *In the upper left window, the variable X is plotted horizontally, and the variable Y is plotted vertically, and the X Scale is from -1 (left) to 2.5 (right) and the Y Scale is from -3 (bottom) to 1.5 (top). In the three remaining windows, the positive approximate Lyapunov exponent is plotted versus the time for 3 different initial points. In all of these three windows, the X Scale is from 0 to 100,000 and the Y Scale is from 0.495 to 0.525.*

The number of Pre-Iterates before plotting is 1000. Different initial points will yield different plots, but almost every initial point in the basin of the attractor should yield the same asymptotic values for the Lyapunov exponents.

5.3 EXAMPLES

In Chapter 2, we have already presented some examples of computing Lyapunov exponents. In Example 2-2b the computation of the two Lyapunov exponents for the Henon map was discussed. The following examples illustrate how to plot approximate Lyapunov exponents versus time (number of iterates).

Example 5-1a: Plotting chaotic trajectories

To illustrate plotting chaotic trajectories and approximate Lyapunov exponents, we first use the pulsed Rotor map.

Plot a chaotic trajectory of the pulsed Rotor map (R)

$$(X,Y) \rightarrow (X + Y \bmod(2\pi), \ Y + 1.08*\sin(X+Y))$$

where the variable X is plotted horizontally, and the variable Y is plotted vertically. The X Scale is from -3.14159 to 3.14159, and the Y Scale is from -3.14159 to 3.14159.

1. Set the parameter RHO to be 1.08:

$$RHO < Enter > \quad 1.08 < Enter >$$

The parameter C1 does not need to be reset, since it is 1 by default.

2. Set the number of Pre-Iterates to 1000:

$$PI < Enter > \quad 1000 < Enter >$$

3. Initialize:
$$I < Enter >$$

4. Set the number of Lyapunov exponents to be computed to be 2:

$$L < Enter > \quad 2 < Enter >$$

5. Enter "plot Trajectory" command T:

$$T < Enter > \ < Enter >$$

and a trajectory will be plotted. The resulting picture consisting of the chaotic trajectory is given in Figure 5-6a. After 5,000,000 iterates (that is, 5,000,000 dots have been plotted), the Lyapunov exponents are 0.133745 and -0.133745.

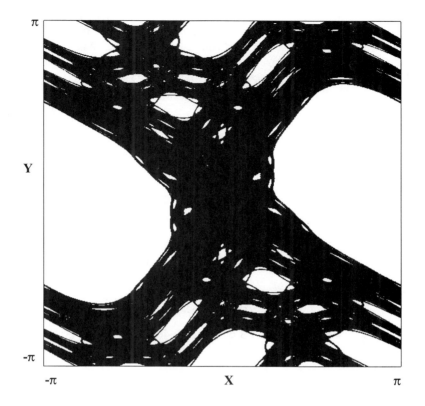

Figure 5-6a: Chaotic trajectory of an area preserving map
A trajectory is plotted for the pulsed Rotor map (R)
$$(X,Y) \rightarrow ((X + Y) \bmod(2\pi), \ Y + 1.08*\sin(X+Y))$$
where both X and Y are computed mod(2π).

The number of Pre-Iterates before plotting is 1000. The number of plotted DOTS of the trajectory is 5,000,000. The approximate Lyapunov exponents (at time 5,000,000) are 0.133745 and -0.133745. The current estimate for the Lyapunov dimension of the attractor is 2. In the figure, the chaotic set seems to have nonzero area. Therefore, the Lyapunov dimension is 2. The Lyapunov dimension will always be 2.0 for an area preserving map in the plane.

Topic of discussion. *The Jacobian of the map for these parameters is identically 1. What does this fact imply? Using blow-ups where necessary, describe the trajectories in the "islands", the white regions.*

Note that the sum of the Lyapunov exponents is 0 since the map is area preserving. Therefore, the Lyapunov dimension of the attractor is 2. In Figure 5-6a, the chaotic set seems to have nonzero area. When this area is indeed positive, the Lyapunov dimension is 2. Since one Lyapunov exponent is clearly positive, the trajectory is chaotic.

Example 5-1b: Approximate Lyapunov exponents
Plot approximate Lyapunov exponents (versus time) of the pulsed Rotor map with the same parameters as in Example 5-1a, where the variable time (number of the iterate) is plotted horizontally, and the variable approximate Lyapunov exponents are plotted vertically. The X Scale is from 0 to 10,000, and the Y Scale is from -0.4 to 0.4.

6. Set the X Scale to be from 0 to 10,000, and set the Y Scale to be from -0.4 to 0.4 (chosen so as include the values of the Lyapunov exponents computed above):

$$\textbf{XS} <Enter> \quad \textbf{0 \ 10000} <Enter>$$
$$\textbf{YS} <Enter> \quad \textbf{-0.4 \ 0.4} <Enter>$$

7. Turn the toggle PT for plotting Lyapunov exponents on:

$$\textbf{PT} <Enter>$$

8. Enter "plot Trajectory" command T:

$$\textbf{T} <Enter> <Enter>$$

and the approximate Lyapunov exponents will be plotted. A resulting picture consisting of the two plotted approximate Lyapunov exponents versus time is given in Figure 5-6b.

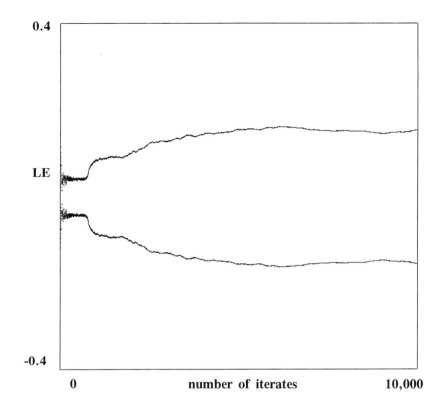

0.4

LE

-0.4

0   number of iterates   10,000

Figure 5-6b: Approximate Lyapunov exponents

The approximate Lyapunov exponents are plotted for the Rotor map (R)

$$(X,Y) \;\rightarrow\; ((X + Y)\, mod(2\pi),\; Y + 1.08*sin(X+Y))$$

The variable time (number of iterates) is plotted horizontally, and the approximate Lyapunov exponents (LE) are plotted vertically. The number of Pre-Iterates before plotting is 1000. Different initial points will yield different plots, but almost every initial point in the basin of the attractor should yield the same asymptotic values for the Lyapunov exponents.

Topic of discussion. *Why is the sum of the Lyapunov exponents zero?*

5.4 EXERCISES
When an exercise does not specify the X or Y Scale, use the default values.

Exercise 5-1. Plot a trajectory and compute the Lyapunov exponents for the Quasiperiodicity map (Q)
$$(X,Y) \rightarrow (X + C1 + RHO*P1, Y + C2 + RHO*P2)$$
(X and Y are computed mod 1) where P1 and P2 are each periodic with period 1 in both X and Y, for the following parameter values:
a. RHO = -1.000, C1 = 0.400, C2 = 0.300;
b. RHO = 0.605, C1 = 0.415, C2 = 0.735;
c. RHO = 0.550, C1 = 0.440, C2 = 0.330. See also Figure 5-1.

Exercise 5-2. Plot the approximate Lyapunov exponents versus the time (number of iterates) of the trajectory with initial condition y = (0,1) for the Rotor map (R)
$$(X,Y) \rightarrow (X + Y \bmod(2\pi), Y + 1.08*\sin(X+Y))$$
Set the X Scale to be from 0 to 4000 and the Y Scale to run from -0.4 to 0.4. See also Figure 5-2 and Figure 5-6.

Exercise 5-3. Plot the approximate Lyapunov exponents versus the time (number of iterates) of the trajectory with initial condition y = (0,1) for the Henon map (H)
$$(X,Y) \rightarrow (1.4 - X*X + 0.3*Y, X)$$
Set the X Scale to run from 0 to 10000 and the Y Scale to run from 0.4 to 0.44. See also Figure 5-3.

Exercise 5-4. Plot the approximate Lyapunov exponents versus the time of the trajectory with initial condition y = (-0.5,-0.5) for the Tinkerbell map (T)
$$(X,Y) \rightarrow (X*X - Y*Y + 0.9*X - 0.6013Y, 2*X*Y + 2*X + 0.5*Y)$$
Set the X Scale to run from 0 to 10000 and the Y Scale to run from 0.18 to 0.22. See also Figure 5-4.

Exercise 5-5. Plot the approximate Lyapunov exponents versus the time of the trajectory with initial condition y = (1,-2) for the Ikeda map (I)
$$Z \rightarrow 1 + 0.9*Z*\exp\{i[0.4 - 6/(1 + |Z*Z|)]\}$$
where Z = X + iY. Set the X Scale to run from 0 to 10000 and the Y Scale to run from 0.49 to 0.53. See also Figure 5-5.

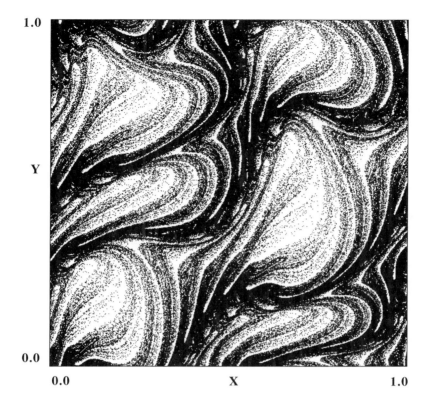

Figure 5-7: Chaotic trajectory

A trajectory is plotted for the Quasiperiodicity map (Q)

$$(X, Y) \rightarrow (X + 0.48 + 0.56*P1, \quad Y + 0.32 + 0.56*P2)$$

where X and Y are computed mod 1 and where P1 and P2 are each periodic functions of X and Y with period 1 in both X and Y.

The number of Pre-Iterates PI before plotting is 1000, the number of plotted DOTS of the trajectory is 1,000,000. The approximate Lyapunov exponents (at time 1,000,000) are 0.012 and -0.014. After 200,000 iterates the Lyapunov dimension of the attractor was estimated at 1.89, and after 1,000,000 it was 1.88.

The map Q is invertible for RHO = 0.56.

5.5 REFERENCES related to DYNAMICS

Computation and theory develop together. This program was written in part to make accessible the tools we have developed and used in the papers of the Maryland Chaos Group. These papers were interwoven with this program in that as the methods were developed, such as the computation of the Lyapunov exponents of trajectories and the Lyapunov dimension of attractors, they were incorporated into the program. Those papers below marked with (*) are of a more general nature and are appropriate for general readings.

An expository paper on dimension, making considerable use of the Generalized Baker map (GB) to provide easily computable examples. The map had been introduced in an earlier expository paper (see E. Ott, Strange attractors and chaotic motion of dynamical systems, Rev. Modern Phys. 53 (1981), 655-671).
    * J.D. Farmer, E. Ott and J.A. Yorke, The dimension of chaotic attractors, Physica 7D (1983), 153-180

The following paper introduced into the journals the ideas of the Lyapunov dimension (sometimes called the Kaplan-Yorke dimension). An earlier meeting proceedings paper in 1979 had announced these ideas. This paper shows a bifurcation diagram plotting Lyapunov exponents, illustrating a phenomenon where pairs of exponents are equal over a range of parameters.
    P. Frederickson, J.L. Kaplan, E.D. Yorke, and J.A. Yorke, The Lyapunov dimension of strange attractors, J. Diff. Equations 49 (1983), 185-207

The following paper on the Lyapunov dimension shows how the attractor can topologically be a torus and yet be so wrinkled that the dimension of the torus is greater than 2.0 (but less than 3). It also computes the box-counting dimension of the Weierstrass nowhere differentiable curve.
    J.L. Kaplan, J. Mallet-Paret, and J.A. Yorke, The Lyapunov dimension of a nowhere differentiable attracting torus, Ergodic Theory & Dynamical Systems 4 (1984), 261-281

CHAPTER 6

BIFURCATION DIAGRAMS

6.1 INTRODUCTION and the METHODS

In the fifties, Myrberg (1958, 1959, 1963) discovered infinite cascades of period doubling bifurcations. The word **"bifurcation"** means a sudden qualitative change in the nature of a solution, as a parameter is varied. The parameter value at which a bifurcation occurs, is called a **bifurcation parameter value**. He found that as a parameter was varied, a fixed point attractor could bifurcate into an attracting period 2 orbit, which could again double to an attracting period 4 orbit, followed rapidly by an infinite sequence, period 8, 16, etc. Period doubling bifurcation will be described more carefully later in this section. Unfortunately, he did not have a computer to produce the pictures of these phenomena. In the seventies the first computer generated bifurcation diagrams appeared in the literature. The concept of **bifurcation diagram** includes a number of ways of plotting a phase variable on one axis and a parameter on another. The pages that follow explain the procedures for making a variety of bifurcation diagrams. What you want to display determines the kind of bifurcation diagram you require. As we will see in a moment, with the program it is rather easy to produce detailed bifurcation diagrams. These bifurcation diagrams show many sudden qualitative changes, that is, many bifurcations, in the chaotic attractor as well as in the periodic orbits.

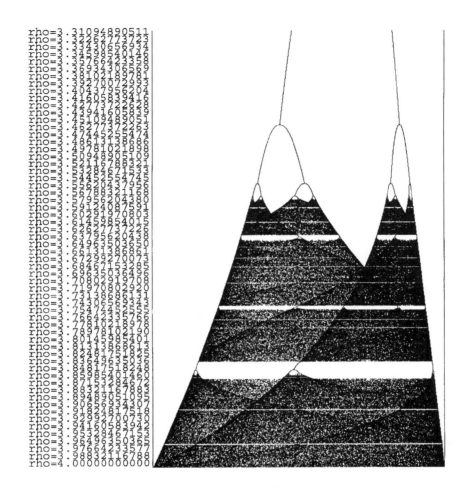

Figure 6-1a: Bifurcation diagram sent directly to the printer

In this bifurcation diagram for the logistic map (LOG)

$$X \;\rightarrow\; RHO*X*(1\text{-}X)$$

the parameter RHO range (BIFR) is from 3.3 to 4.

The diagram was produced using command BIF. The settings are:
y1 = default, BIFI = 0, BIFP = ON, BIFPI = 200, BIFD = 700, and BIFV = 960.

Topic of discussion. *What causes the dark lines within the solid area?*

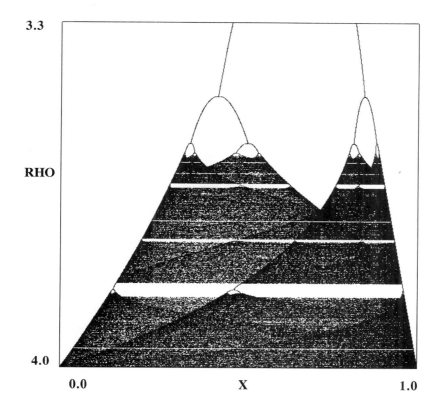

3.3

RHO

4.0

0.0   **X**   **1.0**

Figure 6-1b: Bifurcation diagram created on the screen

This bifurcation diagram for the logistic map is similar to the one in Figure 6-1a, but is created on the screen. The diagram was produced using command BIFS.

There are two different ways of plotting a bifurcation diagram. If the bifurcation diagram is sent directly to the printer (command BIF), then you can set the length of the diagram and you have an option of printing the parameter values (BIFP = ON); see Figure 6-1a. If you create the bifurcation diagram on the screen, then you can set the size of the picture (commands HIGH and WIDE), but parameter values cannot be printed.

Frequently, in bifurcation diagrams the parameter is varied along the horizontal axis, while one of the space coordinates is on the vertical axis. You can obtain such a graphical representation by rotating the picture three times 90 degrees clockwise; see also Figure 6-1c.

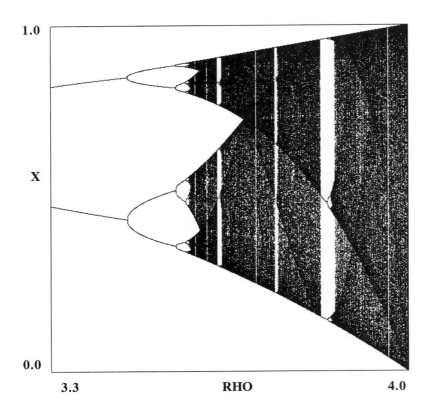

Figure 6-1c: Bifurcation diagram created on the screen and rotated
This bifurcation diagram for the logistic map is similar to the one in Figure 6-1b, but is rotated. The picture is 720 by 720. Square picture can be rotated, while pictures of 640 by 480 (size of VGA screen) cannot be rotated.

After the diagram was created as shown in Figure 6-1b, the command ROT was invoked three times resulting in a rotation of 90 degrees counterclockwise. Each time when the command ROT is executed, it will rotate the picture 90 degrees clockwise.

An alternative way to get this picture is the following. Start with the picture of Figure 6-1b. Flip the picture horizontally (command FLIPH), rotate it (command ROT) and flip it again horizontally (FLIPH). Instead of flipping horizontally, you also may flip the picture vertically (command FLIPV). Hence, entering consecutively the three commands FLIPV, ROT, and FLIPV results in the same picture.

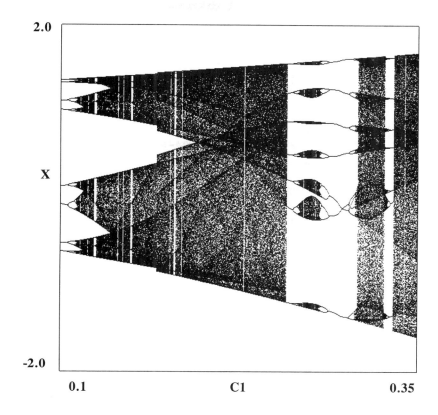

2.0

X

-2.0

0.1   C1   0.35

Figure 6-2: Bifurcation diagram with "bubbles"
In this bifurcation diagram for the Henon map (H)
$$(X, Y) \;\; \rightarrow \;\; (1.25 - X*X + C1*Y, \; X)$$
the parameter C1 range (BIFR) is from 0.1 to 0.35. The diagram was produced using command BIFS. The settings are: y1 = (0,1), BIFI = 0, BIFPI = 200, BIFD = 700, and BIFV = 720.

After the diagram was created, the command ROT was invoked three times rotating the picture 90 degrees counter-clockwise. Each time when the command ROT is executed, it will rotate the picture 90 degrees clockwise.

See also the diagram in Figure 2-8 that was directly sent to the printer using command BIF, where BIFP = ON and BIFV = 880.

Topic of discussion. *What are the horizontal white lines or strips?*

Plotting of bifurcation diagrams

The diagram for a process is constructed as follows. For the process LOG, the LOGistic map, we might select a range of RHO to be, say 3.3 to 4. For the minimum value of RHO and some initial value (x(0),y(0)), when we calculate the first 60 points of the trajectory without plotting, we say there are 60 **pre-iterates**. Then plot the specified coordinate for the next 200 points of the trajectory. Increase RHO slightly, say by 0.01. It may or may not be advisable to reinitialize y whenever the parameter is increased. If it is **not** reinitialized, we say we **follow the attractor** and we take for the initial value the last point which was plotted using the previous RHO value. Then calculate 60 points of this trajectory without plotting and plot the next 200 points. Increase RHO again, and continue increasing until parameter RHO assumes the maximum value. The minimum and maximum value of RHO are set using the "BIFurcation Range" command BIFR. We carry this example out as follows:

dynamics *< Enter >*

LOG *< Enter >*

BIFR *< Enter >* **3.3 4** *< Enter >*

BIFS *< Enter > < Enter >*

The parameter RHO is the parameter that will be varied by default, but parameters like C_1, C_2, ... C_9 can be varied by invoking the "PaRaMeter" command PRM. The number of pre-iterates can be set by using the "BIFurcation Pre-Iterates" command BIFPI, and the number of dots plotted can be set by the "BIFurcation Dots" command BIFD. However, the current default values are BIFPI = 60 and BIFD = 200, which permits the user to sketch the diagram more quickly. The preiterates 200 and computed trajectory of length 700 can be chosen to get quite a reasonable picture of an attractor. Seeing that the diagram looks reasonable, we now make a more densely plotted picture as follows (make sure the printer is turned on):

BIFPI *< Enter >* **200** *< Enter >*

BIFD *< Enter >* **700** *< Enter >*

BIFV *< Enter >* **960** *< Enter >*

BIF *< Enter > < Enter >*

and select the printer you have from the menu that appears on the screen. The resulting bifurcation diagram is shown in Figure 6-1. But what do we mean by an attractor?

Attractors

Several (nonequivalent) definitions of the notion "attractor" appear in the literature. For a discussion of this topic, see for example the articles by Ruelle (1981) and by Milnor (1985). Let F be a map from the m-dimensional space to itself. We say a compact set A in the m-dimensional space is an **attractor** for F if it satisfies the following properties: (1) Invariance: F(A) = A (we say A is **invariant**), (2) Density: A contains an initial point whose trajectory travels throughout A, that is, "it is **dense in A**", and (3) Stability and attraction: initial points starting close to A have trajectories that stay close to A and tend asymptotically to A.

We call an attractor a **chaotic attractor** if it contains a chaotic trajectory (a trajectory with a positive Lyapunov exponent); see Chapter 5 for a discussion of "chaotic trajectory". Chaotic attractors generally may have very complicated structure, except for one dimensional maps. The term **strange attractor**, introduced by Ruelle and Takens (1971), was defined to mean an attractor with a very complicated geometric structure. For most purposes the term chaotic attractor (emphasizing complicated dynamics) and strange attractor (emphasizing complicated geometry) are synonymous. For a discussion of the topic "strange attractor" versus "chaotic attractor" see for example Grebogi, Ott, Pelikan and Yorke (1984). For the prevalent existence of a chaotic attractor in the quadratic family (LOG) and the Henon family (H), see Jakobson (1981), and Benedicks and Carleson (1991), respectively.

The computer allows only a finite number of iterates and we have a limited resolution available for plotting. We cannot test all points near the attractor, so the computer is an imperfect tool. Nonetheless we will often ignore these difficulties and speak of seeing attractors on the computer screen.

Initializing one or more trajectories for each parameter value

Suppose we see an attractor, say a stable periodic orbit, somewhere in the diagram. If we do not reinitialize the trajectory for each parameter value, one may expect that in the diagram we will see the attractor as long as it exists, because after increasing RHO we choose the initial value to be the last point which was plotted. Hence, once the trajectory is close to an attractor, as the parameter is increased, this attractor is followed as long as it exists. This is the procedure used when the "BIFurcation Initializations" command BIFI is set to 0 (which is the default value); that is, the trajectory will not be initialized for each new parameter value. In this case, we say "follow the attractor" as the parameter is varied to indicate that the trajectory is not reinitialized. In contrast, when BIFI is set to 1, it is reinitialized with the initial condition of the storage vector y1. See command BIFI for other options.

Coexistence of attractors

A natural method for investigating the coexistence of different attractors is to produce a collection of bifurcation diagrams, each using a different initial condition. It is most effective to reinitialize after each parameter change. Hence, in such bifurcation diagrams an attractor is not followed, but all attractors corresponding to the different initial conditions of the parameter, can be seen. Set BIFI to some value, say 10, then for each parameter value 10 initial conditions on the line segment [ya,yb] are used. When a trajectory is iterated N times prior to plotting, we say it is **pre-iterated** N times. Each of these 10 initial conditions is pre-iterated BIFPI times and then its next BIFD dots are plotted. Hence, the bifurcation diagram shows BIFI (not necessarily different) trajectories when BIFI is set to a positive value. If the bifurcation diagram is plotted on the screen and the point to be plotted is out of range, a dot is plotted on the left (right) edge of the screen if the dot is to the left (right) of the screen.

Bifurcations and results from a bifurcation diagram

Let $F(\,\cdot\,;m)$ denote a differentiable map depending on a parameter m. Assume that $F(\,\cdot\,;m)$ has a fixed point attractor p(m) for m < m*, and p(m) is an unstable fixed point for m > m*, and at m = m* an attracting period two orbit emerges from the fixed point p(m) when m is increased. The local bifurcation that occurs at m = m* is called a **period-doubling bifurcation**. Similarly, the bifurcation that occurs if an attracting period two orbit collapses to a fixed point yielding an attracting fixed point when m is increased, is called a **period-halving bifurcation**; see Figure 6-3. The unstable fixed point that is involved in these bifurcations has the property that one eigenvalue of DF(p(m);m) is less than -1 for m different from m*, and it is called a **flip saddle**; see also Chapter 11. Another familiar type of local bifurcation is when at m = m* a pair of fixed points are born, one of which is attracting when m is increased slightly, and the other one is unstable and its Jacobian matrix has an eigenvalue greater than 1. This bifurcation is called a **saddle-node bifurcation**; see also Figure 6-3. The unstable fixed point that is involved in the saddle node bifurcation is a **regular saddle**; see also Chapter 11. The fixed point in the bifurcations above can be replaced by a periodic point in this discussion, and similar bifurcations occur. For more details on local bifurcations including Hopf bifurcation, and global bifurcations such as homoclinic and heteroclinic bifurcations, see for example the books by Iooss (1979) and Guckenheimer and Holmes (1983).

When a bifurcation diagram is plotted, several phenomena can be observed: existence of a simple attractor with low period, coexistence of attractors, existence of a chaotic attractor (to be verified by the computation of Lyapunov exponents), and various bifurcations, such as period-doubling bifurcation, period-halving bifurcation, homoclinic bifurcation, heteroclinic bifurcation, and border-collision bifurcation. All these phenomena have to be verified in the phase space, since bifurcation diagrams show only what happens to one coordinate of phase space; that is, they show a projection of the dynamics.

Summary for plotting bifurcation diagrams
- Set the X Scale (command XS).
- Set the PaRaMeter to be varied (command PRM).
- Set the BIFurcation Range (command BIFR).
- Set the number of BIFurcation Pre-Iterates (command BIFPI).
- Set the number of BIFurcation Dots (command BIFD).
- Set the number of BIFurcation Values (command BIFV).
- Set the number of BIFurcations Initializations (command BIFI); if BIFI > 1, then set vectors ya and yb appropriately.
- Plot the BIFurcation diagram (on the Screen) (commands BIFS, BIF).

Plotting of Lyapunov exponents versus a parameter
To approximate the values of the Lyapunov exponents the positive integer n has to be chosen suitably large. For each parameter value, approximations of the Lyapunov exponents can be plotted as a function of n or time. Using the features of plotting bifurcation diagrams, approximate Lyapunov exponents can be plotted as functions of the parameter. A **Lyapunov exponent bifurcation diagram** shows the long term values of Lyapunov exponents of the system, as some parameter is varied. Feit (1979) was the first to publish the Lyapunov exponent bifurcation diagram for the Henon map (H) for C1 = 0.3 and RHO varies between 0 and 1.4; see Example 6-3. Shaw (1981) introduced the Lyapunov exponent bifurcation diagram for the logistic map (LOG); see also Figure 6-4a.

The Lyapunov exponent bifurcation diagrams are constructed similarly to the bifurcation diagrams, but the preiteration time (command BIFPI) and the number of approximate Lyapunov exponents to be plotted (command BIFD) should be larger, due to the slow convergence of approximate Lyapunov exponents. For example, for each value of RHO (or any other parameter) and some initial value, set the BIFurcation Pre-Iterates BIFPI = 10,000 to calculate the first 10,000 points of the orbit without plotting the approximate Lyapunov exponents. Then set BIFurcation Dots BIFD = 1000 to plot the next 1000 approximate Lyapunov exponents. Note that the current default values are BIFPI = 60 and BIFD = 200.

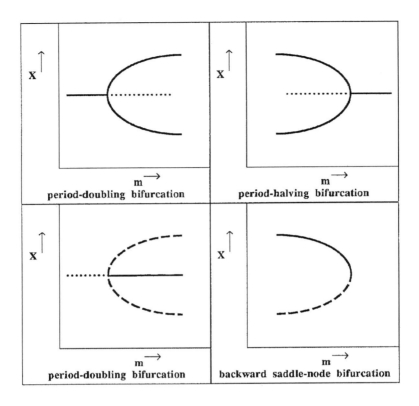

Figure 6-3: Local bifurcations

This schematic picture shows some possible local bifurcations. The solid curves indicate attracting periodic points, the dashed curves indicate regular saddle periodic points, and the dotted lines flip saddle periodic points.

In the upper left window, at the bifurcation value an attracting fixed point becomes a flip-saddle and a period-2 orbit emerges (as the bifurcation parameter m increases); this bifurcation is a period-doubling bifurcation. In the upper right window, the picture is reversed; this bifurcation is a period-halving bifurcation. The lower left window shows an reversed period-doubling bifurcation; it is called "reversed" because the period two orbit is unstable.

In the lower right window, at the bifurcation value two fixed points merge. One is an attracting fixed point, and the other one is a regular saddle. This bifurcation is a saddle-node bifurcation.

Note. The larger values above have been chosen to improve the picture of the Lyapunov exponents. The consecutive dots in a Lyapunov exponent bifurcation diagram can be connected (command CON) as they are plotted using BIFS.

To plot approximate Lyapunov exponents plotted versus a parameter, set the number of Lyapunov exponents $L > 0$, set the horizontal scale to be the range of the Lyapunov exponents, and the vertical scale to the range of the parameter to be varied. The variable PRM contains the name of the parameter that is plotted. By default, PRM is RHO but can be reset equal to the names C1, ..., C9 or PHI using "PaRaMeter" command PRM. In some cases, for example the one dimensional quadratic map, one of the Lyapunov exponents might assume a huge negative value. Therefore, if this might occur, the Screen Diameter SD has to be set to a higher value. When this all has been done, enter (for example) the "BIFurcation Screen" command BIFS, and the approximate Lyapunov exponents are plotted on the screen.

You cannot conclude that what you plot are the actual final values of the Lyapunov exponents. A good procedure is to plot as many dots as the number of preiterates, that is, BIFPI = BIFD. The pictures give an indication as to how much variation the Lyapunov exponent has. This rule of thumb is not always followed here. The reader is encouraged to recompute these diagrams with a variety of BIFD and BIFPI, both equal.

Summary for plotting of Lyapunov exponents versus parameter
- Set the number of exponents to be computed (command L).
- Set the horizontal axis for the range of the Lyapunov exponents (command XS).
- Set the PaRaMeter to be varied (command PRM).
- Set the vertical axis for the range of the parameter to be varied (command BIFR).
- Set the number of parameter values (command BIFV).
- Set the length of the transient time interval (command BIFPI).
- Set the length of the time interval for the approximate Lyapunov exponents to be plotted (command BIFD).
- Set the Screen Diameter to a higher value (command SD).
- **(optional)** Connect the consecutive dots (command CON).
- Plot the Lyapunov exponents (command BIFS or command BIF).

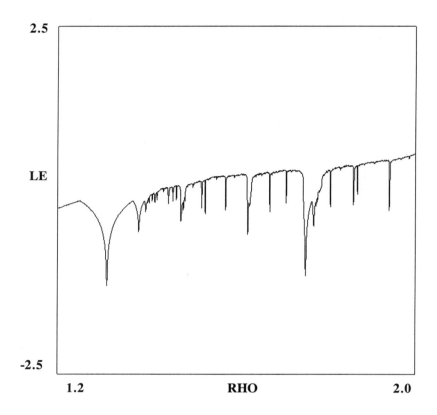

Figure 6-4a: Lyapunov exponent bifurcation diagram

A Lyapunov exponent bifurcation diagram for the one dimensional map (H)

$$(X, Y) \rightarrow (RHO - X*X, X)$$

The diagram was produced using command BIFS. After the picture was created, it was rotated three times 90 degrees clockwise by invoking command ROT. The settings are: y1 = (0.5,0.5), BIFI = 0, BIFPI = 10000, BIFD = 10000, BIFV = 720, and CON = ON.

The value of 10000 for BIFPI and BIFD is very conservative. It is interesting to experiment with much smaller values like BIFPI = 1000, and BIFD = 1, or with large values of BIFV.

Topic of discussion. *The top of the curve looks smooth. Is it really? Then there is a sudden change in direction. What causes this change, and what feature in Figure 6-4b corresponds to it?*

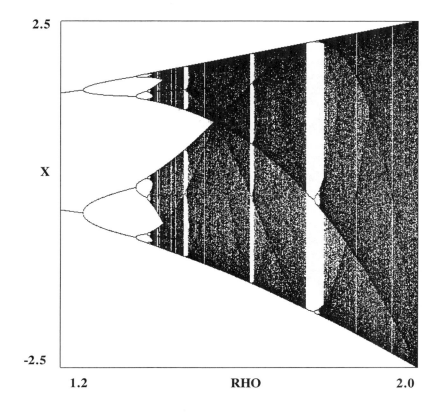

2.5

X

-2.5

1.2   RHO   2.0

Figure 6-4b: Bifurcation diagram
This bifurcation diagram for the one dimensional map (H)
$$(X, Y) \;\rightarrow\; (RHO - X*X, \; X)$$
again studies the parameter RHO where the range (BIFR) is from 1.2 to 2.
    The diagram was produced using command BIFS. After it was completed, it was rotated three times 90 degrees clockwise (ROT). The settings are: y1 = (0.5,0.5), BIFI = 0, BIFPI = 200, BIFD = 700, and BIFV = 720.

    Topic of discussion. *At rho = 1.5 there are 2 black intervals. These merge as rho increases to about 1.55. How can you check whether this is a real merger or just the effect of projecting (X,Y) space onto the X-axis? If they merge, do they really meet at one point? Describe the dynamics where they meet.*

6.2 BIFURCATION DIAGRAM MENU BIFM

Recall that a bifurcation diagram is a picture that shows how the attractor changes as a parameter is slowly varied. Each horizontal line of the picture shows one trajectory for a single parameter value. The trajectory is projected onto one dimension. In the picture, one sees only one coordinate of y plotted horizontally, versus the parameter, plotted vertically. Enter the command BIFurcation diagram Menu BIFM:

BIFM < *enter* >

and the following menu appears on the screen.

BIFM	– –	BIFURCATION DIAGRAM MENU
BIFS	–	make diagram on the Screen
BIF	–	make diagram on the printer
BIFD	–	iterates (Dots) for each rho: 200
BIFI	–	re-Initialize y for each rho: 0
		(BIFI = 0 means y is not re-initialized for new rho)
BIFP	–	Print parameter values on paper (for BIF only): ON
BIFPI	–	PreIterates for each rho: 60
BIFR	–	the Range of parameter rho: from NOT SET to NOT SET
BIFV	–	number of Values of rho (about 720 per page): 480
		The core has 720 horizontal lines; the screen has 480.
PRM	–	PaRaMeter to be varied: rho
		Allowed settings for PRM: C1 RHO)

Useful commands in other menus:
SAME	–	core to be SAME as screen (useful for printing BIFS)
XCO	–	X COordinate: y[0]
XS	–	X Scale: from -2.5 to 2.5

BIFM

Bifurcation diagrams can be made directly on an Epson (or compatible) printer using command BIF or on the screen using BIFS. The printer diagrams can be of arbitrary length, one inch or 20 feet long on an Epson (or compatible) printer. On an HP laser printer the diagrams must fit on one page. A "bifurcation diagram" is a picture that shows what happens to the

---

trajectories as a parameter is slowly varied. The default parameter to be varied is RHO, but it can be changed to another using "PaRaMeter" command PRM. For bifurcation diagrams on the printer, only the coordinate the user chooses for horizontal plotting matters, and that coordinate is called the X COordinate (see command XCO).

The current status of relevant parameters can be seen by viewing the menu, which you call using command BIFM.

Plotting the vertical scales (a warning)

If you use a command like B1 that plots a box with the tic marks, you should call the command **while** the picture is being plotted. Then B1 will use whatever you have set for BIFR on the vertical scale. This scale is usually different from the Y Scale which is set using command YS. If you call B1 **after** the picture is finished plotted, the tic marks will correspond to the Y Scale YS and **not** to BIFR.

Commands for plotting bifurcation diagrams

The following commands are for changing the variables that are listed in the BIFurcation diagram Menu above.

BIF

The command BIF tells the program to create a bifurcation diagram on the printer. In the default case, command BIF instructs the program to calculate the X coordinates of the attractor for 480 different values of RHO. (The number of RHO values can be changed using "number of BIFurcation parameter Values" command BIFV.) For each of these RHO values, the trajectory is first computed for BIFPI = 60 pre-iterates without plotting anything. Then for the next BIFD = 200 iterates the X coordinate is plotted, or rather stored for printing. When the program has computed the results for eight values of RHO, the results are sent to the printer, with one pass of the printer head plotting the eight lines. The only command the user must invoke prior to BIF is BIFR to set the range (scale) of the parameter. When the routine is called, y is automatically initialized to be y1.

HP Printers are restricted to a single page diagram.

BIFS

The "plot BIFurcation diagram on the Screen" command BIFS creates a bifurcation diagram on the screen using as many lines (that is, values of the parameter) as there are horizontal lines on the screen. This routine is slower than BIF. When printing the picture using "Print Picture" command PP, you will get a picture of the core memory. Core memory usually has more lines than there are on the screen. Therefore if you plan to print the picture using command PP, then before making the picture you should invoke command SAME first or use command BIFV. Otherwise, the core will have more horizontal lines than the screen, and the bifurcation diagram will be computed with just enough lines for the screen. Hence a number of the horizontal lines in the core picture will be blank, yielding a picture with tiny gaps. If the command SAME has not been entered, the refreshed picture (command R) may look a little strange. See command BIFF.

L & BIFS or L & BIF

The approximate Lyapunov exponents can be plotted as functions of the parameter. To get the approximate Lyapunov exponents plotted, the number of Lyapunov exponents L must be positive (not 0), the horizontal axis has to be set to be the range of the Lyapunov exponents (command XS), and the vertical scale has to be set to be the range of the parameter to be varied (command BIFR). The parameter RHO is the parameter that will be varied by default, but parameters like C1, C2, ... C9 can be varied by invoking the "PaRaMeter" command PRM.

Commands BIF and BIFS will plot approximate Lyapunov exponents horizontally whenever $L > 0$.

BIFD

The "BIFurcation Dots" command BIFD is used to set the number of iterates or dots at each parameter value that the program will attempt to plot.

BIFF

By default BIFF is ON. The "BIFurcation Flag" command BIFF works with BIFS. When BIFF is ON, it tells BIFS to use BIFV values of RHO (or PRM) instead of the default, which is the number of horizontal lines on the computer screen. In some cases such as Lyapunov exponent bifurcation diagrams, you may wish to use much higher values of BIFV (perhaps 2000) to see the transitory changes. The command BIFF is **not listed in the menus**.

BIFI

The "BIFurcation Initialization" command BIFI is used for initializing at each parameter value when a bifurcation diagram is made. The default is for the program to initialize only at the beginning of the plot and then NOT re-initialize for each value of the parameter. The user can instruct the program to initialize for each value of the parameter named in PRM using command BIFI.

If BIFI is 1, then at each new parameter value, y is set equal to y1 (the position of the small cross).

If BIFI is greater than 1, then the vectors ya and yb must be set, and BIFI distinct, equally spaced points, including ya and yb, are selected on the line segment from ya to yb. If for example BIFI = 2, then just ya and yb are selected. The selected points are then used as initial points for each parameter value when creating bifurcation diagrams using either BIF or BIFS. That means that for each initial point, the process is pre-iterated BIFPI times and then BIFD iterates are then plotted. BIFS plots the results of the different initial points in different colors (up to the number of available colors).

BIFP

The "print BIFurcation Parameter" command BIFP is a toggle. When it is ON, the command BIF creates a picture on the printer and prints values of the parameter on the left side of the page. The default value of BIFP is ON. It will print the value unless command BIFP is set to be OFF. The menu BIFM tells whether BIFP is "ON" or "OFF". The value is printed once every eight parameter values (16 for HP printers) and is the RHO or PRM value of the last of these values. For example, the first value printed is the PRM value of the 8th or 16th parameter value for which computations are made.

BIFPI

The "BIFurcation Pre-Iterates" command BIFPI is used to set the number of pre-iterates at each parameter value before plotting any points.

BIFR for BIF and BIFS

The "BIFurcation Range" command BIFR is needed to set the Range or scale of the parameter for bifurcation diagrams. Do not use YS to set the Y Scale. The default values are both NOT SET.

BIFV

The "number of BIFurcation parameter Values" command BIFV sets the number of horizontal lines that command BIF will use, that is the number of parameter values (about 720 per page).

EDGE

The command EDGE instructs the program what to do when it is plotting a trajectory and it encounters a point that is outside the screen area but is within SD diameters of the screen. In particular, when making bifurcation diagrams, the trajectory sometimes lies outside the screen area. Recall that if the program is plotting a trajectory and encounters a point that is more than SD screen diameters from the screen, it pauses, does not plot anything, and prints a warning in order to avoid a possible overflow that would crash the program. EDGE instructs how to handle the intermediate cases.

If EDGE is OFF, the default case, such a point is not plotted.

If EDGE is ON, such a point is plotted on the edge of the screen. If for example the X coordinate of a point lies to the left of the screen, a point is plotted on the left edge of the screen with the correct Y coordinate; if the Y coordinate is for example above the screen, the point is plotted on the top edge of the screen, using the correct X coordinate; if the X coordinate is to the left of the screen and the Y coordinate is above the screen, the point is plotted in the upper left corner of the screen.

The command EDGE is a toggle. If you enter command EDGE, EDGE will be turned OFF if it is ON, and vice versa.

For bifurcation diagrams on the screen (command BIFS), EDGE is automatically ON. Similarly for the orbit following commands FO and FOB. While the commands BIFS, FO, and FOB are running, calling EDGE turns off edge plotting. The command EDGE is **not listed in the menus**.

FF

The "Form Feed" command advances the paper to the next sheet. The command FF **is not listed in the menus**.

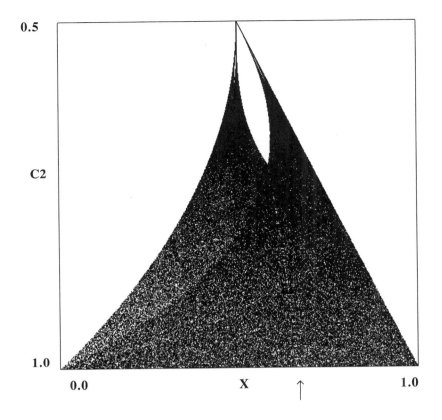

Figure 6-5: Bifurcation diagram

In this bifurcation diagram for the tent map (TT)
$$X \rightarrow (C2/0.5)*X \text{ if } X \leq 0.5$$
$$X \rightarrow (C2/0.5)*[1 - X] \text{ if } X > 0.5$$
the parameter C2 range (BIFR) is from 0.5 to 1.

The diagram was produced using command BIFS. The settings are:
$y1 = (0.5,1)$, *BIFI = 0, BIFPI = 200, BIFD = 700, and BIFV = 720.*

Topic of discussion. *Explain the nearly vertical line or curve of white dots meeting the X-axis at a point shown by the arrow.*

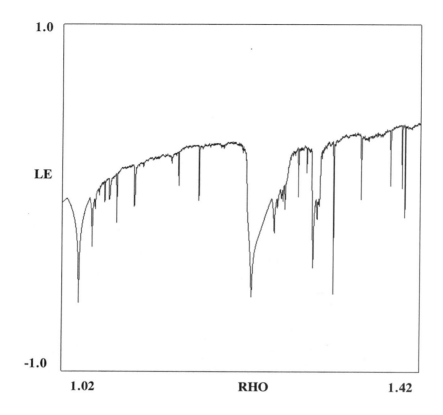

Figure 6-6a: Lyapunov exponent bifurcation diagram

This figure shows a Lyapunov exponent bifurcation diagram for the Henon map
$$(X, Y) \;\; \rightarrow \;\; (RHO - X*X + 0.3*Y, \; X)$$
The diagram was produced using command BIFS. After the picture was created, it was rotated 90 degrees counter-clockwise. The settings are: y1 = (0,1), BIFI = 0, BIFPI = 10000, BIFD = 10000, BIFV = 720 and CON = ON.

Compare this figure with Figure 6-12a where a broader range (BIFR) is chosen.

Topic of discussion. *What causes the above spikes? How far left do the actual spikes go?*

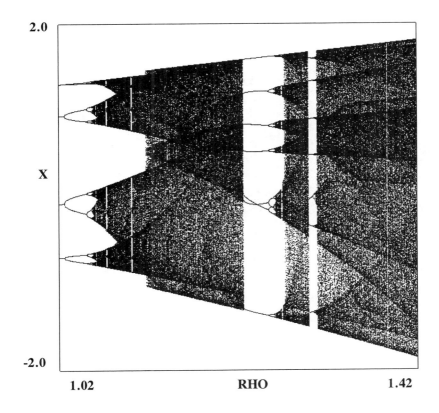

2.0

X

-2.0

1.02   **RHO**   1.42

Figure 6-6b: Bifurcation diagram

This figure shows a bifurcation diagram for the Henon map (H)

$$(X, Y) \rightarrow (RHO - X*X + 0.3*Y, X)$$

where the parameter RHO range (BIFR) is from 1.02 to 1.42.

The diagram was produced using command BIFS. After the picture was created, the command ROT was invoked three times. The settings are:

y1 = (0,1), BIFI = 0, BIFPI = 200, BIFD = 700, and BIFV = 720.

Topic of discussion. *At RHO approximately 1.25 two curves appear to cross. What are these curves? Do they really meet in (X, Y) space or is this a result of projecting (X, Y) space onto the X-axis?*

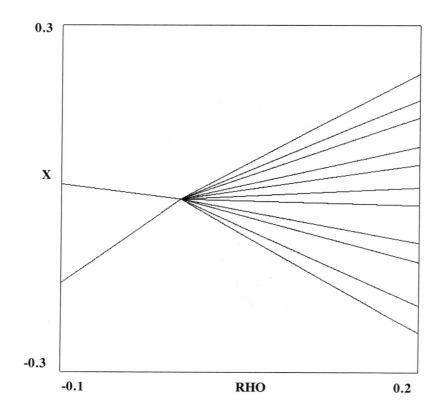

Figure 6-7: Bifurcation diagram

In this bifurcation diagram for the piecewise linear map (PL)

$$(X, Y) \ \rightarrow \ (-1.25*X + Y + RHO, -0.0435*X) \ if \ X \leq 0$$
$$(X, Y) \ \rightarrow \ (-2*X + Y + RHO, -2.175*X) \ if \ X > 0$$

the parameter RHO range (BIFR) is from -0.1 to 0.2. The bifurcation that occurs at RHO = 0, is called a (period-2 to period-11) **border-collision bifurcation***; see also Figures 6-16 and 6-17. The lines converge at X = Y = RHO = 0.0. The diagram was produced using command BIFS. After the picture was completed, it was rotated 90 degrees counter-clockwise (ROT). The settings are:*

y1 = (0,0), BIFI = 0, BIFPI = 200, BIFD = 700, and BIFV = 720.

Topic of discussion. *Why are these curves straight lines?*

6.3 EXAMPLES

In Chapter 2, we presented two elementary examples of plotting bifurcation diagrams. In Example 2-7 the bifurcation diagram of the one dimensional quadratic map (Figure 2-7), and in Example 2-8 a bifurcation diagram for the Henon map was plotted (Figure 2-8 and Figure 6-2). We also presented an example of a Lyapunov exponent bifurcation diagram, see Figure 2-23.

Example 6-1: Bifurcation diagrams

To illustrate plotting of bifurcation diagrams, we first use the Henon map (H)

$$(X,Y) \rightarrow (RHO - X*X + 0.5*Y, X),$$

where RHO varies between 0.8 and 1.06. We deal with the following objectives.

A. Plot a bifurcation diagram of the Henon map for each of the following.
- Follow the development of the attractor using initial value $y1 = (0,1)$.
- Contrast that picture with pictures made initializing at each parameter value for the following values of $y1$:
  $y1 = (1,0)$, $y1 = (1,1)$, $y1 = (0.9,1.1)$ and $y1 = (1.3,0.7)$.
- Make a diagram that initializes at each parameter value with 8 initial conditions.

B. Discuss the coexistence of attractors.
C. Discuss the types of bifurcations that you observe in the bifurcation diagram.

1. Set the parameter C1 equal to 0.5:

C1 <*Enter*>   **0.5** <*Enter*>

2. The parameter RHO is the parameter that will be varied in the default setting. Hence, PRM is already set and the command PRM is not invoked.

Set the range of the parameter RHO to run from 0.8 and 1.06:

BIFR <*Enter*>

0.8 1.06 <*Enter*>

3. Continue with Example 6-1a, Example 6-1b, or Example 6-1c.

Example 6-1a. Bifurcation diagram: Follow the attractor

3a. Now, Set the Vector y1 to be (0,1):

$$SV < Enter >$$

$$SV1 < Enter >$$

$$SV10 < Enter > \quad 0 < Enter >$$

$$SV11 < Enter > \quad 1 < Enter >$$

Notice that if you are using the menu, you should now respond with "OK" twice by hitting <*Enter*> twice.

4a. To create the bifurcation diagram on the screen, enter the "plot BIFurcation diagram on Screen" command BIFS:

$$BIFS < Enter > \quad < Enter >$$

The resulting picture will be similar to Figure 6-8. To reproduce this figure on paper do the following. Make sure the printer is turned on, and enter the "BIFurcation" command BIF:

$$BIF < Enter > \quad < Enter >$$

The bifurcation diagram in Figure 6-8 exhibits a jump from one attractor to another attractor as RHO increases; the jump occurs for RHO ≈ 0.995. Also notice the small loops, so-called "simple bubbles", which appear about 4/5 of the way down the diagram. These represent a period-doubling followed by a period-halving bifurcation. Since the attractor is followed (i.e., BIFI = 0), the jump from a presumably 2-piece chaotic attractor to a simple attractor with period 8 might indicate the coexistence of at least two attractors for an interval of parameter values. See for this coexistence of attractors Example 7-1 in Chapter 7, and Example 6-1c.

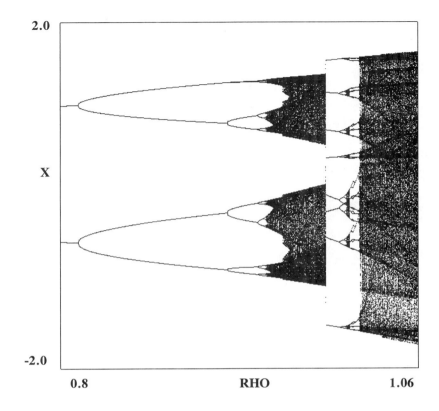

2.0

X

-2.0

0.8   RHO   1.06

Figure 6-8: An apparently discontinuous bifurcation diagram
In this bifurcation diagram for the Henon map (H)

$$(X, Y) \rightarrow (RHO - X*X + 0.5*Y, X)$$

the parameter RHO range (BIFR) is from 0.8 to 1.06. The diagram was produced using command BIFS with the following settings:
y1 = (0,1), BIFI = 0, BIFPI = 200, BIFD = 700, and BIFV = 720.

After the picture was completed, the command ROT was invoked three times resulting in a rotation over 90 degrees counter-clockwise.

Topic of discussion. *What causes this discontinuity? How would the plot change if the BIFR values were reversed to be from 1.06 to 0.8, aside from being upside down?*

Example 6-1b. **Bifurcation diagram: Initialize at each parameter value**

3b. Now, Set the Vector y1 to be (1,0):

$$SV < Enter >$$

$$\textbf{SV1} < Enter >$$

$$\textbf{SV10} < Enter > \quad 1 < Enter >$$

$$\textbf{SV11} < Enter > \quad 0 < Enter >$$

Notice that if you are using the menu, you should now respond with "OK" twice by hitting *< Enter >* twice.

4b. In order to reinitialize at each parameter value, set BIFI to 1:

$$\textbf{BIFI} < Enter > \quad 1 < Enter >$$

5b. Set the number of PreIterates at each parameter value to 200:

$$\textbf{BIFPI} < Enter > \quad 200 < Enter >$$

6b. Plot the BIFurcation diagram on the Screen:

$$\textbf{BIFS} < Enter > \ < Enter >$$

The resulting picture will be similar to the upper left window in Figure 6-9. The bifurcation diagram in the upper left window of Figure 6-9 exhibits a jump from one attractor to another attractor, and so-called simple bubbles. Comparing the diagram of the upper left window in Figure 6-9 with Figure 6-8, a new simple attractor and subsequent bifurcations are observable. Hence, from these two diagrams the coexistence of attractors for an interval of parameter values is obvious.

Similarly, plot the diagrams when the trajectory is initialized at each parameter value where the initializing vector is selected to be y1 = (1,1), y1 = (0.9,1.1) and y1 = (1.3,0.7) respectively. The corresponding bifurcation diagrams are presented respectively in the upper right window, the lower left window, and the lower right window of Figure 6-9.

Conclusion: When plotting bifurcation diagrams, different initial conditions may result in pictures that are quite different. In addition, the diagrams depend dramatically on whether the attractor is followed (that is, BIFI = 0) or is reinitialized at each parameter value (i.e., BIFI = 1).

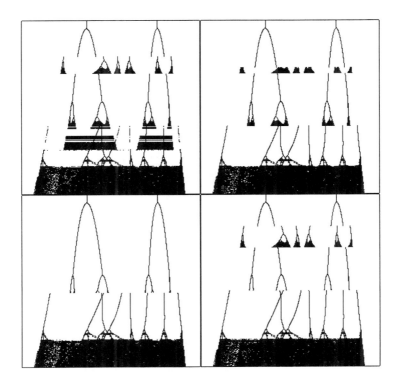

Figure 6-9: Bifurcation diagrams with different initial conditions

These bifurcation diagrams use the same map, parameter value, and scales as in Figure 6-8. Here we set BIFI = 1 so that at each parameter value the trajectory is initialized using y1 as initial condition.

The diagram was produced by using BIFS. The settings are: BIFI = 1, BIFPI = 1000, BIFD = 700, and BIFV = 720.

In the upper left window, y1 = (1,0); in the upper right window, y1 = (1,1); in the lower left window, y1 = (0.9,1.1); and in the lower right window, y1 = (1.3,0.7).

Topic of discussion. *Can you explain what might cause the difference in these 4 diagrams?*

Example 6-1c. Bifurcation diagram: Initialize at each parameter value with 8 initial conditions

3c. Set the Vector ya to be (-1,-1) and Set the Vector yb to be (1,1):

SV < *Enter* >

SVA < *Enter* >

SVA0 < *Enter* > -1 < *Enter* >

SVA1 < *Enter* > -1 < *Enter* >

SVB < *Enter* >

SVB0 < *Enter* > 1 < *Enter* >

SVB1 < *Enter* > 1 < *Enter* >

Notice that if you are using the menu, you should now respond with "OK" twice by hitting < *Enter* > twice.

4c. In order to reinitialize at each parameter value with 8 initial conditions, enter the command BIFI and the value 8:

BIFI < *Enter* > 8 < *Enter* >

5c. Set the number of PreIterates at each parameter value to 200:

BIFPI < *Enter* > 200 < *Enter* >

6c. Since the results from 8 different initial points will be plotted, plot only 100 dots from each:

BIFD < *Enter* > 100 < *Enter* >

7c. Plot the BIFurcation diagram on the Screen:

BIFS < *Enter* > < *Enter* >

The resulting picture will be similar to Figure 6-10. The bifurcation diagram in Figure 6-10 exhibits coexistence of attractors and period halving bifurcation.

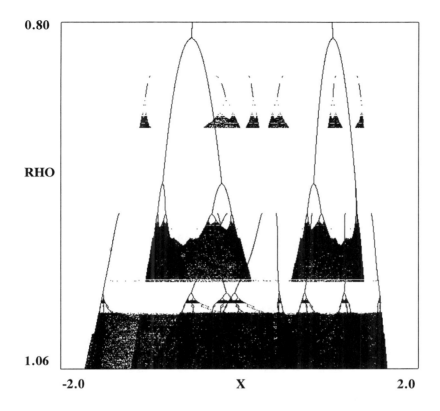

Figure 6-10: Bifurcation diagram with multiple initial conditions
This bifurcation diagram uses the same map, parameter value, and scales as in Figure 6-9. At each parameter value the trajectory is initialized using 8 equally spaced initial conditions on line segment [ya,yb]. The differences in the settings are: ya = (-1,-1), yb = (1,1), BIFI = 8, and BIFD = 300. The picture incorporates all the features displayed in Figure 6-9. It does this by using 8 initial conditions.

Topic of discussion. *Can you find any features here that are in none of the 4 diagrams in Figure 6-9, or are there any absent here that occur somewhere in Figure 6-9?*

Example 6-2. **Blow-up of bifurcation diagram**

In this example, a blow-up (that is, a detailed enlargement) will be made of the diagram in Figure 6-2. Consider the Henon map (H)

$$(X,Y) \;\rightarrow\; (1.25 - X*X + C1*Y, \; X)$$

where C1 varies between 0.26 and 0.29. Set the X Scale to be from 1.15 to 1.3, and plot a bifurcation diagram of the Henon map by following the attractor (BIFI = 0) with initial condition y1 = (0,0).

    1. Set the parameter RHO to be 1.25:

RHO < Enter >   1.25 < Enter >

    2. The parameter C1 is the parameter that will be varied. Set PRM to be C1:

PRM < Enter >   C1 < Enter >

    3. Set the parameter range of C1 to vary from 0.26 to 0.29:

BIFR < Enter >   0.26 0.29 < Enter >

    4. Set the initial condition y1 to be (0,0) using the "Set Vector" command SV:

SV < Enter >
SV1 < Enter >
SV10 < Enter >   0 < Enter >
SV11 < Enter >   0 < Enter >

Notice that if you are using the menu, you should now respond with "OK" twice by hitting < <i>Enter</i> > twice.

    5. Set the X Scale XS to range from 1.15 to 1.3:

XS < Enter >   1.15 1.3 < Enter >

    6. The Screen Diameters SD has to be increased to perhaps 20 since the X Scale has been reduced:

SD < Enter >   20 < Enter >

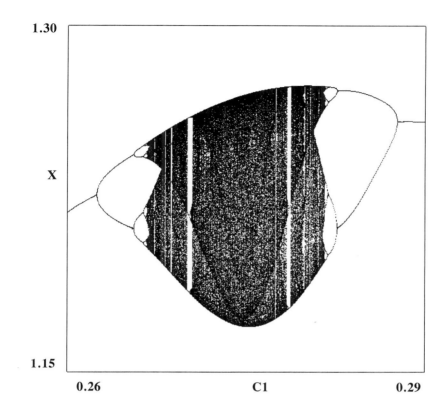

Figure 6-11: A bubble in a bifurcation diagram

This is a bifurcation diagram for the Henon map (H)

$$(X, Y) \;\; \rightarrow \;\; (1.25 - X*X + C1*Y, \; X)$$

It is a blow-up of Figure 6-2. The diagram has been produced using command BIFS. The settings are:

y1 = (0,0), BIFI = 0, BIFPI = 200, BIFD = 3500, and BIFV = 720.

7a. Plot the BIFurcation diagram on the Screen:

BIFS < *Enter* > < *Enter* >

The resulting picture will be similar to Figure 6-11.

7b. To produce a more detailed copy of this figure on paper, turn the printer on and enter the command BIF:

BIF < *Enter* > < *Enter* >

Example 6-3. Lyapunov exponent bifurcation diagram.
Plot the Lyapunov exponent bifurcation diagram for the Henon map

$$(X,Y) \rightarrow (RHO - X*X + 0.3*Y, X),$$

where RHO varies between 0 and 1.4. The first Lyapunov exponent is plotted horizontally, and the parameter RHO is plotted vertically. The X Scale is from -1 to 1.

   1. Set the parameter C1 to be 0.3:

$$\textbf{C1} <Enter> \quad \textbf{0.3} <Enter>$$

   2. Set the number of Lyapunov exponents to be computed to 1:

$$\textbf{L} <Enter> \quad \textbf{1} <Enter>$$

   3. Set the X Scale to range from -1 to 1:

$$\textbf{XS} <Enter> \quad \textbf{-1 1} <Enter>$$

   4. The "PaRaMeter" command need not be invoked, since PRM is RHO by default. Set the BIFurcation parameter Range of RHO to vary from 0 to 1.4:

$$\textbf{BIFR} <Enter> \quad \textbf{0 1.4} <Enter>$$

   5. Set the length of the transient time interval of the exponents to be 500:

$$\textbf{BIFPI} <Enter> \quad \textbf{500} <Enter>$$

   6. Set the number of Dots to be plotted to 500:

$$\textbf{BIFD} <Enter> \quad \textbf{500} <Enter>$$

   7. The default value of initialization vector y1 is (0,2). Set the initial condition y1 using the "CENTer" command CENT:

$$\textbf{CENT} <Enter>$$

   8 Connect the dots of the diagram:

$$\textbf{CON} <Enter>$$

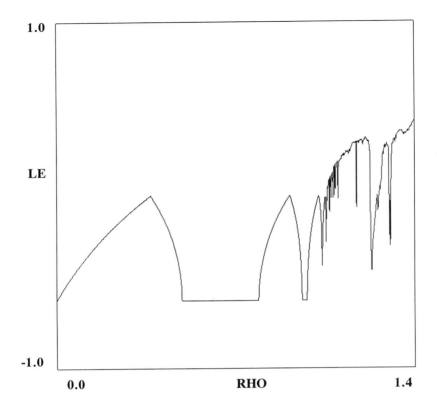

Figure 6-12a: Lyapunov exponent bifurcation diagram

This is a Lyapunov exponent bifurcation diagram for the Henon map (H)

$$(X, Y) \rightarrow (RHO - X*X + 0.3*Y, X)$$

The first Lyapunov exponent (LE) is plotted vertically. The number of Pre-Iterates before plotting is 10,000.

The diagram was produced using command BIFS. After it was completed, it was rotated using the command ROT. The settings are: y1 = (0.5,0.5), BIFI = 0, BIFPI = 10000, BIFD = 700, BIFV = 720, and CON = ON.

Compare this figure with Figure 6-6a where a narrower range (BIFR) is chosen. Compare it also with the bifurcation diagram in Figure 6-12b.

Topic of discussion. *What features can you see in this diagram, and what aspects of the dynamics might explain them?*

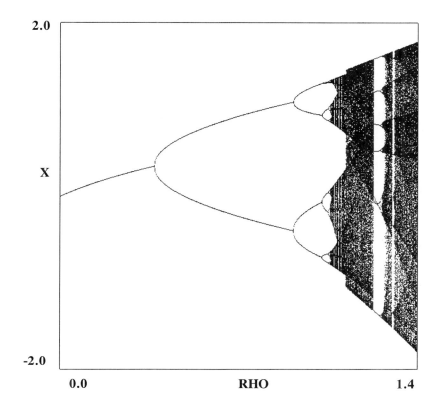

Figure 6-12b: A bifurcation diagram with a period-doubling cascade
This is a bifurcation diagram for the Henon map (H)
$$(X,Y) \; \rightarrow \; (RHO - X*X + 0.3*Y, \; X)$$
The diagram was produced using BIFS; it was rotated 90 degrees counter-clockwise invoking command ROT three times. The settings are:
$y1 = (0.5,0.5), \; BIFI = 0, \; BIFPI = 200, \; BIFD = 700, \; and \; BIFV = 720.$

Topic of discussion. *In the middle of the figure there is an interval of parameters having a stable period 2 orbit. Measure the length $L(2)$ of that interval of parameters. Next measure the lengths $L(4)$ and $L(8)$ of the adjacent intervals of parameters in which there are stable period 4 and 8 orbits respectively. How close are the ratios $L(2)/L(4)$ and $L(4)/L(8)$ to the Feigenbaum number 4.669...; see Feigenbaum (1978, 1979). Do the same for Figure 6-1, and compare the results. How accurate are your measurements of these lengths?*

Note. If only one Lyapunov exponent is plotted, the resulting Lyapunov exponent bifurcation diagram is of higher quality when consecutive dots are connected. However, when two (or more) Lyapunov exponents are plotted, and if the dots on the diagram are connected, a great deal of the information will be lost. Therefore, we recommend connecting the dots in the diagram in the case of plotting one Lyapunov exponent. The command CON is only for screen plots and so cannot be used with BIF.

9. Plot the Lyapunov exponent bifurcation diagram:

BIFS *< Enter > < Enter >*

The resulting picture will be similar to Figure 6-12a. Print this figure:

PP *< Enter >*

Note. Another method to get a high quality picture for plotting Lyapunov exponents without using CON, is to set BIFV to be, say 360,000. This method has been applied in Example 2-23 with the CIRCle map.

6.4 EXERCISES

In the exercises below, we use the following notions. The notion "follow the attractor" means the trajectory is initialized, i.e., y is set to y1, only at the **first** parameter value; the initial condition at a parameter value is the last computed point of the previous value. In other words, BIFI = 0. The resulting bifurcation diagram might depend on the choice of the initial condition y1.

The notion "initialize at each parameter value" means reinitialize the trajectory to the storage vector y1 for every parameter value (that is, use the command BIFI to set BIFI = 1). The resulting bifurcation diagram may exhibit jumps from one attractor to another attractor without a warning. In such cases, attractors may coexist for several of the parameter values. However, this has to be verified. One method for doing so is to produce bifurcation diagrams by "following the attractor" for different initial conditions, or by "initializing at each parameter value with BIFI initial conditions" where BIFI > 1.

The phrase "initialize at each parameter value with 8 initial conditions" means that for each parameter value the trajectories of 8 points on the line segment joining ya with yb are computed (that is, use the command BIFI and set BIFI = 8). This calculation takes 8 times as long as when BIFI = 0 or BIFI = 1, since there is 8 times as much computation. These 8 initial conditions include both ya and yb and are equally spaced on the line segment joining ya with yb. On a color screen, the results of the different initial conditions are plotted in different colors. The resulting bifurcation diagram may exhibit discontinuities. Presumably for several parameter values, there is coexistence of attractors.

In some cases when the trajectory of the initial condition is going to infinity, it might be helpful to plot the basin of the attractor infinity first. (See Chapter 7 for plotting basins.) An initial condition, whose trajectory is bounded, can be chosen in the uncolored region by moving the small cross on the screen.

In the exercises Exercise 6-1 through Exercise 6-10, the phrase "Plot bifurcation diagrams" means do the following either on the screen or on a printer:

a. Plot 3 bifurcation diagrams of the map:
   - Follow the attractor.
   - Initialize at each parameter value.
   - Initialize at each parameter value with 8 initial conditions.

b. Discuss the coexistence of attractors.

c. Discuss the bifurcations and other features that can be observed in the bifurcation diagram.

Actual values of BIFD and BIFPI should be chosen to be smaller for slower computers in order to get results in a reasonable time.

Exercise 6-1. Plot bifurcation diagrams of the LOGistic map (LOG)

$$X \ \rightarrow \ RHO*X*(1-X)$$

where the parameter RHO varies between 3 and 4. See also Figure 6-1. Are all the bifurcations alike?

Exercise 6-2. Plot bifurcation diagrams of the Henon map (H)

$$(X,Y) \ \rightarrow \ (1.25 - X*X + C1*Y, X)$$

where the parameter C1 varies between 0.1 and 0.38. See also Figure 6-2, and the blow-up in Figure 6-11.

Exercise 6-3. Plot both a bifurcation diagram and a Lyapunov exponent bifurcation diagram of the one dimensional quadratic map (H)

$$(X,Y) \ \rightarrow \ (RHO - X*X, X)$$

where the parameter RHO varies between 1.2 and 2. Use BIFPI = BIFD, for a range of values. How fast are the Lyapunov exponents converging? See also Figure 6-4.

Exercise 6-4. Plot bifurcation diagrams of the Henon map (H)

$$(X,Y) \ \rightarrow \ (RHO - X*X + 0.5*Y, X)$$

where the parameter RHO varies between 0.86 and 1.06. See also Figure 6-8, Figure 6-9, and Figure 6-10.

Exercise 6-5. Plot both a bifurcation diagram and a Lyapunov exponent bifurcation diagram of the Henon map (H)

$$(X,Y) \ \rightarrow \ (RHO - X*X + 0.3*Y, X)$$

where the parameter RHO varies between 1.02 and 1.42. Use BIFPI = BIFD, for a range of values. How fast are the Lyapunov exponents converging? See also Figure 6-7.

Exercise 6-6. Plot a bifurcation diagram of the TenT map (TT)

$$X \rightarrow (C2/C1)*X \text{ if } X \leq C1$$
$$X \rightarrow (C2/C1)*[2*C1 - X] \text{ if } X > C1$$

where the parameter C2 varies between 0.5 and 1, for the following values of C1: (a) C1 = 0.5; (b) C1 = 0.3. See also Figure 6-5.

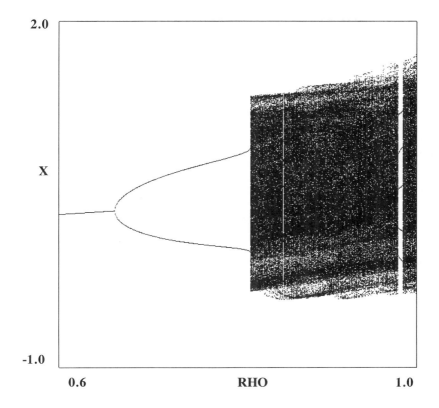

Figure 6-13: A suddenly appearing chaotic attractor

This is a bifurcation diagram for the Ikeda map (I)

$$Z \rightarrow RHO + 0.9*Z*exp\{i[0.4 - 6/(1 + |Z*Z|)]\}$$

The diagram was produced using command BIFS. After it was created, it was rotated (ROT). The settings are: $y1 = (0,1)$, $BIFI = 0$, $BIFPI = 200$, $BIFD = 700$, and $BIFV = 720$.

Topic of discussion. *The upper half of the figure shows stable period 1 and period 2 attractors. The lower part is dominated by a strange attractor. What causes the sudden transition?*

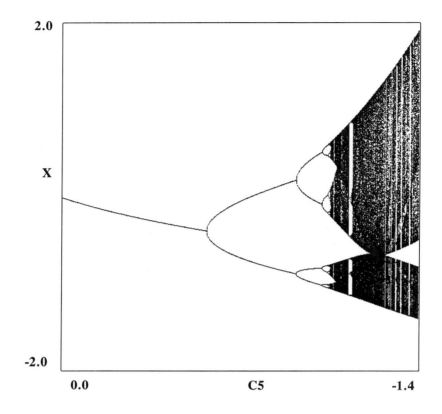

2.0

X

-2.0

0.0   C5   -1.4

Figure 6-14: Bifurcation diagram
This is a bifurcation diagram for the complex Cubic map (C)
$$Z \rightarrow -0.5*Z*Z*Z + Z*Z + C5$$
where Z = X + iY, and the parameter C5 range BIFR is from 0 to -1.4.
   The diagram was produced using command BIFS; after it was created it was rotated using command ROT The settings are: y1 = (0,0), BIFI = 0, BIFPI = 200, BIFD = 700, and BIFV = 720.

   Topic of discussion. *What aspect of the dynamics causes the lower left region to be darker than the lower right, which appears more grey than black?*

Exercise 6-7. Plot both a bifurcation diagram and a Lyapunov exponent bifurcation diagram of the Henon map (H)

$$(X,Y) \rightarrow (RHO - X*X + 0.3*Y, X)$$

where the parameter RHO varies between 0 and 1.4. Use BIFPI = BIFD, for a range of values. How fast are the Lyapunov exponents converging? See also Figure 6-12.

Exercise 6-8. Plot bifurcation diagrams of the Ikeda map (I)

$$Z \rightarrow RHO + 0.9*Z*exp\{i[0.4 - 6/(1 + |Z*Z|)]\}$$

where $Z = X + iY$, and the parameter RHO varies between 0 and 1. See Figure 6-13.

Exercise 6-9. Plot bifurcation diagrams of the complex Cubic map (C)

$$Z \rightarrow RHO*Z*Z*Z + C1*Z*Z + C5$$

where $Z = X + iY$, for the following parameter values and bifurcation parameter range:
(a) C1 = -1, C5 = 1, and RHO varies between 1 and -1.05.
(b) RHO = -0.1, C5 = -1, and C1 varies between 0.2 and 1.8.
(c) RHO = -0.5, C1 = 1, and C5 varies between 0 and -1.4. See also Figure 6-14.

Exercise 6-10. Plot bifurcation diagrams of the Piecewise Linear map

$$(X,Y) \rightarrow (C1*X + Y + RHO, C2*X) \text{ if } X \leq 0$$
$$(X,Y) \rightarrow (C3*X + Y + RHO, C4*X) \text{ if } X > 0$$

where RHO varies between -0.1 and 0.2 for the following choices of the parameters C1, C2, C3, and C4.
(a) C1 = -1.4, C2 = -0.1, C3 = -3, C4 = -4 (period-2 to period-3);
(b) C1 = -1.25, C2 = -0.0435, C3 = -2, C4 = -2.175 (see Figure 6-6);
(c) C1 = -1.25, C2 = -0.042, C3 = -2, C4 = -2.1 (period-2 to chaotic);
(d) C1 = -1.36, C2 = -0.12, C3 = -2, C4 = -2 (period-2 to chaotic).

* **Exercise 6-11.** Describe the details you see in Figure 6-12a. Explain what might cause these features.

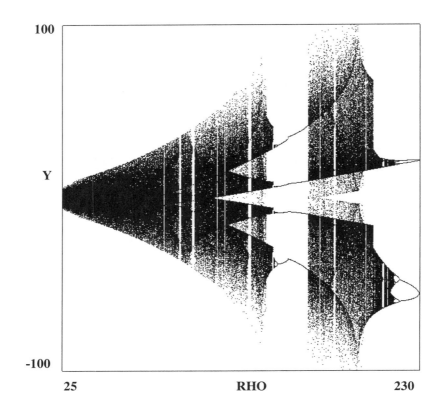

Figure 6-15: Bifurcation diagram

This is a bifurcation diagram for the Poincare Return map of the Lorenz differential equations (LPR)

$$X' = 10*(Y-X), \quad Y' = RHO*X - Y - X*Z, \quad Z' = X*Y - 2.66667*Z$$

using the plane z = rho - 1. A dot is plotted in the (X,Y) plane when a trajectory crosses downward past Z = RHO - 1.

The diagram was produced using commands BIFS and ROT. The settings are: y1 = (0,1,0), BIFI = 0, BIFPI = 200, BIFD = 700, and BIFV = 720.

Topic of discussion. *What happens to cause the change in direction of the curves when RHO is approximately 150?*

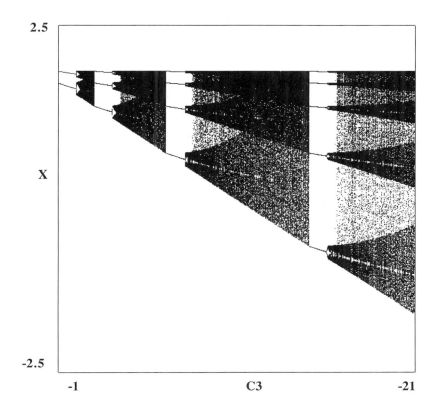

Figure 6-16: Bifurcation diagram

In this bifurcation diagram for a one-dimensional skew tent map (which is the piecewise linear map (PL) with C2 = C4 = 0)

$$X \rightarrow 05*X + RHO \quad if \; X \leq 0$$
$$X \rightarrow C3*X + RHO \quad if \; X > 0$$

the parameter C3 range (BIFR) is from -1 to -21 and the parameter RHO = 0.1. The diagram was produced using command BIFS. After it was completed, it was rotated 90 degrees counter-clockwise. The settings are: y1 = (0,0), BIFI = 0, BIFPI = 200, BIFD = 700, and BIFV = 720. If you select C3 = -3.5 and if you create a bifurcation diagram in which RHO varies from -0.1 to 0.1, then at RHO = 0, there is a bifurcation from a fixed point attractor to a period-3 attractor (this bifurcation is called a (period-1 to period-3) **border-collision bifurcation** *(see also Figure 6-7 and Figure 6-17).*

Topic of discussion. *Which border-collision bifurcations do occur when RHO is varied from -0.1 to 0.1 if C3 = -7.5 0r C3 = -15.5?*

6.5 REFERENCES related to DYNAMICS

This program was written in part to make accessible the tools we have developed and used in our papers. These papers are interwoven with this program in that as the methods were developed, such as the plots of bifurcation diagrams. A number of these papers display figures made with this program. Those papers marked below with (*) are of a more general nature and are appropriate for general readings.

The following paper provides an example of a one dimensional one-hump map with negative Schwarzian derivative, that exhibits both period-doubling and period-halving bifurcation when the parameter is increased.

H.E. Nusse and J.A. Yorke. Period halving for $x[n+1] = MF(x[n])$ where F has negative Schwarzian derivative. Physics Lett. 127A (1988), 328-334

The following papers show how irregular bifurcation diagrams are for maps in the plane, when examined in sufficient detail. The first of these arose from numerical studies of bifurcation diagrams using *Dynamics*.

I. Kan, H. Koçak and J.A. Yorke, Antimonotonicity: Concurrent creation and annihilation of periodic orbits, Annals of Mathematics 136 (1992), 219-252

H.E. Nusse and L. Tedeschini-Lalli, Wild hyperbolic sets, yet no chance for the coexistence of infinitely many KLUS-simple Newhouse attracting sets. Commun. Math. Phys. 144 (1992), 429-442

Mitchell Feigenbaum showed that when period doubling cascades occur, they have a certain universal regularity. The following paper addresses the question of why the cascades must occur as chaos develops.

* K.T. Alligood, E.D. Yorke, and J.A. Yorke, Why period-doubling cascades occur: periodic orbit creation followed by stability shedding, Physica D 28 (1987), 197-205

The following paper examines a local bifurcation phenomenon that has a global impact. For this paper, *Dynamics* has shown to be a fruitful tool.

H.E. Nusse, E. Ott and J.A. Yorke, Saddle-node bifurcations on fractal basin boundaries Phys. Rev. Letters 75 (1995), 2482-2485

This video shows a variety of views (views which are difficult or impossible to see on a PC) of the forced damped pendulum (P), the Ikeda map (I), and the double well Duffing equation (D).

F. Varosi and J.A. Yorke. Chaos and Fractals in simple physical systems as revealed by the computer, Univ. of Maryland, 1991, 55 minutes.

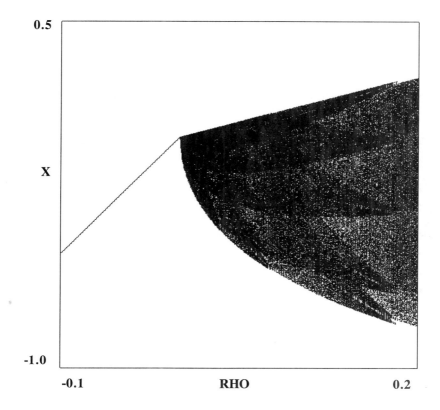

Figure 6-17: Bifurcation diagram

In this bifurcation diagram for the Nordmark map (N)

$$(X, Y) \rightarrow (X + Y + RHO, -0.2*X) \quad \text{if } X \leq 0$$
$$(X, Y) \rightarrow (-2*sqrt(X) + Y + RHO, -0.2*X) \text{ if } X > 0$$

the parameter RHO range (BIFR) is from -0.1 to 0.2. The bifurcation that occurs at RHO = 0, is called a **border-collision bifurcation** *from a fixed point attractor to a chaotic attractor. See also Figures 6-7 and 6-16. The diagram was produced using command BIFS. After the picture was created, it was rotated 90 degrees counter-clockwise (ROT). The settings are:*

$y1 = (0,0)$, BIFI = 0, BIFPI = 200, BIFD = 700, and BIFV = 720.

Border-collision bifurcations

The following papers examine bifurcation phenomena for maps that are piecewise smooth and depend continuously on a parameter. For these papers, *Dynamics* has shown to be a very fruitful tool.

* H.E. Nusse and J.A. Yorke, Border-collision bifurcation including "period two to period three" for piecewise smooth systems. Physica D 57 (1992), 39-57

H.E. Nusse, E. Ott and J.A. Yorke), Border-collision bifurcations: a possible explanation for observed bifurcation phenomena, Phys. Rev. E 49 (1994), 1073-1076

W. Chin, E. Ott, H.E. Nusse, and C. Grebogi, Grazing bifurcations in impact oscillators, Physical Review E 50 (1994), 4427-4444

H.E. Nusse and J.A. Yorke, Border-collision bifurcations for piecewise smooth one-dimensional maps, Intern. Journal of Bifurcation and Chaos 5 (1995), 189-207

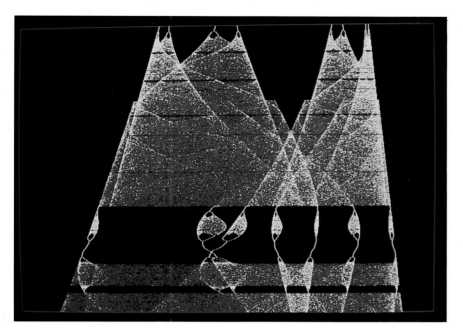

Color Figure 1: Bifurcation diagram with "bubbles" for the Henon map. For more details, see Figure 2-8 and Figure 6-2.

Color Figure 2: The Mandelbrot set. For more details, see Figure 2-14.

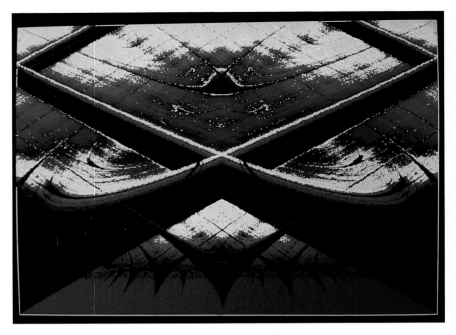

Color Figure 3: Arnol'd tongues and the chaotic parameter set for the circle map. For more details, see Figure 2-24a.

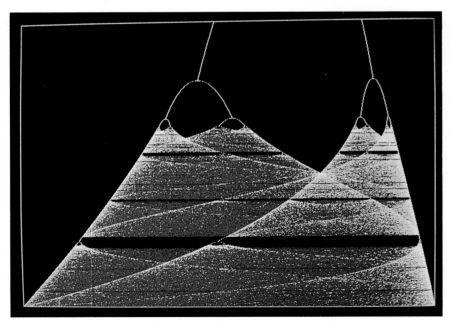

Color Figure 4: Bifurcation diagram for the logistic map. For more details, see Figure 6-1.

Color Figure 5: Generalized basin of attraction and a chaotic attractor for the Henon map. For more details, see Figure 7-2.

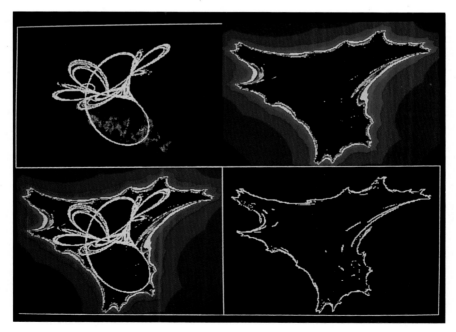

Color Figure 6: Chaotic attractor, basin of infinity, and a Basin boundary Straddle Trajectory for the Tinkerbell map. For more details, see Figure 8-2.

Color Figure 7: Stable and unstable manifolds of a fixed point for the Henon map. For more details, see Figure 9-2.

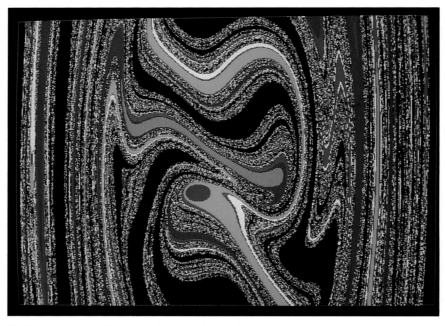

Color Figure 8: Basin of attraction of the time-2π map of the GoodwiN Differential equation. For more details, see Figure 12-3.

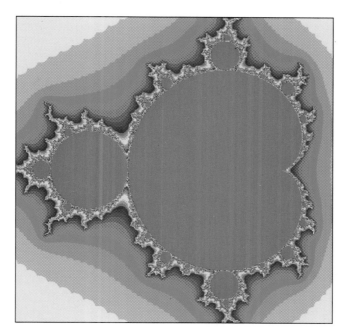

Color Figure 9: The Mandelbrot set. For more details, see Figure 2-14.

Color Figure 10: Arnol'd tongues and the chaotic parameter set for the circle map. For more details, see Figure 2-24a.

Color Figure 11: Computer art: Fractals and Dynamics. For more details, see Figure 4-3.

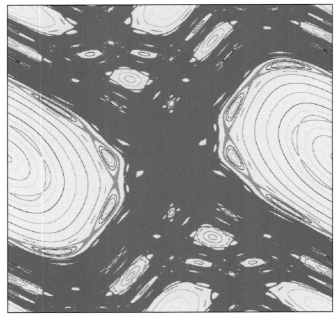

Color Figure 12: Different trajectories for an area preserving map. For more details, see Figure 5-6a.

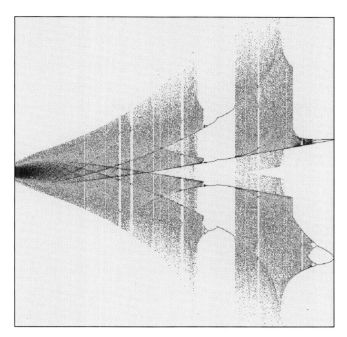

Color Figure 13: Bifurcation diagram for the Poincare Return map of the Lorenz differential equations. For more details, see Figure 6-15.

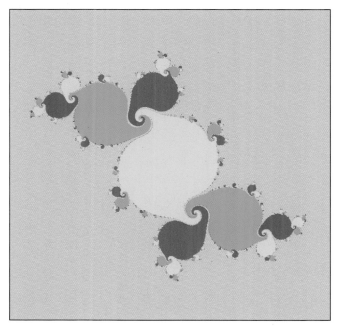

Color Figure 14: Basins of attraction of three fixed point attractors for the third iterate of the complex Cubic map. For more details, see Figure 7-5.

Color Figure 15: Three basins of attraction (Wada basins) for the Pendulum differential equation. For more details, see Figure 7-9a.

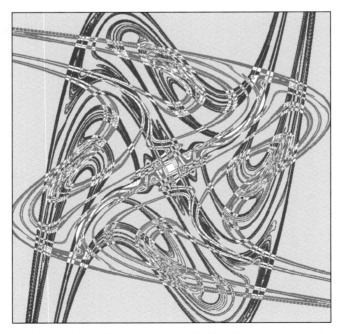

Color Figure 16: Stable and unstable manifolds of a fixed point for the Own-Holmes map. For more details, see Figures 13-1a and 13-1b.

CHAPTER 7

BASINS OF ATTRACTION

7.1 INTRODUCTION and the METHODS

Computing basins of attraction results in a new understanding of the behavior of dynamical systems. The **basin** of attraction of an attracting set is the set of all the initial conditions in the phase space whose trajectories go to that attracting set. Consult Chapter 6 for a definition of the notion "attractor". The main objective of this chapter is to describe the theory and practice of plotting a basin of an attractor. A **generalized attractor** is the union of finitely many attractors, and a **generalized basin** is the basin of a generalized attractor.

In *Dynamics* there are two basic commands for plotting basins. The "Basins and Attractors" command BA plots all the basins and attractors automatically. It is very easy to use. The "BASin" command BAS is often more difficult to use, but it is also more widely applicable. BAS examines a grid of points on the screen. Each grid point represents a different way of initializing a trajectory. Each screen axis represents a parameter like RHO or C1 or a coordinate of y like y[0] or y[1]. BAS colors the grid point to show the behavior of that trajectory. Depending on which options you choose, the coloring can reflect which attractor the trajectory goes to, or whether it diverges, or whether it has a positive Lyapunov exponent, or the period of the attractor if it is periodic, or how fast it approaches the attractor. There are also variants of these two basic commands. For example, BA2 is a variant of BA in higher dimensions and BAN is another variant of BA for plotting quasi-Newton basins. The command BAS can be used in conjunction with command L and command BP for computing "Chaos plots" and "Period plots" in the parameter space. For details, see below.

COMPUTATION OF BASINS: THE BA METHOD

The BA routine must automatically determine how many attractors there are and where they are, whether the attractors are periodic or chaotic. It does so rather simply by keeping track of what is plotted on the screen. No auxiliary information is stored. It cannot detect attractors that are outside the screen's area. The routine BA creates a grid of boxes covering the screen. The default is a 100×100 grid, though this can be changed. See the commands in the "Basin Resolution Menu" BRM in Section 7.2. We refer to the method for plotting basins (and attractors) using command BA as the **BA method**.

We developed the routine BA specifically for *Dynamics*. Boxes are colored using even numbered colors to indicate they contain points of attractors. Colors 2, 4, and 6 for example are green, red, and brownish/yellow. Basins are colored with odd numbered colors 1, 3, 5, ..., and in particular the basin of infinity uses color 1 (dark blue). See Section 3.7 for colors and color numbers. There may be no points that use color 1 since there may be no points whose trajectories diverge. If $2n$ is the even color number for an attractor, its basin is allocated color $2n+1$. While a grid box can contain points of different basins, its color can only reflect one of these. Hopefully you have selected a fine enough grid, such that if a box contains a point of an attractor, the box will not contain points of any other attractor.

Initially all grid boxes are uncolored, but as the routine BA proceeds all grid boxes will eventually be colored. The BA routine selects an uncolored grid box and tests the box by examining the trajectory that starts at the center of the grid box. The trajectory may or may not pass through grid boxes that have previously been colored, but what it encounters will determine how the initial grid box is colored. The use of the BA method is probably better for color screens than for monochrome screens. However, on monochrome screens the basins and attractors will be different shades of grey.

We first describe the simplest version of the BA method, when the BA Parameter BAP is 1. The routine BA selects an uncolored box (which we call box A here). Let C_1 denote the first positive even number not currently used for coloring any grid boxes. This color C_1 will be used for the next attractor located. Box A is tentatively colored with color C_1+1. The routine starts iterating the point at the center of box A, and as long as the resulting trajectory only passes through uncolored boxes, those boxes are tentatively colored color C_1+1, and the routine BA continues the iteration yielding a longer trajectory. As the routine continues iterating, either the trajectory must eventually encounter a previously colored box or the trajectory must diverge. There are several rules as to how the routine BA colors boxes as the trajectory is followed.

(1) The trajectory passes through some box a second time encountering a box colored C_1+1. When MC (default value of MC is 60) consecutive points of the trajectory are in boxes that had previously been colored with color C_1+1, the routine switches so that it colors boxes encountered after this with color C_1. The routine BA continues the iteration. Now each box that the trajectory encounters that is colored with color C_1+1 is changed to color C_1, and uncolored boxes that the trajectory passes through are colored C_1. These are tentatively being identified as boxes containing points of an attractor.

(2) The trajectory "diverges" (see "Diverging Trajectories" below). Then all the boxes colored with color C_1 or C_1+1 are changed to color 1, and the routine BA stops iterating this trajectory.

(3) While routine BA is still plotting uncolored boxes using color C_1+1, the trajectory encounters a box colored C_2, where the number C_2 is odd and $C_2 < C_1$. Then the routine terminates iteration and all boxes colored C_1+1 are changed to color C_2. (If the above event happens while the routine is using color C_1, then no box is changed in color, and iteration continues. The routine BA must be able to deal with cases where a bounded attractor can be quite close to points in the basin of other attractors.)

(4) The trajectory encounters a box colored C_3, where C_3 is even and $C_3 < C_1$. Then the iteration is terminated and all points colored C_1 and C_1+1 are changed to the basin color C_3+1.

If as the routine BA is computing a trajectory, neither of the situations (2), (3) nor (4) occur, then (1) must eventually occur, and eventually the point will come where there are MC consecutive iterates during which no uncolored grid box is encountered nor is any box colored with color C_1+1. At this point, the routine BA stops the computation of that trajectory. This may require many thousands of iterates of the initial condition **if the attractor is chaotic**, but at this point the great majority of the boxes containing points of the attractor will probably have been colored C_1. Any boxes missed will eventually be colored, and if the grid is fine enough, it will be colored C_1+1, the color of the basin of this attractor. Future steps will not change the color of any boxes already colored at this point.

Next, the routine selects a previously uncolored box and the whole process is repeated, and repeated, until there are no uncolored boxes.

The algorithms for both the methods of plotting basins are designed so that (assuming that the basins of the process are open sets) as the grid becomes finer and finer, the accuracy should become arbitrarily high. For the routine BA, this means for each basin B(i) plotted there will be an actual basin A(i) of the process such that the sum of the areas [B(i) minus A(i)] will go to 0 as the grid size goes to 0. The sum is over all the basins, and it is assumed in this calculation that no two A(i) are equal. No theorem to this effect exists yet, but it is the goal in designing basin algorithms.

There are considerable uncertainties in the algorithm above. Method BAS is more reliable, but it often demands that the user has considerable skill and patience. Since the BA method is automatic, it is preferable to the BAS method, if it is reliable. In fact, it can produce strange features in the pictures it creates. A major weakness of this routine (that can be cured) is that it too easily concludes that a point is in the wrong basin; if the routine BA starts at an initial point x_0, its first iterate might be in a colored box (say color C_2). This means that box contains a point $y(0)$ whose trajectory encounters a colored box, colored C_2 or C_2-1. That does not mean that the trajectory of x_0 would also encounter that color if the box containing $x(1)$ was not colored. It might go to another attractor for example. Nonetheless, by rule 3, this trajectory is terminated and colored as if it is in the C_2 basin. This is the main reason that this method as described above produces pictures with some features that are strange, comparing with the much more reliable BAS method. The method in the program is the "modified BA method", which reduces the effect of the weakness.

The modified BA method

The modified method introduces the "Basins and Attractors Precision" number BAP. The BA method corresponds to BAP = 1. In this modified BA method, only case (3) is handled differently. We replace the first sentence of case (3) by:

(3') The trajectory encounters boxes colored C_2 on BAP consecutive iterates, where C_2 is odd and $C_2 < C_1$, and the routine BA is still using color C_1+1 for plotting uncolored boxes. Then we proceed as before.

The default value of BAP is 10, and in the typical cases tested, it produces pictures differing only slightly from those produced using BAS. The bigger BAP is, the more accurate the pictures are but the slower BA runs. Even ignoring the setup time required for BAS, the computation time for BA is usually shorter than for BAS. Consider MC = 600 for more precision.

Since the entire screen is colored, printing the picture would result in a solid black plot. After saving the picture with command TD, you can erase colors using the Ek commands, k = 1, 2, ... and then print.

The first box tested is the box that contains y1.

Diverging trajectories

The point at infinity may be an attractor. We cannot test rigorously if the trajectory of a point is going to infinity, so we say a trajectory **diverges** or **is diverging** if it leaves the screen area and either goes to the left or right of the screen by more than SD widths of the screen, or goes above or below the screen area by more than SD screen heights. SD has the default value 1. The **basin of infinity** is the set of initial points whose trajectories are diverging.

BA for Bifurcation Diagrams

BA permits you to choose the XCOordinate to be a parameter like RHO or C1. (The YCOordinate must be a phase space variable.) Then BA yields pictures that look like bifurcation diagrams. Use XS the set the X Scale through which the parameter is to vary. This method may be also used as an alternative method to creating bifurcation diagrams using at least two initial conditions. Make sure the range of YS is not too small. You need not set PRM to specify which parameter is being varied. Eeach trajectory is first initialized using y1 and then the YCOordinate is set, so depending on your choice of y1, you may or may not detect various attractors.

Plotting of basins in higher dimensions

The routine BA (Basins and Attractors) is designed for 2 dimensional phase space. When the phase space has dimension higher than two, you can use the command BA2 instead of BA to plot basins automatically. In such cases it makes no sense to see if a basin is hit; only attractors count, and since the attractor lies in a higher dimensional space, a single encounter (when projected on to the plane of the screen) does not guarantee that the trajectory is going to that attractor. The basins are plotted for the plane through the vector y1; change its coordinates to translate the plane; set XCO and YCO to select the plane. The attractors are unlikely to lie in the plane being studied, and they are projected onto the plane. Use MC > 100 to avoid plotting errors. The algorithm is the following.

If the initial point has not been colored,
(1) iterate BAP times or 10 times whichever is more;
(2) iterate up to MC times to see if attractor(s) have been hit BAP times in a row; if so, declare that the initial point is in the basin of the last attractor hit; this rule also holds in stage 3;
(3) search for a new attractor; color uncolored boxes using the highest color available = numCoreColors -1. Call this color X. When MC points in a row (ignoring lower colors) are of this color X, assuming an old attractor is not encountered, go to stage 4

(4)   then start coloring using a new attractor color until it is encountered MC times in a row; color all previously encountered boxes except attractor boxes. If the last attractor color cannot be increased, then use the last attractor color; all further attractors will be colored numCoreColors - 4 which is assumed to be even;
(5)   erase the X colored boxes and color the initial point the appropriate basin color if it has not been colored an attractor color.

Exception:
   if at any point the trajectory diverges, then erase all boxes colored and color the initial point color 1

Note: Basins of attractors of color 2n use color 2n+1.

Many basins and too few colors?
   Both BA and BA2 allocate two colors for each basin and attractor pair. Sometimes 16 colors are insufficient. In rare cases 5000 coexisting attractors have been seen; see Feudel, Grebogi, Hunt and Yorke (1996). In such cases you can increase the number of color planes using the SIZE command, allowing up to 65,536 colors. While there only 16 colors will be displayed on the screen, you can use the command CNT to determine what colors are being used, and how many points (pixels) have been colored with specific colors. You can also use command COL which in DOS will provide 256 colors on the screen and in the core picture.

Blow-ups: when BA cannot be used
   The BA routine determines from the screen picture where the attractors are. If you wish to create a blow-up picture of a small region, BA will not work unless part of each attractor is on the screen. In such cases, you should use BAS, as explained in the next section. BA can be used to see where the attractors are and is a help in setting the parameters for BAS. Then change the scales and use BAS to make the blow-up picture.

Summary for plotting basins of attraction utilizing BA
● Plot basin of attraction (command BA).

Summary for plotting bifurcation diagrams utilizing BA
● Select parameter to be varied (command XCO).
● Set the range of the bifurcation parameter (command XS).
● Set the Y Scale (command YS).
● Plot the bifurcation diagram and corresponding basins (command BA).

COMPUTATION OF BASINS: THE BAS METHOD

See Figures 2-10 and 2-20 for examples using the ideas of this section. The BAS method divides the basins into two groups, first the (generalized) basin of attraction A whose points will be plotted, and second the (generalized) basin B whose points will not be plotted. Also, the routine BAS will not plot boxes lying outside A or B. The routine BAS considers a 100 × 100 grid of boxes covering the screen. (See the commands in the "Basin Resolution Menu" BRM in Section 7.2 to change the grid size.) It tests each grid box. If the center of the grid box is in basin A, then the grid box is plotted, that is, it is colored; otherwise the grid box is left uncolored.

In the default case, basin A consists of points whose trajectories are diverging and nothing else, while basin B contains nothing. Hence, BAS will plot a grid box if the trajectory of its center is diverging. Otherwise the grid box is not plotted. If you wish to plot just the basin of infinity, you need to do nothing to set basin A and B. This default setting was used in Figure 2-10. If you do set basin B, the speed of computation can be doubled. But there are also cases where it is essential to specify both basins. For the Henon map with parameter values in Figure 7-2, there are three attractors: infinity, an attracting period 8 orbit, and a 2-piece chaotic attractor. In that figure we plot the points in the basin of infinity and points in the basin of the period 8 orbit. In other words, a grid box is plotted if the trajectory of its center is diverging or if it is approaching the stable period 8 orbit, while a grid box is not plotted if the trajectory of its center is approaching the 2-piece chaotic attractor.

Specifying the basins A and B

There are three storage vectors (y2, y3, y4) that can be used to help specify basin A and there are three storage vectors (y5, y6, y7) that can be used to help specify basin B. By default, they are not set and so are ignored. The routine BAS terminates computing the trajectory of the center of a grid box once the trajectory comes "close to" (that is, within a distance RA of) a storage vector that is set, or if it is diverging. If the trajectory of the center of a grid box is not diverging and if it does not come within a distance RA of one of the storage vectors y2, ..., y7 within MC iterates, the final point reached will be stored in the storage vector ye.

The default values for RA and MC are RA = 0.1 and MC = 60.

How to set storage vectors y2, ..., y7 to specify basins A and B

Suppose we are studying a discrete process, a map F (for example, the Henon map) and we have found two periodic attractors A and B, and assume that for our fixed parameter values the system has only these attractors. Say the periods of A and B are a and b respectively. In the simplest case, we may have a = b = 1.

Using "Replace Vector" command RV, we replace storage vector y2 by one of the points of attractor A. An initial point is in the basin of A if the trajectory through the initial point comes within a distance RA of y2. Therefore, the RAdius RA has to be set correctly, see the subsection "How to determine and set the radius RA" below. Similarly, we replace storage vector y5 by one of the points of attractor B, and we say that an initial point is in the basin of B if the trajectory through the initial point comes within a distance RA of y5.

You can also set y2,...,y7 by first moving the mouse arrow to the desired point. Hold the left mouse button down until there is a beep (about 1 second). Use the mouse to select the appropriate command.

Essential region and Exit time

A **region** is an open and bounded set in the phase space. In this discussion, we assume that the phase space is two dimensional. An example of a region is the entire screen area. The **extended screen region** is the rectangular region with the center of the screen as its center and whose width is 1 + 2*SD times the screen width and whose height is 1 + 2*SD times the screen height. When a trajectory leaves the extended screen region, we say it **diverges** or **is diverging**. For k = 2, ..., 7, we say the yk-disk is the disk centered at yk with radius RA if yk is set; otherwise if yk is NOT SET, the yk-disk is empty. The **essential region** is the extended screen region minus the union of the yk-disks, where k = 2, ..., 7; see Figure 7-1. The **exit time T(x)** of a point x under a map F is the minimal nonnegative integer n such that the nth iterate of x under F is not in the essential region. Hence, T(x) = 0 for x outside the essential region, and T(x) ≥ 1 for x in the essential region.

When and Where does the trajectory of a point go?

The "BASin" routine BAS and several other routines repeatedly use the "**When** and **Where** the trajectory of a point goes" routine WW. The WW routine initializes the trajectory at a specified point, a point which depends on which routine is calling it. It then iterates the point up to a maximum number of iterates. It reports 2 numbers, the "When" and "Where" numbers. The "**When**" **number** of a point p is the exit time T(p). The value of the "**Where**" **number** equals the following:

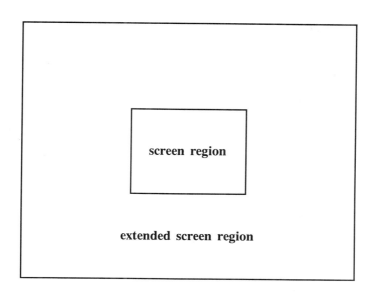

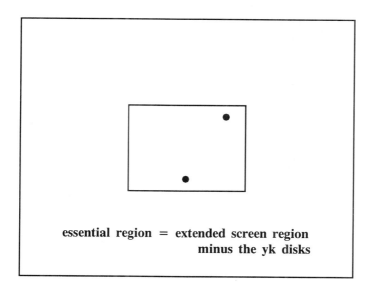

Figure 7-1

1, if the trajectory is found to be diverging
k, if the trajectory first comes within distance RA of the vector yk, where k = 2, ..., 7;
0, if none of the above happens in the allowed number of iterates, that is, MC iterates.

To get the "Where and When" information for the initialization point y1 (marked by small cross), use command WW. For details, see Section 7.2. You can do both using the mouse. Move the mouse arrow to the desired point. Hold the left button down until there is a beep. Use the mouse to select the command **MWW** from the menu.

How to determine and set the radius RA

Let the map F and the two periodic attractors A and B be as above. Assume that for our fixed parameter values the system has only these attractors, and that the periods of A and B are a and b respectively. Write B(x;RA) for the open ball centered at a point x with radius RA. Let p be any point of attractor A, and assume that there exists a ball U centered at p such that the ath iterate of F maps all points of U strictly inside U. Hence, there exists a radius RA > 0 such that the ath iterate of F maps the closure of the ball B(p;RA) into its interior. RA must be chosen so that all the points in B(p;RA) go to attractor A. Frequently it will be necessary to use some higher iterate, say the 3ath iterate of F, mapping a closed ball into its interior.

The program will not automatically set RA to a correct value. Running BA first will often show where the attractors are. You must find a value for RA. This value will depend on the specific map under investigation. The default value of RA is 0.1. If you do not set RA to a correct value, then the routine BAS plots all the initial conditions whose trajectories come within RA of the vector y2. The resulting "basin picture" may be misleading.

The radius RA has to be determined using the "circle and its Iterates" commands OI or OIY (the letter "O" represents the shape of a circle). To speed up the computations, the radius RA should be set as large as possible. A useful criterion is to find a reasonable positive number RA and a positive integer k such that (1) the akth iterate of F maps the closed ball with radius RA centered at y2 into its interior, and (2) the bkth iterate of F maps the closed ball with radius RA centered at y5 into its interior. To obtain the nth iterate of the circle with radius RA centered at y1, invoke the "PeRiod" command PR, and set it to be n. After doing that, enter the "circle and its Iterates" commands OY and OIY. These commands plot circles around those points y2,...,y7 that you have set, and OIY also plots the image of the circles.

Summarizing, RA must be chosen so that all the points go to the same attractor. Use OIY to check if RA has been set correctly. See for more details, Section 7.2 and for an example, see Example 7-1.

Technical Note. The radii for the two storage vectors y2 and y5 can be set independently using the "Radius of Attraction" commands RA2 and RA5. If you do not set RA2 or RA5, the program uses the value RA.

Plotting of basins using BAS

Recall that the routine BAS to plot basins uses the routine WW described above. The number of times that a point p is iterated equals the exit time T(p) but at most MC. Hence, computing stops when the trajectory of the point p leaves the essential region, that is, it diverges or it comes within a distance RA of one of the vectors y2 through y7. If none of the vectors y2 through y7 are set, then the routine BAS plots those grid boxes whose centers will exit the extended screen region within MC iterates.

The rules for what is actually plotted when the "BASin" command BAS is entered, are the following.

If WW returns 2, 3, or 4, then the point p is in basin A and its grid box is plotted.

If WW returns 5, 6, or 7, then the point p is in basin B and its grid box is not plotted.

If WW returns 0, then the trajectory of the point p goes nowhere, and the grid box is not plotted. The final point reached will be stored in storage vector ye.

If WW returns 1, then the trajectory of the point p is diverging. The grid box will be plotted **unless** you have set y2, and in that case it is not plotted. Hence if you want the diverging points to be plotted, then do **not** set the vector y2. If you do **not** want the diverging points to be plotted, then do set y2.

Plotting of basins in higher dimensions

When the phase space has dimension higher than two, we choose a two dimensional plane and plot the intersection of the basin with this plane. In *Dynamics* the plane is always a coordinate plane. That is, two coordinates vary and the rest of the coordinates are fixed. This plane is specified by the vector y1. If for example the phase space is four dimensional, and you choose the phase space coordinates of y1 to be (0.0,1.0,2.0,3.0), and if you specify that the X COordinate and Y COordinates are the coordinate numbers 0 and 2 of the phase space, then the plane will have its coordinates numbered 1 and 3 fixed with values 1.0 and 3.0. This plane contains the vector y1. A series of pictures can show how the basin structure varies when you change the values of the coordinates 1 and 3 of y1.

The routine BAS allows an even more general view when you are interested in points whose trajectories are diverging. You can take a two dimensional slice in the larger space consisting of the parameter and phase space coordinates. The Mandelbrot set is the complement of the basin of infinity for the map $Z \rightarrow Z*Z + C5 + iC6$. Example 2-14 shows the result of BAS after setting the X COordinate to C5 and the Y COordinate to C6.

You can also set one coordinate to be a parameter and the other to be a phase space variable.

Note. The techniques of the BAS method are essential for computing trajectories on basin boundaries as described in Chapter 8.

BAS Plotting scheme

The "BAS Plotting scheme Menu" BASPM, lists commands that set the different colors for plotting basins. The default coloring scheme is that the color used to plot the box is determined by the exit time of the center of the box. The point at the center of a box is checked. If this point exits in 5 iterates, for example, then the entire box is colored with color 5. The color may vary from box to box. It is possible to plot up to 6 basins using a different coloring scheme; see Section 7.2 for details.

The menu BASPM also lists commands that set the different diverging schemes for plotting basins; see Section 7.2 for details.

Summary for plotting basins of attraction utilizing BAS

- Search for the attractors (commands T or BA).
- Replace storage vector(s) by a point on an attractor (command RV).
- Determine a good value for RA (command RA), checking it with command OIY.
- **Optional.** Determine good values for RA2 and RA5 (RA2, RA5).
- Plot basin of attraction (command BAS).

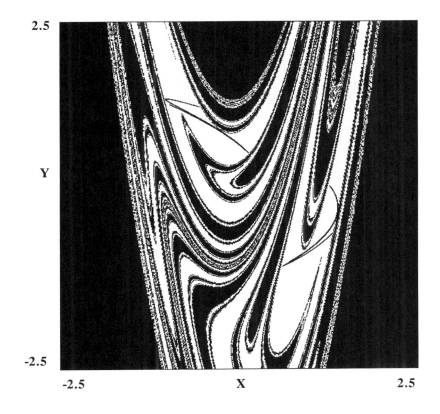

2.5

Y

-2.5

-2.5   X   2.5

Figure 7-2: Basin of attraction

The black area is a generalized basin of the Henon map (H)

$$(X, Y) \;\rightarrow\; (1 - X*X + 0.475*Y, \; X)$$

There are three attractors for this choice of parameters. The routine BAS has plotted the basins of infinity and of a stable period 8 orbit. The white area is the basin of attraction of a 2-piece chaotic attractor. The dots (in the white region) are from the superimposed 2-piece chaotic attractor. Its basin has not been plotted and so is white.

The basin resolution is 720 by 720. The CouNT command CNT reveals that for this figure 43% of the pixels are in the basin of infinity, 31% are in the basin of the period 8 orbit, and 26% are in the basin of the chaotic attractor.

Topic of discussion. *There is a fixed point lying approximately between the two pieces of the chaotic attractor. Where is its unstable manifold in relation to both attractors? What feature corresponds to its stable manifold?*

---

7. Basins of attraction   325

BASIN PICTURE RESOLUTION

Basins can be plotted in different resolutions. The "Basin Resolution Menu" command **BRM** shows commands that set the different resolutions for plotting basins. An "m × n" grid means the screen is divided into a grid m boxes wide and n boxes high. The Low Resolution is a 100 × 100 grid, and the High Resolution is a 720 × 720 grid. The "MEDium Resolution" command RMED sets the resolution for computing the basin to be a 240 × 200 grid. After invoking a resolution command, enter the "plot BASin" command BAS, and a basin will be plotted with the specified resolution. The plotted basin of an attractor A is obtained as follows. For an m × n grid, the routine tests only the center of each box as an initial point and assigns to the grid box a color (respectively, no color) if the trajectory starting at that center point converges to A (respectively, stays away from A). When no resolution command is entered before invoking the command BAS, the resolution of the resulting picture is a 100 × 100 grid, which is the default resolution. Invoking the "Basin Resolution" command **BR**, you can set the resolution to, for example, 1440 × 1440. For more details, see Section 7.2.

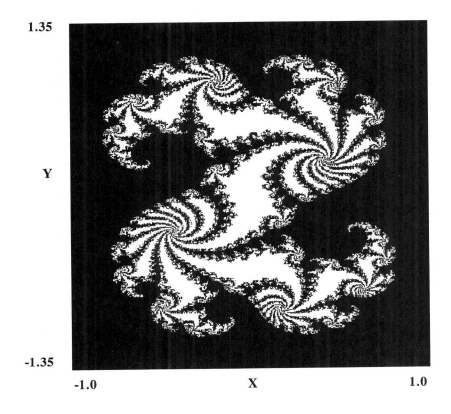

1.35

Y

-1.35

-1.0   X   1.0

Figure 7-3: Basin of attraction of a period 11 attractor

The black area is the basin of infinity of the complex quadratic map (C)
$$Z \rightarrow Z*Z + 0.32 + i0.043$$
where $Z = X + iY$. The white area is the basin of attraction of a stable period 11 orbit.

This can also be plotted after using command COL allowing more colors. Set IPP = 11 so that each point of the attractor becomes a fixed point. Then use command BA.

Topic of discussion. *One of the spirals has a fixed point at the inner tip of the spiral. How are the eigenvalues of that fixed point reflected in the spiral pattern? You can use command RP to find the fixed point, and the Quasi-Newton command Q1 to get the eigenvalues printed on the screen if the small cross is positioned at the fixed point. See also Figure 3-3b. Describe trajectories starting from the tips of other spirals.*

NUMERICAL METHOD FOR COMPUTING NEWTON BASINS

Dynamics utilizes the Newton and quasi-Newton methods for finding period-m points or period-m orbits, where you specify the period m using the PeRiod command PR; see Chapter 10. Since the Newton method is a special case of the quasi-Newton method, we discuss in this section the quasi-Newton method only. For a given point p, the **quasi-Newton basin** of p is the collection of all initial conditions that converge to p under iteration of the quasi-Newton method. The "BAsin Newton method" (command BAN) is a numerical method for plotting the quasi-Newton basins of fixed points or periodic points of a process.

The BAN method is like the BA method except it looks for basins of Quasi-Newton using period PR. It plots the quasi-Newton basins of periodic orbits of period PR. If there are several periodic orbits, it may be best to use 16 color planes (command COLP) and higher resolution (command BRM).

Summary for plotting quasi-Newton basins
● Set the period of the periodic orbit sought (command PR).
● Set the number of color planes (command COLP).
● Plot quasi-Newton basin (command BAN).

NUMERICAL METHODS IN PARAMETER SPACE

This section is aimed at exploring dynamical behavior when varying two parameters such as RHO and C1. See Figures 2-14, 2-24, 2-27, 7-10, and 7-11 for examples using the ideas of this section. The "BASin" command BAS has considerable flexibility. Recall that the Mandelbrot set is the set of parameter values (C5,C6) for which the trajectory of y = (0,0) under iteration of the complex quadratic map $Z \rightarrow Z*Z + C5 + iC6$ is bounded. It is the complement of the set of parameter values (C5,C6) such that the trajectory of (0,0) diverges; see Figure 2-14. The creation of the set of parameter values (RHO,C1) for which the trajectory of y has a positive Lyapunov exponent under iteration of the CIRCle map is an example of a "Chaos plot" in the parameter space; see Figure 2-24. The creation of set of parameter values (C5,C6) in the Mandelbrot set in Figure 2-27 is an example of a "Period plot" in the parameter space.

The "Lyapunov BASin method" (command L and command BAS) and "Basic Period BASin method" (command BP and command BAS) are numerical methods which are useful tools for deriving qualitative dynamics for two varying parameters by creating "Chaos plots" and "Period plots" in a the parameter space. Below we describe the numerical methods for creating Chaos Plots and Period Plots in the parameter space, in which two parameters (for example, RHO and C1) are varied and the other parameters (for example, C2, C3 and C4) are maintained fixed. The resulting pictures are plots in the parameter space. These methods can be used to detect regions in parameter space in which several attractors coexist and to study the prevalence of periodic versus chaotic dynamics.

Chaos plots

Let RHO and C1 be two parameters of a process such as the Henon map. The pair (RHO,C1) is called a **chaotic pair** of parameters (for y) if the trajectory that starts at y is chaotic. Recall that a chaotic trajectory has a positive Lyapunov exponent; see Chapter 5 for the definition of chaotic trajectory. The purpose is to generate a set in the parameter space such that for each point in this set the trajectory through initial state y is chaotic. In other words, we generate the set of parameter values for which the orbit of y converges to a chaotic attractor. A **Chaos plot** shows chaotic pairs of parameters (RHO,C1) for the bifurcation parameters RHO between RHO_{left} and RHO_{right} and C1 between $C1_{left}$ and $C1_{right}$. The objective is to plot (RHO,C1)-values for which the trajectory through the initial condition y has a positive Lyapunov exponent. In this book, Chaos Plots have been constructed as follows. We choose to have a grid of 720 by 720 to be tested and set PI = 1000. BAS now plots a box in the parameter space if the trajectory of the initialization vector y1 has at least one positive Lyapunov exponent provided that the number of exponents being calculated is positive (or more generally, if some Lyapunov exponent is greater than the value of BLM). Use the X COordinate command XCO, the Y COordinate command YCO, the X Scale command XS, and the Y Scale command YS to determine a region in (a two dimensional slice of) the parameter space. The routine BAS runs exactly the same as before, with as many iterates as before, except when it is time to color a box, it checks to see if an exponent is positive. See also "Basin Lyapunov Minimum value" command BLM.

When using BAS with L > 0, you can set the number of Pre-Iterates PI and the Lyapunov exponents will be initialized after PI iterates. If you have PI = 50 and MC = 60, then there will be a total of 110 iterates with the Lyapunov exponents computed for the last 60. Our criterion is that BAS plots the point (small box) in parameter space if at the maximum approximate Lyapunov exponent E is at least BLM = 0.01 after PI = 1000 pre-iterates and MC = 10000 iterates of the initial condition y.

If L = 1, BAS plots when the largest approximate Lyapunov exponent E is at least BLM. If E > BLM, then it depends on the value of E in which color the small box will be plotted. For example, the box is plotted in

color 1 if BLM + 0.00 ≤ E < BLM + 0.05
color 2 if BLM + 0.05 ≤ E < BLM + 0.10

.   .
.   .
.   .

color 12 if BLM + 0.50 ≤ E < BLM + 0.55

If L = 2 and if the phase space is two dimensional, BAS plots color when the largest approximate Lyapunov exponent E is at least BLM. If E > BLM, then the plotting color depends on the number of positive Lyapunov exponents. The small box is plotted in

color 1 if two Lyapunov exponents are positive
color 2 if one Lyapunov exponent is positive

Summary for creating Chaos plots in parameter space
- Set the number of Lyapunov exponents to be computed (command L).
- Set the X COordinate to be a parameter (command XCO).
- Set the horizontal axis for the range of the parameter (command XS).
- Set the Y COordinate to be another parameter (command YCO).
- Set the vertical axis for the range of the parameter (command YS).
- Set the number of PreIterates (command PI).
- Set the number of Maximum Checks (command MC).
- **Optional.** Set the Basin Lyapunov Minimum value (command BLM).
- Create Chaos plot (command BAS).

Period plots

The point (RHO,C1) in parameter space is called a **period-p pair** of parameters (for y) if the trajectory of y under the current process will converge to a stable period-p orbit for these parameter values. The purpose is to generate a set in the parameter space such that for each point in this set the trajectory through initial state y is asymptotically periodic. In other words, we generate the set of parameter values for which the orbit of y converges to a periodic attractor. A **Period-p plot** is the collection of all period-p pairs of parameters (RHO,C1) (for y) for the bifurcation parameters RHO between RHO_{left} and RHO_{right} and C1 between $C1_{left}$ and $C1_{right}$. A **Period plot** is the union of all Period-p plots, where $1 \leq p \leq N$, for some positive integer N. The objective is to plot (RHO,C1)-values for which the trajectory through the initial condition y is approaching a stable period-p orbit; so, generally the trajectory of y has a negative Lyapunov exponent. We construct Period plots as follows. We use a grid of 720 (horizontal) by 720 (vertical) to be tested and set PI = 1000. Our criterion is to plot the point (small box) in parameter space in color p if p is the smallest positive integer for which the trajectory of $y = y_0$ has the following property: after the first MC = m iterates of y_0, the distances $\|y_{m+1+p}-y_{m+1}\|$, $\|y_{m+1+2p}-y_{m+1+p}\|$ and $\|y_{m+1+3p}-y_{m+1+2p}\|$ are all less than ε, where ε is some prescribed positive number.

If the basin coloring scheme is BP (which is BT by default), then L is set to 1 automatically, and BAS now plots a box in the parameter space if either the trajectory of the initialization vector y1 is asymptotically periodic (that is, it converges to a periodic attractor) or the maximum approximate Lyapunov exponent of the trajectory of y1 is greater than the value of BLM; see also Chaos plots above. Use the X COordinate command XCO, the Y COordinate command YCO, the X Scale command XS, and the Y Scale command YS to determine a region in (a two dimensional slice of) the parameter space. The routine BAS runs exactly the same as before, with as many iterates as before, except when it is time to color a box, it checks to see if either the trajectory of y1 is asymptotically periodic or the maximum approximate Lyapunov exponent of the trajectory of y1 is positive. See also "BAS Plotting scheme Menu" BASPM.

When using BAS with the coloring scheme BP, you can set the number of Pre-Iterates PI and the Lyapunov exponents will be initialized after PI iterates. We select N = 12 and $\varepsilon = 0.0001$. A **Period plot** is the union of all Period-p plots, where $1 \leq p \leq 12$. Our criterion is that if the basin coloring scheme is BP, then BAS plots the point (small box) in parameter space if either the trajectory of y converges to a period-p attractor ($1 \leq p \leq 12$) or the maximum approximate Lyapunov exponent E is at least BLM = 0.01 after PI = 1000 pre-iterates and MC = 10000 iterates of the initial condition y.

If the basin coloring scheme is BP, then the color that BAS uses to plot the small box in the parameter space, depends on whether the trajectory of y1 is asymptotically periodic and if this trajectory is asymptotically periodic, the color depends on the period in which color the box will be plotted. For example, the box is plotted in color p if trajectory of y1 converges to period-p attractor ($1 \leq p \leq 12$) and in color 14 if the maximum approximate Lyapunov exponent is at least BLM (which is 0 by default).

After the picture has been created, you can get the Period plot by Erasing color 14.

Summary for creating Period plots in parameter space

- Set the basin coloring scheme to be BP (command BP).
- Set the X COordinate to be a parameter (command XCO).
- Set the horizontal axis for the range of the parameter (command XS).
- Set the Y COordinate to be another parameter (command YCO).
- Set the vertical axis for the range of the parameter (command YS).
- Set the number of PreIterates (command PI).
- Set the number of Maximum Checks (command MC).
- **Optional.** Set the Basin Lyapunov Minimum value (command BLM).
- Create the union of Period plot and Chaos plot (command BAS).
- After the picture has been created, get the Period plot by erasing color number 14 (command E14)

7.2 BASIN OF ATTRACTION MENU BM

In practice, to study the basin of an attractor on the computer, we pick one or more target points and we say that an initial point is in the basin of the target points if the trajectory through the initial point comes close to (that is, within a distance RA of) one of the selected target points. If the target points are chosen on a chaotic attractor, the trajectories will briefly pass within the distance RA of the target point and will then move away, but we terminate the trajectory once the trajectory comes close to a target point. We also have target points to determine a second basin so that if the trajectory comes close to one of those points, computation also stops so as to avoid wasting time.

Enter the "Basin Menu" command (in the default situation)

$$\text{BM} < Enter >$$

and the menu appears on the screen in approximately the following format.

BM	– –	BASIN OF ATTRACTION MENU
BA	–	Basins and Attractors
BAN	–	BAsin for Newton method for points of period PR = 1
BAS	–	plot BASin
MC	–	Maximum Checks per initial point: 60
RA	–	Radius of Attraction for storage vectors: 0.1
WW	–	check Where the trajectory of y1 goes and When it gets within RA of y2, ..., y7 or diverges within MC iterates
		MENUS
BRM	–	Basin Resolution Menu
BASPM	–	BAS Plotting scheme Menu

None of the vectors y2, ..., y7 are set.
BAS will plot only points whose trajectories diverge
since y2 is NOT SET
BAS will plot NO OTHER points since y2, y3, y4 are not set
Current basin resolution: width = 100, height = 100

The point at infinity is automatically a target point and we say an initial point is in its basin if its trajectory leaves the large box that extends SD (Screen Diameter) times the size of the screen in all directions. See command SD.

The defaults do not cover many cases of interest, where we want to see which points go to one attractor and other points go to another, and perhaps all trajectories are bounded.

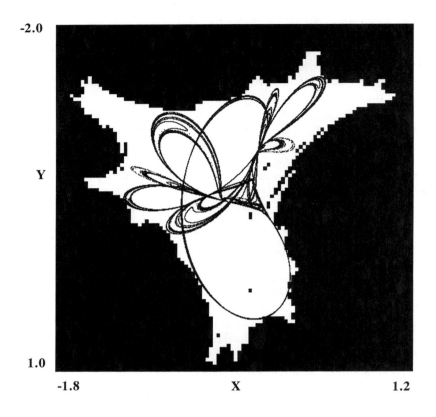

-2.0

Y

1.0

-1.8   **X**   **1.2**

Figure 7-4a: Basin of attraction and chaotic attractor

The black area is the basin of attraction of the attractor at infinity of the Tinkerbell map (T)

(X, Y) → *(X*X - Y*Y + 0.9*X - 0.6013*Y, 2*X*Y + 2.0*X + 0.5*Y)*

*   The basin is plotted in Low Resolution (100 × 100 grid). Black boxes are plotted when the trajectory of an initial point in the box is found to go to infinity. A trajectory on the chaotic attractor is also plotted. The parameters have been chosen specifically so that the attractor is near the boundary of its basin.*

-2.0

Y

1.0

-1.8   X   1.2

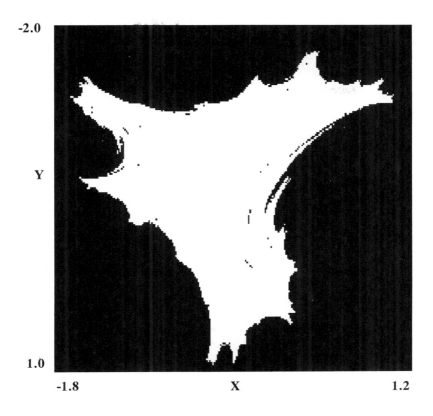

Figure 7-4b: Basin of attraction

This figure is the same as Figure 7-4a except with a higher resolution. The basin is plotted in MEDium Resolution (240 × 200 grid). Black boxes are plotted when the trajectory of an initial point in the box is found to go to infinity. MEDium Resolution is almost as good as High Resolution and it runs about 10 times faster.

Difficult Topic of discussion. *The parameter C2 = -0.6013 has been chosen so that the attractor is very close to the boundary of the basin of infinity. As it decreases further, the attractor will collide with the boundary. Describe what happens at the "crisis" value where they collide.*

BM

The "Basin of attraction Menu" command BM is for calling the menu. The easiest basin command to use is the "Basins and Attractors" command BA.

The second command for plotting basins is the command BAS. The screen area is divided into a grid of boxes. For each box in a grid of small boxes (for example a 100×100 grid) the basin computation (BAS method) determines if the trajectory through the center of that grid box goes near one of the target vectors for the first basin

$$y2, \ y3, \ \text{or} \ y4,$$

before it goes near any of the target points for the second basin,

$$y5, \ y6, \ \text{or} \ y7.$$

If so it colors the grid box and if not, it leaves the spot (box) blank on the screen. (What is an illuminated box on the screen will be a black box on paper.) The program repeats the procedure for all the grid boxes. If the trajectory diverges (as described in command WW), the corresponding grid box may or may not be plotted.

If $y2$ has not been set, the grid box will be colored (plotted) if the trajectory of its center diverges. If $y2$ has been set and the trajectory of the center of the grid box diverges, the box will not be plotted. If the trajectory **goes nowhere**, that is, it has neither diverged nor approached $y2$ through $y7$ in MC iterates, then the initial box is not plotted.

If there are two bounded attractors, set $y5$ to be a point of the other attractor. Then when the trajectory through the center of a grid box comes close to $y2$ or $y5$, the iterating will stop -- and the corresponding grid box will or will not be plotted depending on whether the point it came near was $y2$ or $y5$ -- and then the next grid box will be tested. This might average only 10 or fewer iterates even though the maximum number of iterates is a few thousand. Hence it is important to set $y5$ if there are two bounded attractors in order to speed the computation of the picture.

Note. One of the hardest part about using BAS and BST is the process of choosing the yk's correctly, and part of the problem is knowing what you have done. If you choose the yk's so that their circles of radius RA overlap, then when you retrieve the menu BM, a warning will be plotted stating that some are very close together. This warning does not prevent you from choosing them close together on purpose. The command OY will plot circles centered at the yk's thst are set and so OY will show you your choices.

Figure 7-4c: Basin of attraction
This figure is the same as Figures 7-4a and 7-4b except with a higher resolution. The basin was plotted in High Resolution (960 × 960 grid) using a Unix workstation. Black boxes (which are individual dots) are plotted when the trajectory of an initial point in the box is found to go to infinity. Notice small stray patches of black inside the lower loop of the attractor.

Topic of discussion. *Find the fixed points and the points that map onto the fixed points in one iterate. (Use command II.) How are they all related to the boundary?*

Commands for computation of basins (BA related methods)

BA

The "Basins and Attractors" command BA plots all the basins and the attractors. Grid boxes containing points of attractors are colored using even color numbers. Basins are colored with odd numbered colors, and in particular the basin of infinity uses color 1 (dark blue). If $2n$ is the even color number for an attractor, its basin is allocated color $2n+1$. While a grid box can contain points of several basins and possibly of several attractors, its color can only reflect one of these. Hopefully, when a fine enough grid is selected, so that no box contains points of two different attractors, the box will contain only points of the attractor and of its basin.

Use Color Table command CT to view the color table. Use command Ek possibly with the PATtern command PAT, to erase a color. Then the remaining basins can be printed.

Enter command BA, and the Actions, Hints and Options menu for BA appears and includes the following:

```
ACTIONS, HINTS and OPTIONS for BA

BA2  –  Basins and Attractors if phase space of
            dimension > 2 and if MC > 100
BAC  –  Bright Attractor Colors
BAF  –  tells BST to use the BA picture: OFF
BAP  –  BA Precision = 10 (default value is 10)
BR   –  Basin Resolution (number of boxes)
            width = 100 boxes, height = 100
CB   –  BA, BA2, BAN are Constrained to small Box: OFF
MC   –  Maximum Checks per initial point:  60
```

In the Information window of the Actions, Hints and Options menu for BA, the following text is provided:

```
When picture is complete, use CNT to count pixels
```

BA2

The command BA2 is a version of BA for higher dimensional systems. It plots basins on the coordinate plane through y1; that is, if there are three variables and the X coordinate and Y coordinate are two of them, the third is determined by its value of y1.

BAC

The "Bright Attractor Colors" command BAC is a coloring scheme and is for use with command BA. It effects color numbers 2, 4, 6, and 8, which are used to present attractors. It makes the attractors easier to see. See also Section 3.7 for setting colors.

BAF

This command provides a simplified way of computing Basin Straddle Trajectories. The command BAF is listed in the Straddle Options Menu SOM; see Chapter 8 for its description.

BAP

The "Basins and Attractors Precision" number BAP is used by command BA when plotting the basins and attractors. Higher values produce more accurate pictures but result in slower computations. The default value of BAP is 10. Except in unusual circumstances, this is an excellent value.

CB

The "Constraint Box" command CB instructs the routines BA,BA2 and BAN to Constrain initial points to a small Box, the box that can be created on the screen using arrow keys or mouse. The idea is that you can create a picture with low resolution and then enter the commands CB and CONT (CONTinue) and specify a higher resolution, and then the BA2 command, for example, will refine the picture inside that box.

MC

The "Maximum Checks" command MC is for basin calculations. See "Computations of basins: the BA method" in Section 7.1 for its use with the routine BA.

Commands for computation of basins (BAN related methods)

BAN

The command BAN is a version of BA that plots Basins and Attractors of quasi-Newton method. You must specify a period PR. Then the quasi-Newton method searches for periodic points of that period. BAN computes basins for iterates of the quasi-newton method (unless QUASI is turned OFF).

Enter command BAN, and the Actions, Hints and Options menu for BAN appears and includes the following:

```
ACTIONS, HINTS and OPTIONS for BAN

BAC     –  Bright Attractor Colors
BR      –  Basin Resolution (number of boxes)
               width = 100 boxes, height = 100
CB      –  BA, BA2, BAN are Constrained to small Box:  OFF
COL     –  256 colors (for VGA only)
MC      –  Maximum Checks per initial point:  60
               preferably MC > 400 and even larger if NMS < 0.5
NMS     –  Newton Maximum Step size if set: NOT SET
PR      –  PeRiod of the orbits sought:  1
QUASI   –  use quasi-Newton if ON, else Newton; now:  ON
```

NMS

The "Newton Max Step" command NMS is an optional setting that makes Newton method restrict itself to small steps. If NMS is set, it specifies the largest step allowed. If Newton's method recommends a larger step, it is reduced in size but the direction of the step is unchanged.

When NMS is used with BAN, or RP or RPK, it is useful to increase MC, the Maximum number of steps used, since if small steps are taken, many steps may be required to reach the periodic orbit. For commands RP, RPK, etc., CB restricts the entire search to the small box. It is useful to make Newton Max Step NMS small so the search does not leave the box easily.

QUASI

The command QUASI determines how BAN and the RP-commands operate. QUASI is a toggle. The default setting is ON and it should almost always be left ON. QUASI must be ON for BAN and the RP-commands to use quasi-Newton method. These routines use the Newton method if QUASI is OFF. In other words, if you want BAN to compute Newton basins rather than quasi-Newton basins, turn QUASI in the OFF position. Better results will be obtained when QUASI is ON.

Commands for computation of basins (BAS related methods)

BAS

The "BASin" command BAS has considerable flexibility. It can be used for plotting basins of attractors or Mandelbrot sets. In the former case, the two screen coordinates correspond to coordinates of y. For the Mandelbrot set (Examples 2-14 and 2-27) the coordinates are parameters of the map, namely C5 and C6. In other cases it is useful to have one coordinate of y (say y[1]) plotted vertically and have RHO, for example, plotted horizontally.

For simplicity we start by describing what happens in the default case. The objective is for command BAS to plot the region of small boxes in the screen whose centers appear to be initial points for trajectories that will go to infinity. Since the program does not test every point in the screen, and since it has no rigorous way of testing whether any trajectory will in fact diverge to infinity, it is worthwhile to be clear as to what the program actually does.

The command BAS checks the behavior of trajectories starting at the center of each of the boxes in a 100 by 100 grid of boxes (default). For each box in a 100 by 100 grid boxes, the program first initializes the trajectory, setting **all** coordinates of y equal to those of y1. Then it sets the horizontal and vertical coordinates of y equal to those of the center of the grid box. Then that point is iterated up to a maximum of MC (Maximum Check) times, which has a default time of 60 iterates. After every iterate the program tests to see if the trajectory has left a specified region, which we call the **extended screen region**; see Figure 7-1. In the default case (having SD = 1), the specified region is a rectangle that is three times as wide as the screen and three times as high, with the rectangle centered at the center of the screen. If the trajectory leaves the extended screen region, we say that the trajectory is "**diverging**". If the trajectory "diverges", that is, if it appears to be diverging to infinity (see below), then we say the initial point is in the plotted basin and the corresponding grid box is plotted (that is, the routine draws a small filled-in box, 1/100 of the screen wide and 1/100 of the screen high, centered at the tested point). If on the other hand, after MC iterates, the trajectory is not yet diverging, then the position in the grid is left blank. Then the program goes to the next grid box. After checking all the grid boxes, the program pauses.

On a color screen, the color used to draw the box varies from box to box. If the trajectory diverges, then the program determines the time (number of iterates) required. The color table positions are numbered, and the number of the color table position used to color the boxes that are plotted is the time it has taken for the trajectory to start to diverge. If this time of diverging is larger than the number of color table positions, it cycles back through the color table, skipping color table position 0 which is the background color. See Section 3.7 about color number and color table position number.

The command BAS has considerably flexibility. In conjunction with computing Lyapunov exponents (command L), BAS can be used to create **Chaos plots** in the parameter space. In conjunction with command BP, BAS can be used to create **Period plots** in the parameter space. For more details on Chaos plots and Period plots, see Section 7.1 and command BP.

MC

The "Maximum Checks" command MC is for basin calculations. It allows you to specify the maximum number of iterates that will be attempted to see if an initial point's trajectory will approach one of the points y2, ..., y7, (depending on which of these have been set) or if it is diverging. Sometimes there are attracting points that you don't not know about. If the trajectory is iterated MC times without approaching any of the target points in either attractor, then the final point is stored in storage vector ye for later investigation. You may wish to include such points in one of the two sets of target points. See also "Straddle Maximum Checks" command SMC which is the number of checks used for straddle orbits.

Experience indicates that occasionally quite a few points will take 100 or more iterates so you have to experiment on the maximum number of iterates. Sometimes it might be better to set MC to be a few thousand. If there are two attractors, one of which can be infinity, this is acceptable. If MC is 100 or more and the trajectory goes nowhere, a warning will be reported on the screen "The trajectory has been iterated MC times". To suppress these warnings, set Text Level TL to 1 using command T1.

RA

The "RAdius of attraction" RA is used by several routines, such as the basin routine BAS, the "When and Where" command WW, the basin boundary straddle routines BST and ABST, and sometimes the straddle routine SST. A trajectory is declared to be going to an attractor if it gets within this "RAdius of attraction" distance from any of the points y2, ..., y7 (depending on which of these have been set). See also the subsection "How to determine and set the radius RA".

RA2, RA5

The radii for the two storage vectors y2 and y5 can be set independently using the "Radius of Attraction" commands RA2 and RA5. If you do not set RA2 or RA5, the program uses the value RA. The commands RA2 and RA5 are **not listed in the menus**.

WW

The "When and Where" command WW is essential for basin calculations and for straddle calculations (Chapter 8). The command WW tells whether the trajectory that starts at a specified point is "diverging" or if it approaches one of the test points y2 through y7. This routine is implicitly called repeatedly (usually thousands of times) by basin and straddle routines. *Practice with this command will familiarize you with what those routines are actually doing.* Different calling routines use different initial points. When the user directly calls WW, the routine uses y1, the small cross as its initial point.

More specifically, this routine stops as soon as the trajectory diverges (that is, goes more than SD (Screen Diameters) screen widths or heights from the screen) and it stops as soon as it comes within a distance RA (RAdius of attraction) of any of the points y2 through y7 (ignoring any of those points the user has not set), and if neither of those occurs before MC (Maximum Check) iterates, it stops then anyway. The routine then reports which of these events occurred and how long it took for the event to occur. If it stops only because it has reached MC iterates, it stores in storage vector ye the point where the trajectory stopped. The latter is useful because a system may have an attractor that you did not know about, and this allows you to access a point of the attractor.

It typically takes much longer for the trajectory to be iterated and checked MC times (default value of MC is 60 iterates) than to discover the trajectory has approached some specified point. Therefore, if MC is set to a large value when running basin routines, the vectors y2 through y7 should be set in such a manner that MC iterates is only rarely if ever reached. When the routine is used implicitly, it tests other initial points and does not use the small cross.

When using WW for testing straddle trajectories, the maximum number of iterates is Straddle Maximum Check SMC = 2000 instead of MC. The user may reset MC to be 2000 to simulate what it does for straddle orbits.

Enter command BAS, and the Actions, Hints and Options menu for BAS appears and includes the following:

```
ACTIONS, HINTS and OPTIONS for BAS

BR      –  Basin Resolution (number of boxes)
           width = 100 boxes, height = 100
L       –  number of Lyapunov exponents: L = 0
           If L > 0, BAS plots chaotic basin or creates Chaos plot

BASC    –  Color scheme for BAS with current settings
BLM     –  criterion for Lyapunov exponent to be near 0:  0
BP      –  BAS plots periodic attractor Period and chaotic attractors
MC      –  Maximum Checks per initial point:  60
OY      –  draw circles about all yx's (x = 2,...,7)
OIY     –  draw the circles OY plus their MI iterates
RA      –  Radius of Attraction for storage vectors:  0.1

           MENU
BASPM   –  BAS Plotting scheme Menu
```

BASC

The "BAS Color scheme" command reports the current coloring scheme for BAS, it displays the current color table and provides some additional information on the colors.

BLM

The "Basin Lyapunov Minimum value" command BLM is for setting the lower bound for Lyapunov exponents when using command BAS and the number of Lyapunov exponents to be computed is positive.

When using BAS with $L > 0$, you can set the number of Pre-Iterates PI and the Lyapunov exponents will be initialized after PI iterates. If you have PI = 60 and MC = 90, then there will be a total of 150 iterates with the Lyapunov exponents computed for the last 90. For example, in the case of the CIRCle map (CIRC), if you enter:

$$XCO < Enter > \quad RHO < Enter >$$

$$YCO < Enter > \quad C1 < Enter >$$

$$XS < Enter > \quad 0\ 6.28 < Enter >$$

$$YS < Enter > \quad 0\ 8 < Enter >$$

$$L < Enter > \quad 1 < Enter >$$

$$\mathbf{PI} <Enter> \quad 60 <Enter>$$

$$\mathbf{MC} <Enter> \quad 90 <Enter>$$

$$\mathbf{BAS} <Enter> <Enter> <Enter>$$

you obtain a picture.

If you are interested in (a part of) the chaotic parameter set, then you may wish to set BLM to be 0.1. See also Section 7.1. The default value of BLM is 0.

OY

This command OY is similar to O (see the Circles command O). Command OY plots circles that are centered at all the yk points that have been set, where k = 2,...,7, while routine O (or O1) draws a circle centered at y1.

If y2 is set, the circle centered at y2 (and its iterates for OIY) is plotted in color 2, the circle centered at y3 in color 3, etc. The routine starts by listing the yk's that have been set in the color that is used to plot them.

The radii used are RA, and RA2 and RA5 for y2 and y5 if they have been set. When OY is called, a note appears on the screen informing the user about command OIY, see below.

OIY

The command OIY draws the OY command's circles and their MI iterates. The circle centered at yk and its iterates are plotted in color k, where k = 2,...,7. See also commands OY, O, and OI.

OYK, OIYK

The command OYK is for drawing a circle centered at a storage vector with radius RA.

The command OIYK is for drawing a circle centered at a storage vector with radius RA, and its MI iterates.

Ok, OIk, where k = 1, 2, ..., 7, a, ..., e

The command Ok (where k = 1, 2,...,7, a,...,e) is similar to O. This command draws a circle centered at yk with radius RA, if yk has been set.

The command OIk (where k = 1, 2, ..., 7, a, ..., e) is similar to OI. This command draws a circle centered at yk with radius RA and it MI iterates, if yk has been set. The circle is plotted in the active color (default is color 14), the second iterate is plotted in color 2, the third iterate is plotted in color 3, etc. See also commands O, OI, OY, and OIY.

Basin Resolution Menu

Entering the "Basin Resolution Menu" command BRM

$$\textbf{BRM} < Enter >$$

results in a menu with the following commands

```
╔════════════════════════════════════════════════════════════╗
║  BRM    – –      BASIN RESOLUTION MENU                       ║
║                                                              ║
║  The following commands set the resolution for BAS, BA, BA2. ║
║  'm × n' grid means a grid of m boxes wide and n boxes high. ║
║                                                              ║
║  RB     –  Basin flexible Resolution, now:  100 ×  100  grid ║
║  RL     –  Low Resolution:                  100 ×  100  grid ║
║  RMED   –  MEDium Resolution:               240 ×  200  grid ║
║  RMH    –  Medium-High Resolution:          320 ×  272  grid ║
║  REGA   –  EGA Resolution:                  640 ×  360  grid ║
║  RVGA   –  VGA Resolution:                  640 ×  480  grid ║
║  RH     –  High Resolution (= core grid):   720 ×  720  grid ║
╚════════════════════════════════════════════════════════════╝
```

Commands for basin picture resolutions

The following commands are listed in the Basin Resolution Menu above. If you enter a resolution command and the command BA(S), then the basin will be computed in the corresponding resolution. For example, if the resolution command is REGA, then the basin will be computed in the EGA resolution.

RB

The "BAS Resolution" command RB allows the user to change the grid size; the first number is the number of boxes the grid is wide and the second is the number of boxes the grid is high. The two numbers must be entered with a space but no comma between them. For example, entering the command RB followed by entering 2 integers "m n" results in an m × n grid (that is, a grid which is m pixels wide and n pixels high). If the resolution is chosen to have more boxes than there are points on the screen (or points in the core picture), then each screen (or core) point would correspond to more than 1 grid box. In effect, this allows you to check several points per pixel (point on the screen or in the core). Note however that calculations times increase with resolution and such a calculation can be very long. For example, if the resolution is chosen to be 1000 by 1000

---

when the core is 500 by 500, then each core pixel will be checked 4 times. This allows a more thorough checking of the basin.

Once the resolution is set, it remains fixed until it is changed by the user.

RL

The "Low Resolution" command RL sets the low resolution for the basin computation. **Low resolution** means a 100×100 grid of boxes is used.

REGA

The "EGA Resolution" command REGA sets the EGA resolution for the basin computation. **EGA resolution** means a 640×360 grid is used.

RMED

The "MEDium Resolution" command RMED sets the medium resolution for the basin computation. **Medium resolution** means a 240×200 grid is used. Therefore this resolution has about 6 times as many points and is 6 times slower than low resolution.

RMH

The "Medium-High Resolution" command RMH sets the medium-high resolution for the basin computation. **Medium-high resolution** means a 320×272 grid is used. The width 320 is 1/3 of the Epson printer's horizontal resolution and 272 is 1/2 of the number of vertical lines being used.

RVGA

The "VGA Resolution" command RVGA sets the VGA resolution for the basin computation. **VGA resolution** means a 640×480 grid is used.

RH

The "High Resolution" command RH sets the high resolution for the basin computation. **High resolution** means the same resolution as the core grid, which is usually a 960×544 grid, is used for PC's. If the command SQ has been called, the core will be a 720×720 grid. This is the core resolution used for most pictures in this book. These high resolution computations can be very time consuming, about 52 times as long as low resolution.

BAS Plotting scheme Menu

Entering the "BAS Plotting scheme Menu" command BASPM

BASPM < *Enter* >

results in a menu with the following commands

```
BASPM  – –     BAS PLOTTING SCHEME MENU

               COLORING OPTIONS FOR BAS
BT      –  plot Basin with color number  =  capture_Time mod(16)
BT1     –  plot Basin with color number  =  capture_Time up to 16
BT2     –  BT with color number  =  numColors capture_Time / MC
BN      –  plot Basin with color number  =  Number of ball entered
BMOD    –  plot Basin with color number  =  1+capture_Time(modPR)
BP      –  BAS plots periodic attractor Period and chaotic attractors

               WHEN BAS PLOTS DIVERGENT INITIAL POINTS
BASA    –  All: plot for All diverging trajectories -- default
BASN    –  Not: plot for No diverging trajectory
BAST    –  Top: plot whenever diverging with Y coord  >  0
BASR    –  Right: plot when diverging with X coord  >  0
BASL    –  Left: plot when diverging with X coord  <  0
BASB    –  Bottom: plot when diverging with Y coord  <  0
TV      –  Top+V: plot when inVerse diverges with Y coord  >  0

Color scheme is now BT
Diverging scheme is now BASA
```

Commands for color schemes for plotting basins

BT, BT1, BT2

The "Basin exitTime" command BT is for plotting a basin in different colors. The color plotted is based on the computed exit time minus BMIN. You can set BMIN using command BMIN. The resulting procedure is the default procedure for coloring basins. Invoking this command affects how the computer carries out the command BAS. The algorithm for choosing the color depends on the time required for the trajectory to terminate. It only plots the point if that time is greater than or equal to BMIN, which has a default value of 0. The color number for coloring when BAS plots using scheme BT, is given by

---

$$\text{color number} = (\text{exitTime - BMIN - 1}) \bmod (\#\text{colors - 1}) + 1$$

that is, the color number is the remainder plus 1 when time minus BMIN is divided by the quantity (number of color table positions minus 1). For color screens, #colors minus 1 is usually 15. Set BMIN by using command BMIN. Compare this with command BN.

The "Basin Time 1" command BT1 plots the basins with color equal to exitTime minus BLM, up to the number of colors minus 1. If the escape time exceeds this value, then it still plots the basins with color #colors - 1.

The "Basin Time 2" command BT2 is like BT but now it plots the basin with color proportional to exitTime/MC; if this value is near 0, color 1 is used while if the ratio is near 1, color 15 is used (when 16 colors are available). The color number for coloring when BAS plots using scheme BT2, is given by

$$\text{color number} = 1 + [(\text{exitTime - BMIN})*(\#\text{core colors - 1})]/(\text{MC - BMIN})$$

then if color number is less than 1, reset color number to 1.

BMIN

For this command, see command BT. The default value of BMIN is 0. The command BMIN is **not listed in the menus**.

BMOD

Invoking this command affects how the computer carries out the command BAS. The algorithm for choosing the color then depends on the time required for the trajectory to terminate. Here "time" means the number of times the point has been iterated when its trajectory first comes within a distance RA of one of the vectors y2 through y7. Choose PR less than the number of colors. Then the boxes are colored using

$$\text{color number} = \text{time} \bmod(\text{PR}) + 1$$

that is, the remainder plus 1 when time is divided by PR.

One way to use this command is to choose one attracting periodic point of period PR, and set y2 equal to that point. Select radius RA so that the disk of radius RA centered at y2 contains only points that are attracted to y2. Enter commands BMOD and BAS.

BN

The "Basin Number" command BN is for plotting different basins in different colors. Invoking this command affects how the computer plots when carrying out the command BAS. It causes the computer to choose the plotting color according to where the trajectory goes. If a trajectory terminates by approaching y2, color 2 (color table position 2) is plotted at the initial point. If it terminates by approaching y3, ..., y7, it plots colors 3, ..., color 7, respectively. If the trajectory diverges toward infinity, color 1 is used and if it terminates only because it reaches MC (Maximum Check) iterates, nothing is plotted. These colors denoted by these numbers can be changed using the "Set Color" command SC and the "Set Color Table" command SCT.

BP

The "Basin Period" command BP affects what BAS plots. If the BAsin Coloring scheme is BP, then BAS can be used for finding the period of attractors. This method can be used, for example, when both coordinates are parameters. Then it allows you to search parameter space and see if there are periodic attractors. This command is designed to be used with at least one coordinate being a parameter, though that is not necessary. Use commands CT (Color Table) and CNT (CouNT) to evaluate which colors have been plotted.

Assuming there are 16 colors and that the BAS Coloring scheme is BP:
(1) BAS will plot color n if trajectory of y converges to periodic orbit of period n, where n < 13.
(2) If the trajectory of y diverges, then BAS plots color 15. Then, the initial point (which may be a point in the parameter space) gets color that equals the period encountered.
(3) If the trajectory of y has a positive Lyapunov exponent, BAS plots color 14 for the initial point (actually, if the maximum Lyapunov exponent is greater than BLM (which you can set)).
(4) BAS plots color 13 if the maximum Lyapunov exponent is near zero (the absolute value of the maximum Lyapunov exponent is less than BLM); you must set BLM > 0.
(5) If none of the above four cases occurs (for example, trajectory of y converges to a high period attractor), the routine BAS plots nothing (which means that the initial point gets color 0).

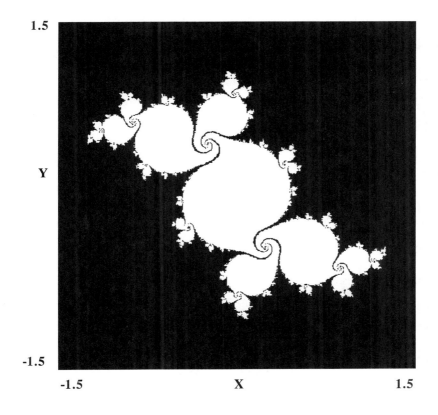

1.5

Y

-1.5

-1.5   X   1.5

Figure 7-5: Basin of attraction

The black area is the basin of infinity of the complex quadratic map (C)

$$Z \rightarrow Z*Z - 0.11 + 0.6557i$$

where $Z = X + iY$. The white area is the basin of attraction of a stable period 3 orbit.

Topic of discussion. *Describe the basins in a small neighborhood of each of the fixed points.*

Set Iterates Per Plot IPP = 3. Then make this picture using command BA with MC = 300. Explain the coloring patterns of the resulting picture. This picture requires that your computer has enough memory to be able to store at least 8 colors.

Commands for plotting basins when trajectories are diverging

BASA

The "BAS All" command BASA is a diverging scheme for the routine BAS. If the diverging scheme is BASA, then BAS plots All diverging trajectories.

BASN

The "BAS Not" command BASN is a diverging scheme for the routine BAS. If the diverging scheme is BASN, then BAS does Not plot diverging trajectories.

BASB, BASL, BASR, BAST

The "BAS Bottom" command BASB is a diverging scheme for the routine BAS. If the diverging scheme is BASB and a trajectory is diverging, then BAS plots whenever y[Yco] < 0 for the first point of the trajectory that is outside the extended screen region (see Figure 7-1).

The "BAS Left" command BASL is a diverging scheme. If the diverging scheme is BASL, then BAS plots if y[Xco] < 0 for the first point of the trajectory that is outside the extended screen region.

The "BAS Right" command BASR is a diverging scheme. If the diverging scheme is BASR, then BAS plots if y[Xco] > 0 for the first point of the trajectory that is outside the extended screen region.

The "BAS Top" command BAST is a diverging scheme. If the diverging scheme is BAST, then BAS plots if y[Xco] > 0 for the first point of the trajectory that is outside the extended screen region.

TV

The "Top + inVert" command TV is a diverging scheme for the routine BAS. It is a combination of BAST and V.

7.3 EXAMPLES

In Chapter 2, we have already presented a few elementary examples of basins. Example 2-9 creates a picture of all the basins and attractors for the Henon map (Figures 2-9a and 2-9b) and Example 2-15 creates a picture of all the basins and attractors for the complex quadratic map (Figure 2-15). Examples 2-10 and 2-20 plot the basin of the attractor at infinity for the Henon map. The Mandelbrot set is discussed in Example 2-14. The following example illustrates how to set the radius RA and how to find attractors with small basins.

Example 7-1: Plotting basins using the routine BAS
To illustrate plotting of basins, we first use the Henon map

$$(X,Y) \rightarrow (1 - X*X + 0.475*Y, X)$$

where the variable X is plotted horizontally, and the variable Y is plotted vertically. The X Scale is from -2.5 to 2.5, and the Y Scale is from -2.5 to 2.5.
  We deal with the following objectives.
● Search for possibly coexisting attractors of the Henon map.
● Plot basins of attraction of the Henon map.

Example 7-1a: Plot the basin of infinity and a chaotic attractor
  1. Set the parameters C1 to be 0.475:

$$C1 <Enter> \quad 0.475 <Enter>$$

Set the parameter RHO to be 1:

$$RHO <Enter> \quad 1 <Enter>$$

  2. The default value for the storage vector y1 is (0,2). Initialize and plot its Trajectory:

$$I <Enter> \quad T <Enter> <Enter>$$

and a message appears on the screen saying that the trajectory is going to infinity. This implies that infinity is an attractor. Therefore, we can plot the basin of infinity first. Hit <Esc> and Initialize:

$$<Esc> \quad I <Enter>$$

Hit <*Esc*> again and plot the BASin of infinity:

BAS <*Enter*> <*Enter*>

A 100 by 100 grid of points will be tested. The program determines if the trajectory through each of those points goes within MC (Maximum Checks) iterates to infinity (that is, goes more than SD diameters from the screen) in which case a small box about the point will be plotted. If instead the trajectory does not go to infinity within MC iterates (default value MC = 60) nothing is plotted. Then the next grid point is tested. Observe that there is a reasonably large blank region on the screen indicating that there is at least one other attractor.

3. The program has the feature that, if the trajectory of a point does not diverge within MC iterates, the final point of this trajectory is stored in ye. That point is likely to be on or near an attractor. Now view the Y Vectors:

YV <*Enter*>

and notice that the vector ye is set (ye = (1.33621,-0.368079). Now set y1 to be equal to ye by using the "Replace Vector" command:

RV <*Enter*> 1 <*Enter*> e <*Enter*>

Clear the screen (and core memory):

C <*Enter*>

Initialize and plot the Trajectory:

I <*Enter*> **T** <*Enter*> <*Enter*>

and a 2-piece chaotic attractor appears on the screen. After a while, store the picture (for later use) on the disk in file H1.PIC. To do this, hit

<*Esc*>

to terminate the routine. Change the Root Name to H1:

RN <*Enter*> **H1** <*Enter*>

followed by the "To Disk" command:

TD <*Enter*> <*Enter*>

and the 2-piece chaotic attractor is stored in H1.PIC.

Example 7-1b: Choose a point on the attractor and set RA so that the circle with RAdius RA is in the basin of that attractor

4. The vector y0 is the point on the attractor which is the latest plotted point on the chaotic attractor (see Example 7-1a). For plotting a basin, set vector y5 to be equal to y0 either with the command **RV50** or:

$$RV < Enter > \quad 5 < Enter > \quad 0 < Enter >$$

Check that the default value of RA, which is 0.1, suffices for the chaotic attractor. To do this, you can set the vector y1 to be equal to y5, which is the latest plotted point on the attractor, either with the command **RV15** or by using the "Replace Vector" command:

$$RV < Enter > \quad 1 < Enter > \quad 5 < Enter >$$

or you can move the small cross to a point on the attractor. The latter option is often more convenient. For example, move the small cross to y1 = y0 = (0.763034, -0.967787), and use the "circle" command O:

$$O < Enter > \ < Enter >$$

Hit $< Esc >$ to terminate the routine O, and enter command BAS again. Observe that the disk centered at y1 with radius RA = 0.1 is in the blank region.

An extra test can be made to be sure that the disk of radius RA centered at the point y1 is in fact in the basin of the attractor. Set MI to 10 for example using "Maximum number of Iterates" command:

$$MI < Enter > \quad 10 < Enter >$$

Now plot the circle and its Iterates:

$$OI < Enter > \ < Enter >$$

A circle of radius RA about y1 and the first 10 iterates of points on the circle will be plotted. These iterates should be spread out along the attractor. If it seems that some of the trajectories starting on the circle are leaving the vicinity of the attractor, then either RA should be made smaller, or preferably, y1 should be replaced by another point near the attractor, for retesting. Verify that RA = 0.1, is a good value. To stop the circle drawing routine, hit

$$< Esc >$$

and the program returns to the menu. Clear the screen: $C < Enter >$.

Example 7-1c: Find the remaining period 8 attractor and store it
From now on, to be specific, we assume that y5 has been replaced by
(0.763034,-0.967787) and that the vector y2 is not set (see Example 7-1b).

5. Recall that, if the trajectory of a point is not diverging or does
not come within a distance RA of y5 within MC iterates while plotting a
basin, the final point of this trajectory is stored in ye. Therefore, plot
a basin so that ye gets set. Now invoke the "BASin" command:

$$\textbf{BAS} < Enter > < Enter >$$

only the basin of the attractor at infinity is plotted. Now view the Y
Vectors:

$$YV < Enter >$$

and notice that the vector ye is set (ye = (0.244295,-0.285432). Now set y1
to be equal to ye either with the command **RV1e** or:

$$\textbf{RV} < Enter > \quad 1 < Enter > \quad e < Enter >$$

Clear the core picture:

$$\textbf{CC} < Enter >$$

Initialize and plot a Trajectory:

$$\textbf{I} < Enter > \quad \textbf{T} < Enter > < Enter >$$

and a stable period 8 orbit appears on the screen. After a while, hit

$$< Esc >$$

which results in the menu on the screen, and store the picture (for later
use) on the disk in file H2.PIC:

$$\textbf{RN} < Enter > \quad \textbf{H2} < Enter >$$
$$\textbf{TD} < Enter > < Enter >$$

and the stable period 8 orbit is stored in H2.PIC.

Note. Instead of **RN** < *Enter* > **H2** < *Enter* > **TD** < *Enter* > < *Enter* > you also may
enter **TD** < *Enter* >, select in menu **RN**, and enter **H2** < *Enter* > < *Enter* >.
(The second < *Enter* > is for actually saving the picture.)

Example 7-1d: Plot the basin of infinity together with the basin of the period 8 orbit

6. Set the value for RA (RAdius of attraction). To do this, set first the vector y1 to be equal to y0, which is the latest plotted point on the period 8 attractor (see Example 7-1c):

RV <*Enter*>   1 <*Enter*>   0 <*Enter*>

For example, y1 = y0 = (0.244295, -0.285432). The default value for RA is RA = 0.1, so it may be good to change the X Scale and the Y Scale to a length of 0.5 with y1 at the center (approximately). For example, set the X Scale to be from 0 to 0.5, and the Y Scale to be from -0.5 to 0:

XS <*Enter*>   0 0.5 <*Enter*>

YS <*Enter*>   -0.5 0 <*Enter*>

Now set PR (PeRiod) to be 8, since the period of the periodic attractor is 8:

PR <*Enter*>   8 <*Enter*>

Try first the "plot circle and its Iterates" command OI for the default value RA = 0.1:

OI <*Enter*> <*Enter*>

and observe that the value is (much) too large. Now set RA to be 0.05:

RA <*Enter*>   0.05 <*Enter*>

Clear the core picture:

CC <*Enter*>

and observe the value 0.05 is still too large as OI continues to work. Decrease the value of RA to 0.01:

RA <*Enter*>   0.01 <*Enter*>

Clear the core picture:

CC <*Enter*>

and observe the value 0.01 might be somewhat too large. Change the X Scale and Y Scale again. For example, set the X Scale to run from 0.2 to 0.3 and the Y Scale to run from -0.33 to -0.23:

XS <*Enter*> **0.2 0.3** <*Enter*>

YS <*Enter*> **-0.33 -0.23** <*Enter*>

Before decreasing the value further, first try to increase the value of PR to, say 16:

PR <*Enter*> **16** <*Enter*>

Clear the core picture:

CC <*Enter*>

and observe that the ball with radius 0.01 centered at y1 is mapped into its interior. Therefore, the choice RA = 0.01 suffices for the period 8 attractor.

7. To have the original setting, leave the program, and read in the stored picture H2.PIC:

Q <*Enter*>

QX <*Enter*>

dynamics h2.pic <*Enter*>

The vector y0 is the point on the attractor which is the latest plotted point of the period 8 orbit (see Example 7-1c). For plotting a basin, set vector y3 to be equal to y0:

RV <*Enter*> **3** <*Enter*> **0** <*Enter*>

Set RA to be 0.01:

RA <*Enter*> **0.01** <*Enter*>

Although RA has been set to 0.01, it might be worthwhile to check if RA can be set to a larger value for the chaotic attractor. (The program has the feature that RA2 and RA5 can be set independently.) Recall that the vector y5 was set to be a point on the chaotic attractor, namely y5 = (0.763034,-0.967787). Now set RA5 to be 0.1:

$$RA5 < Enter > \quad 0.1 < Enter >$$

Note that the vector y2 is not set, so when invoking the "plot BASin" command BAS, the generalized basin of the attractors infinity and the period 8 orbit is plotted. Now plot this generalized basin:

$$BAS < Enter > < Enter >$$

and verify that the vector ye is again set (ye = (0.30688,-0.193635)) and its trajectory is approaching the period 8 attractor. Store the data (for later use) in the file H3.DD by using the "Dump Data" command:

$$RN < Enter > \quad H3 < Enter >$$
$$DD < Enter > < Enter >$$

8. The trajectory ending at ye does not come within a distance RA of y3 in MC iterates, and it does not come within distance RA5 of y5, nor is it diverging within MC iterates. Now first set MC to a greater value to increase the number of iterates to be checked. Increase the value of MC to 1000:

$$MC < Enter > \quad 1000 < Enter >$$

Clear the screen and core memory

$$C < Enter >$$

and again plot the generalized basin:

$$BAS < Enter > < Enter >$$

which should cause the generalized basin of infinity and the period 8 orbit to be plotted. A 100 by 100 grid of points will be tested. The program determines if the trajectory through each of those points goes to infinity (that is, goes more than SD diameters from the screen -- see command SD (Screen Diameter)). If it does, then a small box about the point will be plotted. Also, if the trajectory through each of those points comes within a distance of RA = 0.01 of the period 8 orbit, a box about the point will be plotted. If instead the trajectory goes near the point y5 (within a distance RA5 = 0.1) the iteration of that trajectory is terminated and nothing is plotted. Then the next grid point is tested. When the basin computation is finished, enter the "To Disk" command:

$$\mathbf{TD} < Enter > \; < Enter >$$

to store the picture in file H3.PIC. Verify that the value of ye has not been changed.

The resulting picture is similar to Figure 7-2. Make sure the printer is on and Print the Picture:

$$\mathbf{PP} < Enter >$$

and select your printer.

Note. The resolution of the picture that appears on the screen, is low. By first invoking a command like RH (High Resolution), the command BAS generates a picture of high resolution.

Example 7-1e: Plot the 3 basins separately in windows

Instead of plotting a generalized basin, one also may plot the three basins separately in windows as follows. Read in the data file H3.DD:

dynamics h3.dd < *Enter* >

To plot the basin of infinity in the upper left window, set the vector y6 to be equal to y3, and then set the vector y3 to be equal to y4, which is "NOT SET" so that y3 will be NOT SET:

RV < *Enter* >   **6** < *Enter* >   **3** < *Enter* >

RV < *Enter* >   **3** < *Enter* >   **4** < *Enter* >

(Alternatively you could use the short-hand commands **RV63** and **RV34**.) Enter the "Y Vectors" command YV to check that the vector y3 is not set, while both the vectors y5 and y6 are set.

Open the upper left window (window W1):

OW < *Enter* >

OW1 < *Enter* > < *Enter* >

Enter the "plot BASin" command:

BAS < *Enter* > < *Enter* >

and only the basin of infinity will be plotted.

To plot the basin of the period 8 orbit, set the vector y2 to be equal to y6, and then set the vector y6 to be equal to y4. Enter the "Y Vectors" command YV to check that the vectors y3 and y4 are not set, while both the vectors y2 and y5 are set. Enter the command y6 to verify that y6 is not set. Open the upper right window (window W2), and when the "plot BASin" command BAS is entered, only the basin of the period 8 orbit will be plotted.

To plot the basin of the 2-piece chaotic attractor, set the vector y6 to be equal to y2, and then set the vector y2 to be equal to y5, and then set the vector y5 to be equal to y6. Enter the "Y Vectors" command YV to check that the vectors y3 and y4 are not set, while both the vectors y2 and y5 are set. Since y2 has been replaced by y5, the radius of attraction has to be adapted. Set RA2 to be 0.1, and set RA5 to be 0.01. Open the lower left window (window W3), and when the "plot BASin" command BAS is entered, only the basin of the chaotic attractor will be plotted.

The resulting basin of attraction pictures (in High Resolution) and a picture of the 2-piece chaotic attractor are given in Figure 7-6.

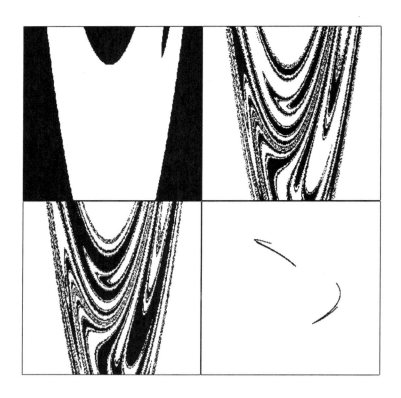

Figure 7-6: Basin of attraction

The black area in each of the two upper windows W1 and W2, and the lower left window W3 is a basin of the Henon map (H)

$$(X, Y) \;\longrightarrow\; (1 - X*X + 0.475*Y, \; X)$$

The attractors corresponding to the black area in W1, W2, and W3 are the attractor at infinity, a stable period 8 orbit, and a two piece chaotic attractor, respectively. The white area is these windows is the generalized basin of attraction of the two remaining attractors. The 2-piece chaotic attractor is shown in window W4. The variable X is plotted horizontally, and the variable Y is plotted vertically. In all the four windows, the X Scale is from -2.5 (left) to 2.5 (right) and the Y Scale is from -2.5 (bottom) to 2.5 (top).

To plot the basin of the period 8 orbit, set the vector y2 to be equal to y6, and then set the vector y6 to be equal to y4. Enter the "Y Vectors" command YV to check that the vectors y3 and y4 are not set, while both the vectors y2 and y5 are set. Enter the command y6 to verify that y6 is not set. Open the upper right window (window W2), and when the "plot BASin" command BAS is entered, only the basin of the period 8 orbit will be plotted.

To plot the basin of the 2-piece chaotic attractor, set the vector y6 to be equal to y2, and then set the vector y2 to be equal to y5, and then set the vector y5 to be equal to y6. Enter the "Y Vectors" command YV to check that the vectors y3 and y4 are not set, while both the vectors y2 and y5 are set. Since y2 has been replaced by y5, the radius of attraction has to be adapted. Set RA2 to be 0.1, and set RA5 to be 0.01. Open the lower left window (window W3), and when the "plot BASin" command BAS is entered, only the basin of the chaotic attractor will be plotted.

The resulting basin of attraction pictures (in High Resolution) and a picture of the 2-piece chaotic attractor are given in Figure 7-6.

Example 7-1f: Plot the 3 basins using coloring scheme BN

Instead of plotting a generalized basin, one also may plot the three basins separately as follows. Read in the data file H3.DD:

$$\text{dynamics h3.dd} < Enter >$$

Recall that y3 is a point on the period 8 attractor and y5 is a point on the chaotic attractor. Fetch the "BAS Plotting scheme Menu":

$$BASPM < Enter >$$

and select the coloring option BN:

$$BN < Enter >$$

Enter the "plot BASin" command:

$$BAS < Enter > < Enter >$$

and the basin of infinity will be plotted in color 1, the basin of the period 8 attractor will be plotted in color 3, and the basin of the chaotic attractor will be plotted in color 5.

After the picture is completed, you may wish to save it to disk. To have a picture of the basin of the period 8 attractor, erase color 1 (command E1) and erase color 5 (command E5).

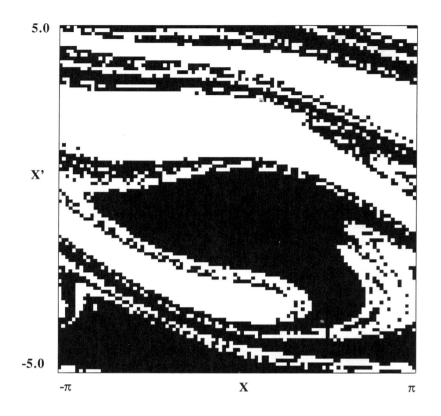

Figure 7-7a: Basin of attraction

This figure shows two basins of attraction of the (second order differential equation) forced damped Pendulum (P)

$$X'' + 0.2*X' + sin\ X = 2*cos(t)$$

Both, the black and white area are each the basin of attraction of a stable fixed point of the time-2π map. First, the vectors y2 and y5 were set at the location of these attracting fixed points in the black and white area. To be specific, these vectors are y5 = A = (-0.473, 2.037) and y2 = B = (-0.478, -0.608). Thereafter the basin was plotted using Low Resolution.

Note. *Command RET is a toggle that allows you to switch to plot trajectories in continuous time when using command T instead of one dot each cycle.*

Topic of discussion. *Describe the behavior of the trajectories that are attracting fixed points in the picture. When RHO is decreased to 1.7 for example, another type of fixed point attractor becomes more important. Describe its behavior.*

---

364   Dynamics: Numerical Explorations

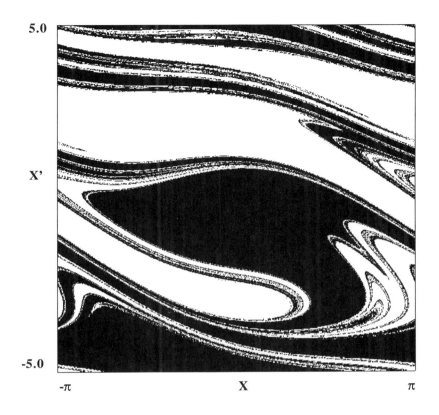

Figure 7-7b: Basin of attraction

This figure is the same as Figure 7-7a except in higher resolution. The basin was plotted using High Resolution. This picture takes a long time to make.

Topic of discussion. *Plot this picture in color using BAS, not necessarily in high resolution, and explain the incredibly complicated color pattern. Techniques in later chapters, e.g., the SST routine, will be helpful, but are not necessary to begin the discussion.*

Example 7-2: Basins of the forced damped Pendulum
    In this example, plotting of basins of attraction is illustrated by
using the forced damped Pendulum differential equation (P)

$$X''(t) + 0.2*X'(t) + \sin X(t) = 2*\cos(t)$$

where the variable X is plotted horizontally, and the variable Y = X' is
plotted vertically. The X Scale is from -3.14159 to 3.14159, and the Y
Scale is from -5 to 5. Search for possibly coexisting attractors of the
pendulum, and plot a basin of attraction for each of these attractors. Note
that this process requires much more computer time than the previous
example.

    1. Set the parameter RHO = 2 (the other parameters are the default
values), and change the Y Scale to from -5 to 5.

    2. Enter the "plot Trajectory" command T, and a trajectory converges
to a stable fixed point. Hence, the time-2π map has at least one stable
fixed point.

    3. Check with the "circle and its Iterates" command OI that the
default value RA = 0.1 is a good value for the stable fixed point, that is,
the points on the circle of radius RA all go to the same attractor.

    4. Use the "Replace Vector" command RV to replace the vector y2 by the
stable fixed point.

    5. Verify that there is a second stable fixed point for the time-2π
map, and that RA = 0.1 suffices.

    6. Set y5 to be equal to this second fixed point.

    7. Plot basin of attraction.
    For this equation, the time-2π map has attracting fixed points, namely
A = (-0.473,2.037) and B = (-0.478, -0.608). The resulting pictures in two
different resolutions are given in Figure 7-7.

    Note. You may wish to create a basin picture for the equation above
using the BA routine and compare the resulting pictures.
    You may set MC to be 1000 and compare the two resulting pictures using
the routine BA for MC = 60 (default) and MC = 1000 (command CNT to CouNT
the pixels).

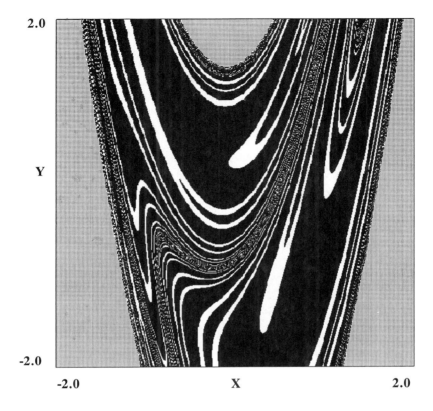

Figure 7-8: Basin of attraction
The black area is the basin of an attracting period 4 orbit of the Henon map (H)

$$(X, Y) \rightarrow (0.855 - X*X + 0.5*Y, X)$$

The white area is the basin of a stable period 6 orbit and the grey area is the basin of infinity.

Topic of discussion. *The fixed points are saddles and they are limiting points of infinitely many bands, black and white converging upon them. Relate one or both of the eigenvalues and eigenvectors of each fixed point to the scaling and positioning of the bands near the fixed point.*

7.4 EXERCISES

In the exercises below, in some cases it might be helpful to plot the basin of the attractor at infinity first. To find initial conditions whose trajectories are bounded and going to an attractor, points can be chosen in the uncolored region by moving the small cross on the screen. When in an exercise the X Scale or Y Scale are not specified, just use the default values.

Exercise 7-1. There are 3 attractors for the Henon map (H)
$$(X,Y) \rightarrow (1 - X*X + 0.475*Y, X).$$
Plot the generalized basin of the attractor at infinity and a stable period 8 orbit, and plot the basin of the stable period 8 orbit. See Figures 7-2 and 7-6.

Exercise 7-2. There are 2 attractors for the complex quadratic map (C)
$$Z \rightarrow Z*Z + 0.32 + i0.043$$
where Z = X + iY. Set the X Scale to run from -1 to 1 and set the Y Scale to run from -1.35 to 1.35. Plot the basin of infinity, and plot the basin of the stable period 11 orbit. See Figure 7-3 and Figure 8-5 (upper left window).

Exercise 7-3. There are 2 attractors for the Tinkerbell map (T)
$$(X,Y) \rightarrow (X*X - Y*Y + 0.9*X - 0.6013*Y, 2*X*Y + 2.0*X + 0.5*Y)$$
Set the X Scale to run from -1.8 to 1.2 and set the Y Scale to run from 1 to -2. Plot the chaotic attractor, and plot the basin of infinity. See Figure 7-4.

Exercise 7-4. There are two stable fixed points attractors of the time-2π map of the forced damped Pendulum equation (P)
$$X''(t) + 0.2*X'(t) + \sin X(t) = 2*\cos(t)$$
Set the Y Scale to run from -5 to 5. Plot the basin of one of those stable fixed points of the time-2π map of the forced damped Pendulum equation. See Figure 7-7.

Exercise 7-5. A period 2 orbit is one of the 2 attractors of the Henon map (H)
$$(X,Y) \rightarrow (1.4 - X*X - 0.3*Y, X)$$
Set the X Scale to run from -3 to 3, and set the Y Scale to run from -3 to 11. Plot the basin of infinity. See Figure 9-4 (upper left window).

Exercise 7-6. There are two attractors for the Henon map (H)

$$(X,Y) \rightarrow (2.66 - X*X + 0.3*Y, X)$$

Set the X Scale to run from -3 to 3, and set the Y Scale to run from -3 to 3. Plot the basin of the stable period 3 orbit. See Figure 9-8 (upper left window).

Exercise 7-7. There are two attractors for the Henon map (H)

$$(X,Y) \rightarrow (RHO - X*X + 0.9*Y, X)$$

Plot the basin of attractor at infinity for the parameter values: RHO = 0.4, RHO = 0.5, and RHO = 0.6. See Figure 9-5a and Figure 9-5b (upper left window).

Exercise 7-8. A stable period 4 orbit and a stable period 6 orbit are the two attractors of the Henon map (H)

$$(X,Y) \rightarrow (0.855 - X*X + 0.5*Y, X)$$

Set the X Scale to run from -2 to 2, and set the Y Scale to run from -2 to 2. Plot the basin of the stable period 4 orbit. See Figure 7-8.

Exercise 7-9. A period 3 orbit is one of the two attractors of the complex map (C)

$$Z \rightarrow Z*Z - 0.11 + i0.6557$$

where Z = X + iY. Set the X Scale to run from -1.5 to 1.5 and set the Y Scale to run from -1.5 to 1.5. Plot the basin of infinity. See Figure 7-5.

Exercise 7-10. There are two attractors of the complex Cubic map (C)

$$Z \rightarrow 0.8*Z*Z*Z - Z*Z + 1$$

where Z = X + iY. Set the X Scale to run from -1.5 to 2 and set the Y Scale to run from -1.1 to 1.1. Plot the basin of the fixed point attractor. See Figure 8-2 (lower right window).

Exercise 7-11. Plot the chaotic attractor and the basin of a fixed point attractor of the Ikeda map (I)

$$Z \rightarrow 1 + 0.9*Z*\exp\{i[0.4 - 6/(1 + |Z*Z|)]\}$$

where Z = X + iY. Set the X Scale to run from -5 to 7, and set the Y Scale to run from -5 to 7.

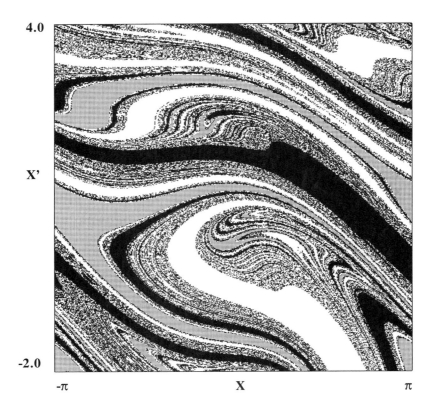

Figure 7-9a: Three basins of attraction (Wada basins)

This figure shows three basins of attraction of the time-2π map of the forced damped Pendulum differential equation

$$X''(t) + 0.2*X'(t) + sin\ X(t) = 1.66*cos(t)$$

The routine BA has plotted the basins of a fixed point attractor, of a period two attractor, and of another period two attractor of the time-2π map of the Pendulum differential equation. The black area is the basin of a period two attractor of the time-2π map of this differential equation. The white area is the basin of attraction of another period two attractor of the time-2π map of the Pendulum equation. We used command E7 (with PAT = 0) to Erase the plotted region using color 7. The grey area is the basin of attraction of a fixed point attractor of the time-2π map of the Pendulum equation, and is obtained by shading the color of the basin of the fixed point attractor using the Erase commands E3 with PATtern PAT = 6.

Each of the basins is a Wada basin, that is, each point that is on the boundary of one of them, is on the boundary of all three. For references, see Section 7-5.

---

370   Dynamics: Numerical Explorations

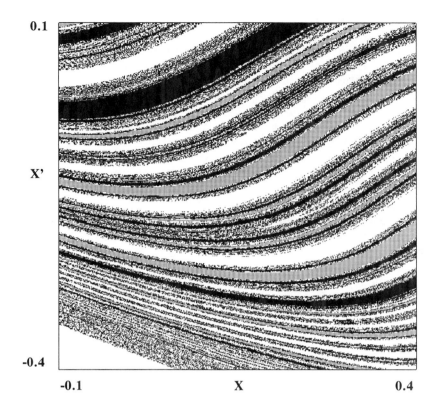

Figure 7-9b: Blow-up of the three basins of attraction

This figure shows a blow-up of the three basins of attraction of the the previous figure. Since the routine BA cannot be used, for example, the fixed point attractor in not in the selected region, the routine BAS was utilized to create this picture.

First, we set storage vector y3 to be the fixed point attractor, (plot a trajectory of a point in the basin of color 3 (the grey basin in Figure 7-9a), hit <Esc> and Replace y3 by y0). Next, set storage point y5 to be a point of the period two attractor in the basin of color 5 (the black basin in Figure 7-9a) and similarly, set storage vector y7 to be a point of the period two attractor in the basin of color 7 (the white basin).

Set the Radius of Attraction appropriately (for example, RA = 0.01). To have each of the basins uni-colored, set the basin coloring scheme and enter BN. Then enter command BAS and the three basins are plotted. After the picture has been finished, erase the colors as described in the caption of Figure 7-9a.

Topic of discussion. What kind of pictures do you expect if more blow-ups are being made?

7.5 REFERENCES related to DYNAMICS

This program was written in part to make accessible the tools we have developed and used in our papers. These papers are interwoven with this program in that as the methods were developed, such as the plots of basin boundaries, they were incorporated into the program and were then developed further in the program to be of greater use in illustrating the concepts for which we were aiming. A number of these papers display figures that were made with this program. Those papers marked with (*) are of a more general nature and are appropriate for general readings.

Basin boundary

Boundaries of basins may be intriguing. We say a point is in the **basin boundary** if it is a limit of points of at least two different basins; see Grebogi, Ott and Yorke (1983). Basin boundaries play an important role in the dynamics of many systems.

Computing the basin of a fixed point attractor of the forced damped Pendulum in Example 7-2, we observe irregularities in the coloring of the basin. In the forced Pendulum Example in Chapter 8 we show that the basin of a fixed point attractor surrounds an invariant set of curves, the trajectories on these curves do not approach the fixed point attractor. Since these curves are surrounded by only one basin, they are not on the basin boundary. Exercise 8-6 provides another example of this phenomenon. See Chapter 8 for more details.

Technical Note. For many systems there exists an **accessible** periodic orbit on the basin boundary, and its stable manifold lies in the basin boundary and can be dense in the basin boundary. Then plotting the stable manifold of the accessible periodic orbit yields a sketch of the basin boundary (or at least a part of the basin boundary). See also Chapter 8 and Chapter 9.

The following paper introduced the idea of fractal basin boundary of physically meaningful systems and shows how the dimension of a basin boundary is intimately related to the problem of knowing in which basin initial points are.

* C. Grebogi, S.W. McDonald, E. Ott and J.A. Yorke, Final state sensitivity: An obstruction of predictability, Phys. Letters 99A (1983), 415-418

The following paper describes why boundaries are sometimes fractal. It introduces the uncertainty dimension as a quantitative measure for final state sensitivity in a system.

* S.W. McDonald, C. Grebogi, E. Ott, and J.A. Yorke, Fractal basin boundaries, Physica D 17 (1985), 125-153

The next paper looks at the role of "basic sets" that lie on basin boundaries. These are invariant sets that determine the fractal dimension of those boundaries. The paper addresses the question of why different regions of a boundary might be expected to have the same dimension.

C. Grebogi, H.E. Nusse, E. Ott, and J.A. Yorke. Basic sets: sets that determine the dimension of basin boundaries. In Dynamical Systems, Proc. of the Special Year at the University of Maryland, Lecture Notes in Math. 1342, 220-250, Springer-Verlag, Berlin etc. 1988

The next paper shows that the box-counting dimension, the uncertainty dimension, and the Hausdorff dimension are all equal for one and two dimensional systems which are uniformly hyperbolic on their basin boundary.

H.E. Nusse and J.A. Yorke, The equality of fractal dimension and uncertainty dimension for certain dynamical systems, Commun. Math. Phys. 150 (1992), 1-21

The next two papers introduced the double rotor example and describe its basin boundary structure. In this example phase space is four dimensional so basin boundary plots show how a two dimensional plane intersects the basins. The basin boundary structure observed in this model is quite different from previously reported examples, though once it was understood why the behavior is different, it was found that similar examples could even occur in one-dimensional maps.

C. Grebogi, E. Kostelich, E. Ott, and J.A. Yorke, Multi-dimensional intertwined basin boundaries and the kicked double rotor, Phys. Lett. 118A (1986), 448-452; see also an erratum: Physics Letters 120A (1987), 497

C. Grebogi, E. Kostelich, E. Ott, and J.A. Yorke, Multi-dimensioned intertwined basin boundaries: basin structure of the kicked double rotor, Physica D 25 (1987), 347-360

The following paper gives the double rotor equations in a particularly nice form, incorporating the corrections in the erratum above.

F. J. Romeiras, C. Grebogi, E. Ott, W. P. Dayawansa, Controlling chaotic dynamical systems, Physica D 58 (1992), 165-192

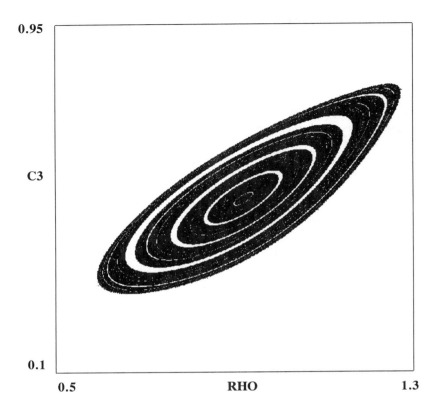

0.95

C3

0.1

0.5　　　　　　　　　　　　　RHO　　　　　　　　　　1.3

Figure 7-10a: Chaos plot
The black area is the resulting Chaos plot of the Cobweb map (CW)
$$X \rightarrow -4*C3*arctan(4x) + (1-C3)*X + 4*RHO*C3$$
*To create this picture, set the parameters C1 = 0.25 and C2 = 4. Next, set the coordinates XCO to be RHO and YCO to be C3, and set the scales XS to run from 0.1 to 1.3 and YS to run from 0.1 to 0.95. The Lyapunov exponent should be computed, so set L = 1. Set the number of Pre-Iterates and the number of Maximum of Checks (for example, PI = 1000 and MC = 1000). Finally enter command BAS. You have to be a little patient, because the computation takes a while. The routine BAS plots a point (small box) at (RHO,C1), if the pair (RHO,C1) is a **chaotic pair** (that is, the trajectory of y1 for that pair has a positive Lyapunov exponent). The resulting picture is a slice in a four dimensional parameter space and is called a **Chaos plot**.*

Topic of discussion. *The Chaos plot seems to consists of a series of nested annuli. What do you think what slices for other values of C1 and C2 look like?*

---

374   Dynamics: Numerical Explorations

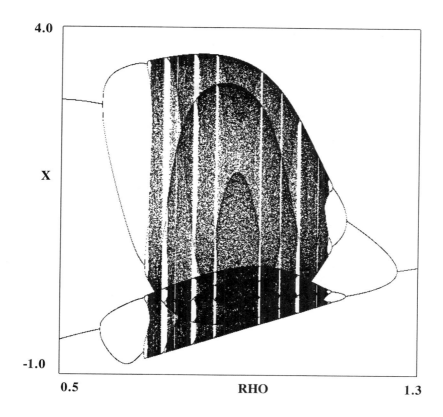

4.0

X

-1.0

0.5   RHO   1.3

Figure 7-10b: Bifurcation diagram
This is a bifurcation diagram for the Cobweb map (CW)

$$X \rightarrow -2*arctan(4x) + 0.5*X + 2*RHO$$

See Chapter 6 for how to create this picture (set the parameters C1 = 0.25, C2 = 4, and C3 = 0.5. (If you continue from the previous picture in Figure 7-10a, then set XC0 to y[0], set the XS to run from -1 to 4.) The picture has been ROTated by applying ROT three times.

In Figure 7-10a, we select C3 = 0.5 and investigate how the dynamics will depend on the parameter RHO. For an interval of parameter values including RHO = 0, there is a period two attractor. If RHO is increased, a period-doubling bifurcation occurs resulting in a period 4 attractor. When RHO is further increased, after infinitely many period-doubling bifurcations, chaotic behavior arises. Finally, after infinitely many period-halving bifurcations, there is a fixed point attractor.

Topic of discussion. *Create the corresponding Lyapunov bifurcation diagram and compare the diagrams with Fig. 7-10a by drawing the horizontal line for C3 = 0.5. What do you observe? What about other values of C3?*

---

7. Basins of attraction   375

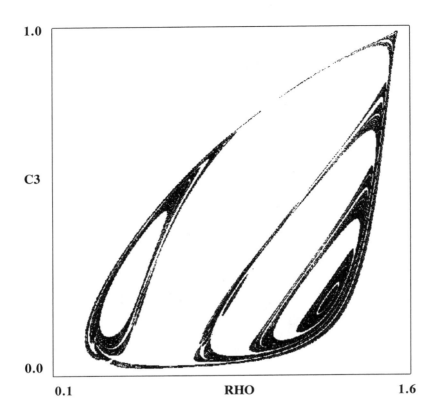

Figure 7-11: Chaos plot
The black area is the resulting Chaos plot of the Cobweb map (CW)
$$X \rightarrow -4*C3*arctan(4x) + (1-C3)*X + 4*RHO*C3$$
To create this picture, set the parameters C1 = 0.05 and C2 = 10. Next, set the coordinates XCO to be RHO and YCO to be C3, and set the scales XS to run from 0.1 to 1.6 and YS to run from 0 to 1. The Lyapunov exponent should be computed, so set L = 1. Set the number of Pre-Iterates and the number of Maximum of Checks (for example, PI = 1000 and MC = 1000). Finally enter command BAS. You have to be a little patient, to watch the picture being finished. The routine BAS plots a point (small box) at (RHO,C1), if the pair (RHO,C1) is a **chaotic pair** *(that is, the trajectory of y1 for that pair has a positive Lyapunov exponent). The resulting picture is a slice in a four dimensional parameter space and is called a* **Chaos plot.**

Topic of discussion. *The Chaos plot seems to consists of a series of nested annuli. What do you think slices for other values of C1 and C2 look like?*

Crises

Crises are sudden changes in chaotic attractors as parameters are varied. In particular as a parameter is varied, the attractor may move and the basin boundary moves until they collide, obliterating both the attractor and the basin, but leaving chaotic saddles that can be observed using the saddle straddle trajectory method.

* C. Grebogi, E. Ott, and J.A. Yorke, Chaotic attractors in crisis, Phys. Rev. Lett. 48 (1982), 1507-1510

* S.W. McDonald, C. Grebogi, E. Ott, and J.A. Yorke, Structure and crises of fractal basin boundaries, Phys. Lett. 107A, (1985) 51-54

C. Grebogi, E. Ott, Ph. Romeiras, and J.A. Yorke, Critical exponents for crisis induced intermittency, Phys. Rev. A, 36 (1987), 5365-5380

B.-S. Park, C. Grebogi, E. Ott, and J.A. Yorke, Scaling of fractal basin boundaries near intermittency transitions to chaos, Phys. Rev. A 40 (1989), 1576-1581

W.L. Ditto, S. Rauseo, R. Cawley, C. Grebogi, G.-H. Hsu, E. Kostelich, E. Ott, H.T. Savage, R. Segnan, M. Spano and J.A. Yorke, Experimental observation of crisis induced intermittency, Phys Rev. Letters 63 (1989), 923-926

H. Bruce Stewart, Y. Ueda, C. Grebogi and J.A. Yorke, Double crises in two-parameter dynamical systems, Phys. Rev. Lett. 75 (1995), 2478-2481

Wada property and Wada basins

The following paper introduces the concept of "Wada property" for basins. Three or more basins have the **Wada property** if every point that is on the boundary of closure of one basin is on the boundary of all basins. In particular, three basins have the Wada property if every point on the boundary of the closure of one of the basins is on the boundary of all three of them. Basins that have the Wada property have fractal boundaries.

J. Kennedy & J.A. Yorke, Basins of Wada, Physica D 51 (1991), 213-225

Whether three or more basins have the Wada property is numerically hard to verify. The following papers introduce "Wada point" and "Wada basin" which mean the following. A point is a boundary point of a basin if every open neighborhood of that point has a nonempty intersection with at least one other basin. A boundary point of a basin is a **Wada point** if every open neighborhood of that point has a nonempty intersection with at least three different basins. A basin is a **Wada basin** if every boundary point of that basin is a Wada point. To be able to verify the existence of Wada basins numericaly, the fundamental notion of "basin cell" has been introduced (see Figure 9-10c for a basin cell). A **cell** is a region with a piecewise smooth boundary, whose edges are alternately pieces of stable and unstable manifolds of a periodic orbit. A **basin cell** is a cell that is a trapping region. Of course, the periodic orbit that generates the cell must be well chosen in order that the cell being a basin cell. Whether a given cell is a basin cell, is in some sense easy to verify numerically by using *Dynamics*. However, the error estimates have still to be carried out.

    * H.E. Nusse and J.A. Yorke, Basins of Attraction, Science 271 (1996), 1376-1380

    * H.E. Nusse and J.A. Yorke, Wada basin boundaries and basin cells, Physica D 90 (1996), 242-261

    H.E. Nusse and J.A. Yorke, The structure of basins of attraction and their trapping regions, Ergodic Th. & Dynamical Systems **17** (1997), 463-481

Chaos plots and Period plots

Dynamics was and is an extremely useful tool for some numerical exploration in the four-dimensional parameter space of a simple economic model. The study of this model was reduced to investigating the CobWeb map. Some examples of both Chaos plots and Period plots are included in the following paper.

    J.A.C. Gallas and H.E. Nusse, Periodicity versus chaos in the dynamics of cobweb models, J. Economic Behavior and Organization 29 (1996), 447-464

CHAPTER 8

STRADDLE TRAJECTORIES

8.1 INTRODUCTION and the METHODS

The ideal behind the *Dynamics* program is that we must be able to visualize what is happening in a dynamical system. That means not just seeing chaotic attractors, but all important elements of dynamics. This chapter presents newly developed techniques for observing trajectories that lie in invariant sets which are not attractors. When trying to understand the global dynamics of a pendulum, a pendulum with friction, one must be familiar with the unstable solution that has the pendulum bob inverted, pointing straight up, as well as the attracting solution. Almost every trajectory starting near this unstable steady state will diverge from it, being pulled down by gravity. Most nonlinear systems have analogues of this unstable state. Sometimes the analogue is a single point, an unstable steady state, and sometimes an invariant chaotic set that is not attracting. We refer to such invariant chaotic sets that are not attracting as chaotic saddles. A **chaotic saddle** is an invariant compact set C that is neither attracting nor repelling and contains a chaotic trajectory which is dense in C. In other words, a chaotic saddle is an invariant compact set C that is neither attracting nor repelling, and there exists a point in C whose trajectory has a positive Lyapunov exponent and travels throughout C. An example of a chaotic saddle is the invariant Cantor set for the Henon map with parameter values RHO = 5, C1 = 0.3; see Example 2-19 which introduces Saddle Straddle Trajectories. Basin boundaries are examples of invariant sets that are not attractors; see Chapter 7 for the definition of basin boundary. They sometimes contain chaotic saddles.

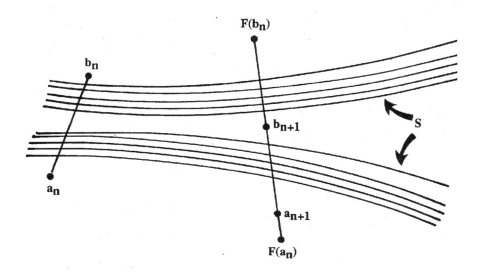

Figure 8-1: Straddle trajectory

Plotting any straddle trajectory consists of three steps which the program cycles through repeatedly:

- *Refine (i.e., shorten) the segment that extends from $F(a_n)$ to $F(b_n)$ until they are at most SDIS apart; the procedure for refining the segment depends on the type of straddle trajectory being created. SDIS has a default value of 0.00000001. The new points are a_{n+1} and b_{n+1}.*

- *Plot a point, say a_{n+1}.*

- *Compute $F(a_{n+1})$ and $F(b_{n+1})$.*

A straddle orbit is **accessible** *(from the "a" side) if the refinement procedure shortens the segment so that there are no points of the set S in the segment from $F(a_n)$ to a_{n+1}.*

We cannot pick an initial point sitting exactly on invariant sets (unless we are especially lucky) since we can only specify about 15 digits for an initial condition. Many invariant sets are quite unstable, and most trajectories starting near the set move rapidly away. The error in the sixteenth digit plus the exponential growth in errors results in a numerical trajectory that moves away from the desired set very quickly. We say **a line segment straddles a set** S if the segment intersects S but the end points of the segment are not in S. We also say a **pair of points** a and b **straddle** S if the segment that connects them straddles S.

Let F denote the map under consideration. F might be the Henon map, or even the single time-step map of a differential equation (using a fixed time step differential equation solver), taking an initial point one fixed time step further along a trajectory. Unlike a normal trajectory which is a sequence of points, a **straddle trajectory** is a sequence of short line segments, denoted $\{(a_n, b_n)\}_{n=0,1,\dots}$, and each of the segments must straddle a specified set S. All the segments (except possibly the initial one) must be shorter than a specified straddle distance which we denote by SDIS. (This length SDIS has a default value of 0.00000001 in *Dynamics*.) Furthermore, we require that $\|a_{n+1} - F(a_n)\|$ and $\|b_{n+1} - F(b_n)\|$ are small for n > 0, on the order of SDIS. See Figure 8-1. Notice that virtually all trajectories that we investigate have numerical errors and we are specifying here that the size is approximately SDIS. Our ability to compute straddle trajectories for an unstable set S depends on the sensitivity to initial data and tendency of trajectories starting near S to move rapidly away from S.

Since we cannot pick an initial point sitting exactly on such an invariant set S that is not attracting, we choose a pair of points ya and yb that are usually not close together, but which nonetheless straddle S. We let the computer repeatedly shorten the segment from ya to yb in such a way that it continues to straddle the invariant set, repeating until the ends are no more than SDIS apart. The new points are a_0 and b_0.

The first question asks how the routine moves them together and knows that the shorter segment still straddles the set. Then the second question is how the routine produces a trajectory once pairs of points are close together. The answer to the first question depends on the kind of straddle trajectory being used, while the answer to the second question is the same for all straddle orbits. There are four kinds of straddle trajectories and they are implemented in this program, but really only two major types: basin boundary straddle trajectories and saddle straddle trajectories.

BASIN BOUNDARY STRADDLE TRAJECTORIES

Refinement procedure for BST, Basin boundary Straddle Trajectory

A basin is the collection of initial points whose trajectories go to a specified attractor. A generalized basin is the collection of initial points whose trajectories go to one of the attractors in a specified collection of attractors. The basin boundary is defined to be the set of all points x in the phase space for which each open neighborhood of x has a nonempty intersection with at least two different basins; see Chapter 7. The **BST method** is a straddle method for finding trajectories on the basin boundary. The Basin boundary Straddle Trajectory is based on a simple idea. It makes use of the fact that we can distinguish between two kinds of trajectories, those that wind up being attracted to one collection of attractors and those that are attracted to other attractors. We can thereby distinguish between two generalized basins. The initial point ya must be in one generalized basin and yb must be in the generalized basin for the remaining attractors. The two generalized basins must include all the basins in region of interest. An example of a generalized basin is the black area in Figure 7-2; the attractors for this generalized basin are infinity and a stable period 8 orbit. More simply, imagine that there are just two attractors (one of which may be the point at infinity) and the two generalized basins are just the basins of these two attractors and ya is in one basin and yb is in the other. When the segment from ya to yb is refined, that is, shortened, the old endpoints ya and yb are moved to the endpoints of the shorter segment. The routine BST then refines the line segment from ya to yb, using what we call the **bisection method**. That is, it bisects the line segment between ya and yb. It chooses a point c between ya and yb, (usually the midpoint), and it determines which basin c is in. If it is in the same basin as ya, then ya is moved to c, while if it is in the same basin as yb, then yb is moved to c. We now have points ya and yb that are closer together than at the start of the refinement procedure and the segment still straddles the basin boundary. This refinement procedure is applied repeatedly until ya and yb are quite close together, that is, ya and yb are less than SDIS apart, say SDIS = 10^{-8}.

Basin boundary Straddle Trajectory

The second question is how to get a trajectory. When ya and yb are less than SDIS apart, plot ya or yb. In Figure 8-1 ya and yb are called

$$a_n \text{ and } b_n.$$

The routine BST now applies the map F and computes the two first iterates

$$F(a_n) \text{ and } F(b_n).$$

Since the trajectories of ya and yb go to different attractors, the iterates may be somewhat further apart. They may have moved a long distance from where they were, but the two first iterates will still be rather close together. They may for example have moved from 0.00000001 apart to 0.0000001 in one iterate. The routine will refine the segment. More precisely, if the new endpoints are more than SDIS apart, the routine BST applies the refinement procedure (bisection method) as described above. If the new endpoints are less than SDIS apart no refinement is carried out. Then plot an endpoint (the two are very close together so they will almost certainly be plotted at the same point on the screen). Repeating this whole process of iterating the map once and applying the refinement procedure, gives a trajectory of successive segments. See Figure 8-1 for illustration. Of course there are small imperfections introduced in the trajectory due to the refinements, but any computer trajectory has numerical errors. These refinement errors are quite small often about SDIS in size.

Select two generalized basins as described in Chapter 7. You must choose them so that *almost every point must be in a basin*. If the trajectory through ya or yb is iterated SMC times without diverging or coming close to one of the points y2, ..., y7, (SMC has a default value of 2000), the BST routine will stop. It is always necessary to set at least one of the vectors y2, ..., y7. The distance RA specifies how close the trajectory must come to one of those points to be recognized as lying in a basin. This is described in detail in Chapter 7. If the trajectories of some points go to infinity, they might constitute one basin. It may be sufficient to select one other point, e.g., y5 on the bounded attractor, if there is just one bounded attractor.

How to plot a Basin boundary Straddle Trajectory. To plot a Basin boundary Straddle Trajectory, first the storage vectors ya and yb have to be set. At any time, their values can be obtained by retrieving the Y Vectors. The values of ya and yb will be changed. Their initial values are at the diagonal corners of the screen (ya at the bottom left and yb at the upper right corner) when the map is first selected. (If they have been moved or the coordinates have been changed, command DIAG puts them at the corners.) To plot a Basin boundary Straddle Trajectory, the vectors ya and yb have to be set such that ya is in one (generalized) basin and yb is in the other. In order to do this, frequently one can just use the "Replace Vector" command RV, to replace ya by y2 (or y3 or y4) and to replace yb by y5 (or y6 or y7), assuming these vectors have been set.

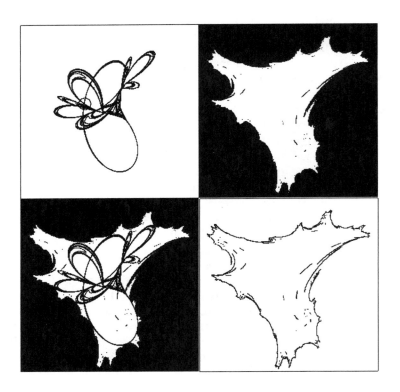

Figure 8-2: Basin boundary Straddle Trajectory

This figure shows the basin of infinity and a Basin boundary Straddle Trajectory of the Tinkerbell map (T)

$$(X, Y) \; \rightarrow \; (X*X - Y*Y + 0.9*X - 0.6013*Y, \; 2*X*Y + 2*X + 0.5*Y)$$

There are two attractors: a bounded chaotic attractor and a point at infinity. In all the four windows, the variable X is plotted horizontally, the variable Y is plotted vertically, and the X Scale is from -1.8 (left) to 1.2 (right) and the Y Scale is from 1 (bottom) to -2 (top).

The upper left window shows the chaotic attractor and the ball with center at y5. The upper right window shows the basin of the attractor at infinity (in black), and the white area is the basin of the chaotic attractor; see also Figure 7-4. The lower left window is a composite of the upper two. The lower right window shows a Basin boundary Straddle Trajectory on the common boundary of the basin of the chaotic attractor and the basin of infinity.

Topic of discussion. *The upper right figure reveals disconnected pieces of the basin of infinity. Are there infinitely many pieces?*

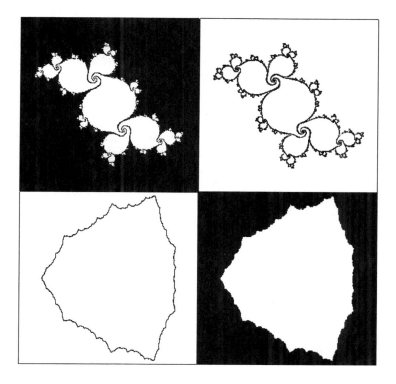

Figure 8-3: Basins and basin boundary straddle trajectories
This figure uses the complex Cubic map (C)
$$Z \rightarrow RHO*Z*Z*Z + C1*Z*Z + C5 + iC6$$
where $Z = X + iY$. The variable X is plotted horizontally, and the variable Y is plotted vertically.

In the upper two windows, RHO = 0, C1 = 1, C5 = -0.11, C6 = 0.6557, and the X Scale is from -1.5 (left) to 1.5 (right) and the Y Scale is from -1.5 (bottom) to 1.5 (top). In the upper left window, the black area is the basin of the attractor at infinity, and the white area is the basin of a period three orbit; see also Figure 7-5. The upper right window shows a Basin boundary Straddle Trajectory on the common boundary of the two basins.

In the lower two windows, RHO = 0.8, C1 = -1, C5 = 1, C6 = 0, and the X Scale is from -1.5 (left) to 2 (right) and the Y Scale is from -1.1 (bottom) to 1.1 (top). In the lower right window, the black area is the basin of the attractor at infinity, and the white area is the basin of a fixed point attractor. The lower left window shows a Basin boundary Straddle Trajectory on the common boundary of the two basins.

After the vectors ya and yb are set correctly, just enter the "Basin boundary Straddle Trajectory" command BST to plot a basin boundary straddle trajectory. Such a numerical trajectory has been produced in Example 2-21 and Example 2-22 in Chapter 2, and the more complicated examples 8-1 and 8-2 in this chapter.

Summary for plotting Basin boundary Straddle Trajectories

- Specify (generalized) basins by setting for each attractor one of the vectors y2, ..., y7 so that the trajectory of almost every initial point either diverges or comes within distance RA of one of these vectors (command SV or command RV).
- Set vectors ya and yb in different (generalized) basins (command RV or command SV).
- Determine the RAdius of attraction (command RA).
- Plot Basin boundary Straddle Trajectory (command BST).

SADDLE STRADDLE TRAJECTORIES

Refinement procedure for SST, Saddle Straddle Trajectory

The routine for computing a Saddle Straddle Trajectory works without having a basin boundary present. It works by evaluating the **escape time** (or **exit time**) for points, which is time that it takes the trajectory starting at a point to leave the essential region. See Chapter 7 for the definition of essential region and exit time. Sometimes we do not have two basins to use and there is no way to divide the diverging trajectories into two classes. Almost all initial points might go to the same attractor, but there can still be a set of points that do not go to the attractor. These points are said to have escape time infinity. In the simplest case, imagine that the trajectories of almost all points diverge. The way the program tests to see if the trajectory of a point is diverging is by seeing if it leaves the extended screen region; see Chapter 7 for the definition of extended screen region. If SD = 0, the extended screen region is the screen. For any point p, the routine SST can compute the escape time for p, the number of iterates of the map F that it takes the trajectory to leave the essential region. See also the discussion of the escape time in the discussion of the "When and Where" routine WW. The escape time is the time reported by WW, and the routine SST ignores where that trajectory goes but keeps track of how long it takes the trajectory to get there.

The **SST method** is a straddle method for finding trajectories on these (chaotic) invariant sets. The saddle straddle method works using segments that straddle the set of points with infinite escape time, that is, the points whose trajectories never leave the essential region. In practice that means that we choose a grid of points on the segment between ya and yb so that at least one of those points has a strictly longer escape time than do ya and yb. An example is the Henon map with parameter values RHO = 4.2 and C1 = 0.3 and the screen region is the region in the plane which is the square centered at the origin with sides of length 6. The exit time will be computable for each initial point. The routine SST then chooses a grid of points on the line joining ya to yb, for example 31 equally spaced points with ya being point 0 and yb being number 30. (That says that DIV = number of DIVisions of the ya-yb segment is 30). Assume ya and yb are chosen so that the maximum escape time of the 31 points is strictly bigger than the escape times for ya and yb. Then the routine SST refines (or shortens) the segment as new endpoints two of the grid points. The two should be chosen so that there is a third grid point between them having a strictly larger escape time than either of the two. The two do **not** have to be chosen so that the grid point with the maximum escape time lies between them. This is a refinement method, a method for shortening segments, and this is repeated until the endpoints are quite close together. This refinement method is called the **PIM triple method**. The refinement requires a third grid point between the new endpoints ya and yb to be a **Proper Interior Maximum**, that is, it is interior to ya and yb, with an escape time strictly greater than that of ya or yb. When one can guarantee there exists a Proper Interior Maximum is outside the scope of this book, but there are theorems that guarantee the procedure will work for hyperbolic sets in the plane if DIV is large enough and the essential region is properly chosen. These results show that under appropriate conditions, that whenever the segment has a proper interior maximum, then in fact there is some point on the segment where the escape time is infinite. See the references in Section 8.5. It also seems to work well for sets that are not hyperbolic, for example, sets that have a tangency between the stable and unstable manifolds of a periodic point.

The SST method requires the map being one dimensionally unstable and will not work for maps that have two positive Lyapunov exponents.

How to plot a Saddle Straddle Trajectory

The second question mentioned above is how to get a trajectory once a refinement procedure is available. To be brief, the Saddle Straddle Trajectory is obtained similarly to the Basin boundary Straddle Trajectory.

To plot a Saddle Straddle Trajectory, first the storage vectors ya and yb have to be set. Their initial values are at the diagonal corners of the screen (ya at the bottom left and yb at the upper right corner) when the map is first selected. These vectors can be set by invoking the "Set Vector" command SV. To plot a Saddle Straddle Trajectory, frequently one can just use the "DIAGonalize" command DIAG, to replace ya by the lower left corner of the screen and to replace yb by the upper right corner of the screen.

Assuming the vectors ya and yb are set correctly, to plot a saddle straddle trajectory, enter the "plot Saddle Straddle Trajectory" command SST. See Example 2-19 in Chapter 2, and Example 8-2 in this chapter for such numerical trajectories. For more examples, see the Figures in this chapter and the articles mentioned in Section 8.5.

Note. In some cases when there are at least two attractors, it may be interesting to study both types of straddle trajectories for different choices of the vectors ya and yb. See the forced damped pendulum example (Example 8-2) in this chapter.

Summary for plotting Saddle Straddle Trajectories
- Specify basins by setting one of y2, ..., y7 for each attractor.
- Set vectors ya and yb (command RV or command SV).
- Determine the RAdius of attraction (command RA).
- Plot Saddle Straddle Trajectory (command SST).

ACCESSIBLE STRADDLE TRAJECTORIES

A straddle trajectory is designed to straddle some invariant set S which may be quite complicated, as it is in the examples in this chapter. Such sets S are **fractals** consisting of an uncountable collection of curves all linked together with countably many gaps intervening. (The meaning of that complicated phraseology is best revealed by pictures of such sets. See Figure 2-19 (stable and unstable manifolds) and the boundaries between basins in figures 2-10 and 2-20.) In this same sense they are similar to the chaotic attractors displayed throughout these pages. A point p in S is **accessible** if there is a continuous arc ending at p such that the arc intersects S only at p. The arc is called an **access arc** for p. In a simple case where S itself consists of an arc (e.g., if S was the basin boundary consisting of a simple arc), all of its points are accessible. For more complicated cases, most points of S are not accessible. The accessible points form an outer envelope that comes arbitrarily close to every point of S. An **accessible trajectory** is a trajectory each of whose points is accessible. Such trajectories often are quite simple. For many examples the accessible trajectories converge to a periodic trajectory which is accessible and is a saddle. For those periodic trajectories, one branch of its local unstable manifold is an accessible arc. Routines ABST and ASST find basin boundary and saddle straddle trajectories that represent accessible trajectories.

The refinement procedures above leave considerable choice. For a straddle trajectory, the segment between ya and yb may contain an infinity of points on the basin boundary (for basin boundary straddle orbits) or with an infinite escape time (for saddle straddle orbits). There will be one such point c which is closest to ya. This point c is accessible from the ya end of the ya-yb segment. The segment from ya to c is an access arc. The accessible procedures aim at choosing a refinement procedure so that the shorter segment is guaranteed to contain c. The endpoints still straddle the point c. Notice that if ya and yb are switched, the accessible trajectory is likely to change completely. The refinement procedures for obtaining accessible straddle trajectories is much more complicated than the refinement procedures for the straddle trajectories presented above. We refer the reader to the literature cited in Section 8.5 of this chapter. The bold may wish to look at the program's source code in the files *ycomb.c* and *ybasins.c*.

When the vectors ya and yb and the RAdius RA are set correctly for the BST or SST routines, use ABST to plot an "**Accessible Basin boundary Straddle Trajectory**", and use ASST to plot an "**Accessible Saddle Straddle Trajectory**". See Figure 8-5 in this chapter for such numerical trajectories. For more examples, see the Figures in this chapter and the article in Section 8.5.

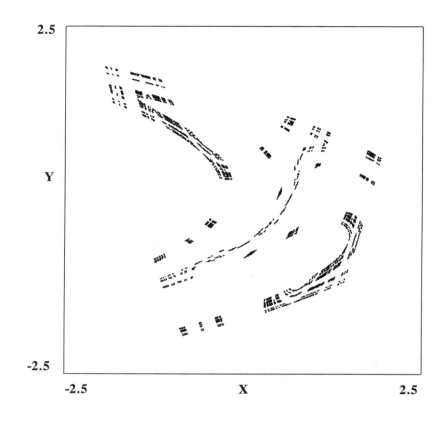

Figure 8-4: Saddle Straddle Trajectory
This figure shows a Saddle Straddle Trajectory of the Henon map (H)
$$(X, Y) \rightarrow (1.2 - X*X + 0.9*Y, X)$$
The only attractor is the point at infinity.

Figure 8-5: Straddle trajectories (ABST and BST)

This figure shows three straddle trajectories of the complex quadratic map

$$Z \rightarrow Z*Z + 0.32 + i0.043$$

where $Z = X + iY$. In all the four windows, the variable X is plotted horizontally and the variable Y is plotted vertically. The X Scale is from -1 (left) to 1 (right) and the Y Scale is from -1.35 (bottom) to 1.35 (top).

The upper left window shows the basin of the attracting period 11 orbit. The upper right window shows the straddle trajectory (ABST) that is accessible from the basin of the periodic attractor. Here ya was chosen in the basin of the periodic attractor, while yb was chosen in the basin of infinity. The lower left window shows the basin boundary straddle trajectory (BST) on the common boundary of the basin of a stable period 11 orbit and the basin of the attractor at infinity. The lower right window shows the straddle trajectory (ABST) that is accessible from the basin of infinity. Here ya was chosen in the basin of infinity while yb was chosen in the basin of the attracting periodic orbit.

LYAPUNOV EXPONENTS and OTHER TRAJECTORY TOOLS

Since straddle trajectories are in fact trajectories, the commands used with trajectories can be used with straddle trajectories. These commands include K, KK, and KKK for plotting crosses and the command IN for INcremental coloring. Straddle trajectories can be single-stepped using the period interrupt $<.>$; they have Lyapunov exponents which are computed when $L > 0$. (Since the straddle orbits do not lie on an attractor, the Lyapunov dimension is not meaningful; see Chapter 5 for more details on the concept of dimension.)

8.2 STRADDLE TRAJECTORY MENU STM

Enter command STM, and the Straddle Trajectory Menu appears on the screen in approximately the following format.

STM – –   STRADDLE TRAJECTORY MENU
BST   – plot Basin boundary Straddle Trajectory using ya and yb SST   – plot Saddle Straddle Trajectory using ya and yb
MENU SOM – Straddle Options Menu
Enter *STM for help

STM

The command STM is for calling the Straddle Trajectory Menu. Straddle trajectories are essentially trajectories that lie in unstable sets, that is, sets that are invariant but are unstable in the sense that trajectories starting near (but not on) the set will diverge from the set. Basin boundaries are such sets. Another example of such a set is a collection of points that do not go to an attractor even though almost all points nearby do go to the same attractor. The first case where there are two attractors is especially simple and reliable. The second case is much more difficult.

Commands for plotting straddle trajectories

BST

The "Basin boundary Straddle Trajectory" command BST is for finding a trajectory on a basin boundary, starting between ya and yb; thus ya and yb must be chosen straddling the basin boundary.

To have a basin boundary, there must be two basins. One basin is determined by

$$y2, \ y3, \ \text{and} \ y4,$$

as described in Chapter 7, and the other basin is described by

$$y5, \ y6, \ \text{and} \ y7.$$

The points ya and yb must be in different basins, that is, one must be in the y2-y3-y4 basin and the other must be in the y5-y6-y7 basin. The default setting for ya is the lower left corner of the screen and yb is the upper right corner of the screen. When they are in different basins, we can say they straddle the basin boundary.

The program can test to find which basin any point is in. It determines which basins ya and yb are in. When these two points are more than a specified distance apart, the program tests the midpoint between them. If the test point is in the same basin as ya, it replaces ya by that test point, while if the test point is in the same basin as yb, it replaces yb by the test point. We still have two points, one in each basin, but now they are only half as far apart. This bisection process is continued until the distance between ya and yb is within the "Straddle DIStance" SDIS of each other, currently 0.00000001. If you do not want to test the midpoint between ya and yb, see BSTF and BSTR for two other choices.

Enter command BST, and the Actions, Hints and Options menu for BST appears and includes the following:

ACTIONS, HINTS and OPTIONS for BST		
ABST	–	compute (from ya's basin) Accessible BST trajectory
BAF	–	tells BST to use the BA picture: OFF
BSTR	–	make BST choose point between ya and yb ad Random: OFF
CB	–	Constrained to small Box: OFF
		if ON, trajectories "diverge" when leaving small box
DIAG	–	set ya = lower left screen corner and
		yb = upper right screen corner
DIV	–	DIVisions of line ya-yb for ABST, SST, ASST, GAME: 30
KKK	–	plot permanent cross at each trajectory point: OFF
OY	–	draw circles about all yk's (k = 2,...,7)
OIY	–	draw the circles OY plus their MI iterates
SDIS	–	subdivide ya-yb until ya and yb are close: 1e-008
SMC	–	Maximum number of Checks for Straddle methods: 2000
TL	–	Text Level currently: T3

The Information window of the Actions, Hints and Options menu for BST contains information on which storage vectors are set. In the default, this window includes:

The following vectors are set:
y1   ya   yb

SST

The "Saddle Straddle Trajectory" command SST is capable of finding and plotting chaotic trajectories that do not lie in a chaotic attractor. The basic idea is to look for a trajectory that does not "exit" from a specified "transient region". This region can have holes, in which case the trajectory could exit the region by entering one of the holes. The region is a rectangle minus the holes centered at any of the points $y2$, ..., $y7$ that have been set. The radius of those holes is RA, the "RAdius of attraction". For example, if the Henon map is studied in parameter ranges that are too large to admit a chaotic attractor, then none of the vectors $y2$, ..., $y7$ should be set, and SST looks for trajectories that do not leave a box that includes the screen. The size of the rectangle is determined by the "Screen Diameter" command SD.

The program has a subroutine (essentially WW) that calculates the exit time, the time it takes the trajectory through a point to leave the box (or a more complicated set if some of the vectors $y2$ through $y7$ are set). This routine is the same one that is used in determining to which basin a trajectory goes, but here only the time it takes is used. See command DIV.

Enter command SST, and the Actions, Hints and Options menu for SST appears and includes the following:

ACTIONS, HINTS and OPTIONS for SST	
ASST	– compute Accessible SST trajectory
CB	– Constrained to small Box: OFF
	if ON, trajectories "diverge" when leaving small box
DIAG	– set ya = lower left screen corner and
	yb = upper right screen corner
DIV	– DIVisions of line ya-yb for ABST, SST, ASST, GAME: 30
KKK	– plot permanent cross at each trajectory point: OFF
OY	– draw circles about all yk's $(k = 2,...,7)$
OIY	– draw the circles OY plus their MI iterates
SDIS	– subdivide ya-yb until ya and yb are close: 1e-008
SMC	– Maximum number of Checks for Straddle methods: 2000
TL	– Text Level currently: T3

The Information window of the Actions, Hints and Options menu for SST contains information on which storage vectors are set. In the default, this window includes:

The following vectors are set:
y1   ya   yb

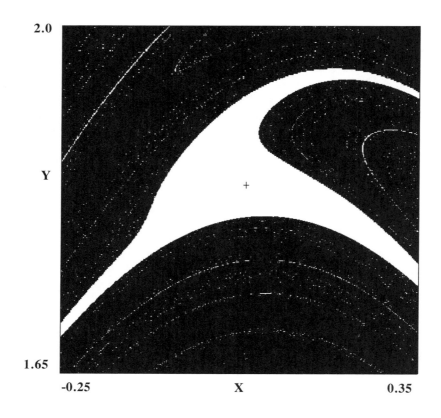

2.0

Y

1.65

-0.25   X   0.35

Figure 8-6a: Basin of attraction of infinity
The black area is the basin of infinity of the Henon map (H)
$$(X,Y) \quad \rightarrow \quad (3.5 - X*X - 0.9*Y, \ X)$$
The white area is the basin of attraction of a period 2 orbit. The cross shows the position of one of the attracting periodic 2 points. The other point is outside the region shown.

Topic of discussion. *The accessible basin boundary straddle trajectory that is accessible from the basin of the period 2 attractor tends to a periodic orbit. Where does this periodic orbit lie? What is its period? Can you suggest why the period is what it is in view of the above picture?*

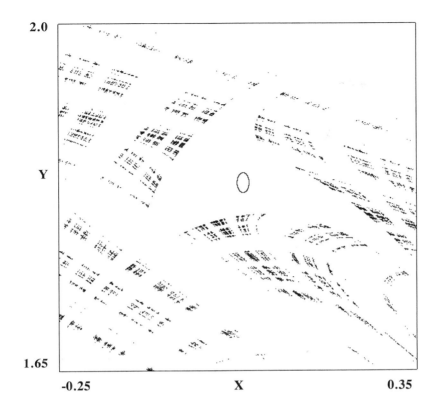

2.0

Y

1.65

-0.25   X   0.35

Figure 8-6b: A trajectory lying on a basin boundary

This figure shows a basin boundary straddle trajectory (BST) on the common boundary of the basin of the attractor at infinity and the basin of a period 2 orbit of the Henon map (H) with the same parameter values used in Figure 8-6a.

A circle has been plotted, centered at one of the attracting stable period 2 points with radius RA.

Topic of discussion. *Notice that the basin boundary straddle trajectory does not travel throughout the boundary in Figure 8-6a. How do you know that the basin boundary is an unbounded set? A quite difficult question is: Is the basin boundary straddle trajectory bounded?*

Straddle Options Menu

The Straddle Trajectory Menu has one submenu, namely the Straddle Options Menu. Enter the "Straddle Options Menu" command

$$\text{SOM} < Enter >$$

and the following menu appears in approximately the following format.

SOM   – –   STRADDLE OPTIONS MENU

ABST   –   compute (from ya's basin) Accessible BST trajectory
ASST   –   compute Accessible SST trajectory
BAF   –   tells BST to use the BA picture: OFF
DIV   –   DIVisions of line ya-yb for ABST, SST, ASST, GAME: 30
GAME   –   like ASST, ABST, and SST, only
                    you choose the new ya and yb each time
SDIS   –   subdivide ya-yb until ya and yb are close: 1e-008
SMC   –   Maximum number of Checks for Straddle methods: 2000

Commands for straddle options

ABST

The "Accessible Basin boundary Straddle Trajectory" command ABST computes a BST trajectory which is accessible from the basin that ya is in.

ASST

The "Accessible Saddle Straddle Trajectory" command ASST computes a Saddle Straddle Trajectory whose initial point is accessible from ya without first encountering other points with infinite escape time.

BAF

This command provides a simplified way of computing Basin Straddle Trajectories. Ordinary, you would have to let the program know where the attractors are by setting the appropriate storage vectors y2, ...,y7. Here you can create a picture on the screen using BA and and if the "BA Flag" BAF is set to ON, the routine BST will do the rest. See Chapter 7 for information on BA.

The command BAF is a toggle of value for plotting Basin Straddle Trajectories (command BST). BST will use the picture created by BA if the toggle "BAF" is ON. The color used for plotting must be more than 11. The BA picture must be in core, and BAF be turned ON. The routine requires the step size X_Step to be set.

If the picture is an old picture created before this feature was added, it will not have that stored, so it will not run.

BSTF

The "Basin boundary Straddle Trajectory Fraction" command BSTF specifies which point is chosen between ya and yb. The line segment from ya to yb is parameterized from 0 to 1, and BSTF can take any value between 0 and 1. The default value is 0.5. This command is **not listed in the menus**.

BSTR

BSTR stands for Basin boundary Straddle Trajectory - Random "midpoint". If this is turned ON (default is OFF), then the point that BST selects between ya and yb as a new candidate for replacing either ya or yb, is selected at random (from a uniform distribution). When BSTR is OFF, the candidate point is the midpoint between ya and yb.

BSTR is a toggle: if it is ON, enter it again and it turns OFF.

DIV

The "DIVisions of line ya-yb" command DIV sets the number of subsegments in the refinement. For commands SST, ASST, ABST, and GAME, the line segment from ya to yb is split into DIV subintervals and an escape time is determined for each of the DIV+1 endpoints.

The refinement algorithms try to find a new ya, yb pair of points from this collection of DIV + 1 points.

Dynamics allows a maximum value for DIV to be 5000, while *Smalldyn* only allows this value to be 200. Typically, the default of DIV = 30 works fine for BST and SST. However, in some cases, the routines ABST and ASST need a higher value for DIV.

GAME

GAME is comparable with ASST, ABST, and SST, only the user chooses the new ya and yb each time from the grid of points examined.

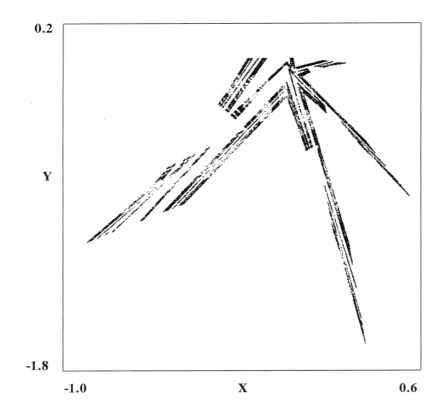

Figure 8-7: Saddle straddle trajectory
This figure shows a Saddle Straddle Trajectory (SST) of the Piecewise Linear map (PL)

$$(X, Y) \; \rightarrow \; (-1.25*X + Y + 0.05, \; 0.18*X) \text{ if } X \leq 0$$
$$(X, Y) \; \rightarrow \; (2*X + Y + 0.05, \; -3*X) \text{ if } X > 0$$

In this figure, the trajectory of almost every initial point diverges toward infinity.

Topic of discussion. *The origin is an unstable fixed point. Where are other periodic orbits located relative to the above plot? The Random Periodic orbit finder RP will plot periodic orbits here for period PR < 30. Command RPK will plot crosses at those orbits so that the periodic points can be distinguished from the SST plotted points.*

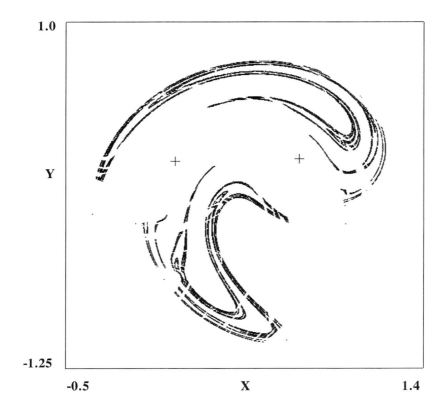

1.0

Y

-1.25

-0.5   X   1.4

Figure 8-8: Saddle straddle trajectory
*A Saddle Straddle Trajectory in the basin of a period 2 attractor of the
Ikeda map (I)*

$$Z \rightarrow 0.75 + 0.9*Z*exp\{i[0.4 - 6/(1 + |Z*Z|)]\}$$

where $Z = X + iY$.

*The crosses show the position of the period 2 attractor. In addition,
there is an attracting fixed point outside the screen area. Command BA
shows the entire screen region is in the basin of the period 2 attractor.*

Topic of discussion. *Here RHO is 0.75. As it increases to 0.82, the
period 2 orbit becomes unstable. Describe the attractor at RHO = 0.82 and
compare it to this figure. What happens to the straddle orbit as RHO
increases to the critical value. What is the basin of the period 2 orbit
just before it becomes unstable?*

SDIS

The "Straddle DIStance" command SDIS is for setting the target distance between points ya and yb. All straddle trajectories use some procedure to refine the line segment between ya and yb, that is, to choose new points ya and/or yb that are on the segment between the current ya and yb. This procedure decreases the distance between ya and yb and is used repeatedly until the distance between them is less than the "Straddle Distance" SDIS.

SMC

The "Straddle Maximum Checks" SMC is the number of iterates used by straddle trajectories to see where a trajectory goes. SMC is the number used by straddle trajectory routines, BST, SST, ABST, ASST, and GAME. By comparison the "Maximum Checks" number MC is used by the basin routine BAS and by WW. SMC has a default value much larger than that of MC because straddle trajectories require the determination of where a trajectory is going. Thus as the routines test where a trajectory goes, it never needs SMC iterates if the straddle method is running correctly.

8.3 EXAMPLES

In Chapter 2, we have already presented a few easy examples of straddle trajectories. Example 2-21 deals with a Basin boundary Straddle Trajectory for the Henon map, Example 2-22 discusses a Basin boundary Straddle Trajectory for the Tinkerbell map, and Example 2-19 concerns a Saddle Straddle Trajectory for the Henon map.

In the examples below, we assume that you are familiar with
- setting parameters to the desired values (command RHO, C1, etc.)
- setting the X Scale and Y Scale (commands XS and YS)
- setting a good value for the Radius of Attraction (command RA) by using the "circle and its Iterates" command OI
- Setting Vectors (command SV) and Replacing Vectors (command RV)
- Clear screen (commands C and CC)

All these features are discussed in detail in Example 7-1 in Section 7.3, and you may benefit from going through that example. As an aid, the commands required are given in parentheses, sometimes with hints of what else must be typed. In all two dimensional examples, you may wish to use command BA to find the locations of basins and attractors before beginning.

Example 8-1: A basin boundary straddle trajectory

To illustrate plotting of a Basin boundary Straddle Trajectory, we first use the Henon map (H)

$$(X,Y) \; \rightarrow \; (0.5 - X*X + 0.9*Y, \; X)$$

where the variable X is plotted horizontally, and the variable Y is plotted vertically. The X Scale is from -1.5 to 1.5, and the Y Scale is from -1.5 to 1.5. Exercise 7-7 dealt with plotting the two basins for this map.

The objective of this example is to plot a basin boundary straddle trajectory on the common boundary of the basin of the attractor at infinity and the basin of a stable period 2 orbit. The steps needed to plot such a straddle trajectory are the following.

1. Set the parameters RHO and C1 to be RHO = 0.5 and C1 = 0.9.

2. Set the X Scale (XS) and Y Scale (YS) to run from -1.5 to 1.5.

3. Search for the attractors and store the gained information. To do this, you may do the following: (a) plot all the basins and attractors (BA); (b) use arrow keys to move the small cross to the light blue region; e); (c) then initialize (I) and plot the trajectory (T). This results in a stable period 2 orbit.

Observe now that there are at least two attractors, namely the attractor at infinity and a stable period 2 orbit. The routine BA finds no other attractors. Replace y5 by a point of the period 2 attractor.

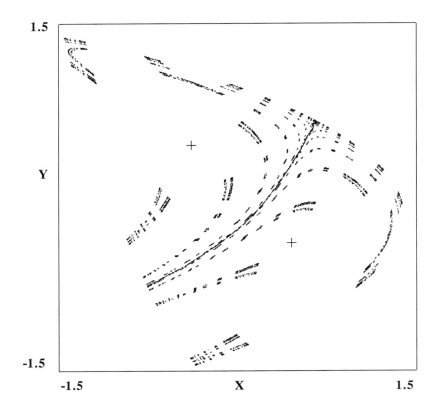

Figure 8-9: A trajectory that lies on the basin boundary

This figure shows a Basin boundary Straddle Trajectory on the common boundary of the basin of the attractor at infinity and the basin of a period 2 orbit of the Henon map (H)

$$(X, Y) \rightarrow (0.5 - X*X + 0.9*Y, X)$$

The crosses show the position of the attracting period 2 orbit.

Topic of discussion. *Find the fixed points of the map. It looks almost as if there is a continuous curve lying in the set plotted out by the Basin boundary Straddle Trajectory. Describe where points on the spine go when iterating the map. What suggests this spine does not contain a continuous curve?*

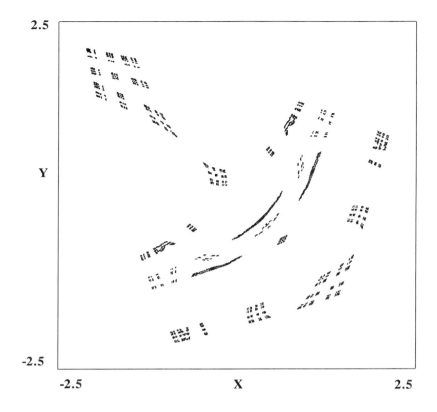

Figure 8-10: A bounded trajectory when almost all trajectories diverge
This figure shows a Saddle Straddle Trajectory of the Henon map (H)

$$(X,Y) \;\longrightarrow\; (1.5 - X*X + Y,\ X)$$

See also Figure 9-2. Notice that the Jacobian = -C1, so C1 = 1 implies a quadrilateral will map to a shape of the same area. If vectors ya, yb, yc, and yd are positioned at the corners of a quadrilateral, the command ABI will draw the quadrilateral and its iterate. Warning: SST and BST both use vectors ya and yb.

Topic of discussion. *This remarkable set seems to consist largely of many copies of a fractal set lying in a (curvilinear) quadrilateral. In fact, all the points lie inside wedge-shaped quadrilaterals. Verify that the edges are related to the stable manifold of one fixed point and the unstable manifold of the other. Command RPK can be used to locate the fixed points. Find out where the larger of these quadrilaterals will map.*

Note. Another way to search for the attractors, is the following: (a) plot the basin of infinity (BAS); then either (b1) use arrow keys to move the small cross to the blank region on the screen **or** (b2) replace vector y1 by the value stored in ye (RV 1 e); (c) then initialize (I) and plot the trajectory (T). This results in a stable period 2 orbit. Besides the two attractors mentioned above, the routine BAS finds no other attractors.

4. Check that the value RA = 0.1 is a good value for the stable period 2 point y5 (command OI with MI = 8).

5. Set the storage vectors ya and yb, one in the basin of infinity and the other in the basin of the period 2 orbit. For example, set vector ya to be (-1.5,-1.5) (SV) which is in the basin of infinity, and replace yb by y5 which is a point of the stable period 2 orbit (RV b 5).

6. Now plot the basin boundary straddle trajectory (BST). The resulting picture is given in Figure 8-9.

Example 8-2a: A basin boundary straddle trajectory
In this example, the plotting of straddle trajectories is illustrated by using the forced damped Pendulum differential equation (P)

$$X''(t) + 0.2*X'(t) + \sin X(t) = 2*\cos(t)$$

where the variable X is plotted horizontally, and the variable Y = X' is plotted vertically. The X Scale is from -3.14159 to 3.14159, and the Y Scale is from -5 to 5. Example 7-2 described the plotting of the two basins. There are two attractors, each a fixed point. Hence, a Basin boundary Straddle Trajectory can be plotted. The steps needed to plot such a basin boundary straddle trajectory are the following.

1. Set the parameter RHO to be 2 (the other parameters are the default values).

2. Set the Y Scale to be from -5 to 5.

3. Search for the attractors of the time-2π map and store the gained information. See Example 7-2 for finding the two fixed point attractors for this map. For this equation, the time-2π map has stable fixed points, viz. A = (-0.473,2.037) and B = (-0.478, -0.608) and has no other attractors. Replace y2 by a stable fixed point and y5 by the other stable fixed point.

4. Check that the value RA = 0.1 is a good value for both the stable fixed point stored in y2 and the one stored in y5 (OI with MI = 10).

5. Set the storage vectors ya and yb. To plot a basin boundary straddle trajectory replace (RV) for example, ya by y2 and yb by y5. The vectors ya and yb are then in different basins.

6. Plot a basin boundary straddle trajectory (BST). The resulting picture is given in Figure 8-11.

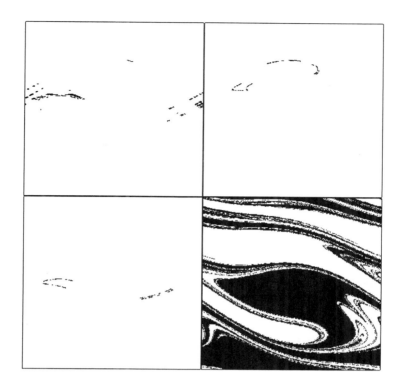

Figure 8-11: Straddle trajectories and basin
This figure shows a variety of straddle trajectories and basins of the forced damped Pendulum differential equation (P)

$$X''(t) + 0.2*X'(t) + \sin X(t) = 2*\cos(t)$$

where the variable X is plotted horizontally, and the variable X' is plotted vertically. The upper left window shows a Basin boundary Straddle Trajectory (BST) of the time-2π map. It is on the common boundary of the two basins. The upper right window and the lower left window each show a Saddle Straddle Trajectory (SST) of the time-2π map. The straddle trajectory in the upper right window is in the white basin, and the straddle trajectory in the lower left window is in the black basin. In the lower right window, the black area and white area are each the basin of attraction of a stable fixed point of the time-2π map; see Figure 7-7.

In all four windows, the X Scale is from -π (left) to π (right) and the Y Scale is from -5 (bottom) to 5 (top).

Topic of discussion. *Describe a nonlinear transformation that takes the points of the upper left window to those of the lower right, and takes one basin onto the other.*

Example 8-2b: A saddle straddle trajectory

On a color screen, transient chaotic behavior is strongly suggested by the irregular speckled color pattern in the plotted basin. This indicates that the basin might have a chaotic saddle embedded in it. This fact was already mentioned in Chapter 7. We now plot a saddle straddle trajectory in each basin. First set the storage vectors ya and yb.

7. To plot a Saddle Straddle Trajectory, set ya and yb, for example, to be the lower left corner and upper right corner of the screen (DIAG).

8. Plot a saddle straddle trajectory (SST). The resulting picture is given in Figure 8-11.

To plot a Saddle Straddle Trajectory, which is in the other basin, set ya to be the upper left corner of the screen, for example, (arrow keys and command RV a 1), and replace yb by the stable fixed point in that basin (command RV b 5).

Example 8-2c: Accessible straddle trajectories

After BST has been plotting for a while, it is instructive to invoke ABST. The resulting Accessible Basin boundary Straddle Trajectory converges rapidly to a period 3 orbit. Similarly, invoking the "plot Accessible Saddle Straddle Trajectory" command ASST (after SST has been running) results in a period 2 orbit.

8.4 EXERCISES

In the exercises below, it is assumed that you know how to plot the basin of any attractor using command BAS. For example, before plotting basin boundary straddle trajectories it may be convenient for you to do the related Exercise in Section 7.4 first. See Chapter 7, for plotting basins. If you have made a "Dump Data" file (DD) for a related exercise in Chapter 7, you can use this file for plotting a basin boundary straddle trajectory with a minor change only. Read this file in when you start the program, and set the vectors ya and yb such that they are in different (generalized) basins. When the X Scale or Y Scale are not specified in an exercise, use the default values.

Exercise 8-1. There are two attractors for the complex quadratic map
$$Z \rightarrow Z*Z + 0.32 + i0.043$$
where $Z = X + iY$, ∞ and a period 11 orbit. Plot a basin boundary straddle trajectory and an accessible basin boundary straddle trajectory on the common boundary of the basin of infinity and the basin of the stable period 11 orbit. Set the X Scale to run from -1 to 1 and set the Y Scale to run from -1.35 to 1.35. Compare your results with a BA basin and attractor plot using MC = 1000 and BAP = 30. This example requires large values of MC and BAP. See Figure 8-5.

Exercise 8-2. Plot a basin boundary straddle trajectory on the common boundary of the basin of infinity and the basin of the chaotic attractor of the Tinkerbell map (T)
$$(X,Y) \rightarrow (X*X - Y*Y + 0.9*X - 0.6013*Y, \ 2*X*Y + 2.0*X + 0.5*Y)$$
Set the X Scale to run from -1.8 to 1.2 and set the Y Scale to run from 1 to -2. See Figure 8-2, lower left window.

Exercise 8-3. Plot a saddle straddle trajectory and an accessible saddle straddle trajectory of the Henon map (H)
$$(X,Y) \rightarrow (1.2 - X*X + 0.9*Y, \ X)$$
See Figure 8-4.

Exercise 8-4. Plot a basin boundary straddle trajectory on the common boundary of the basin of infinity and the basin of the stable period 2 orbit of the Henon map (H)
$$(X,Y) \rightarrow (3.5 - X*X - 0.9*Y, \ X)$$
Set the X Scale to run from -1 to 2.5 and set the Y Scale to run from -1 to 2.5. See Figure 8-6b for a blow-up.

Exercise 8-5. Plot a saddle straddle trajectory of the Piecewise Linear map (PL)

$$(X,Y) \rightarrow (-1.25*X + Y + 0.05, 0.18*X) \text{ if } X \leq 0$$
$$(X,Y) \rightarrow (2*X + Y + 0.05, -3*X) \text{ if } X > 0$$

Set the X Scale to run from -1 to 0.6 and set the Y Scale to run from -1.8 to 0.2. See Figure 8-7.

Exercise 8-6. Plot a saddle straddle trajectory in the basin of a period 2 orbit of the the Ikeda map (I)

$$Z \rightarrow 0.75 + 0.9*Z*\exp\{i[0.4 - 6/(1 + |Z*Z|)]\}$$

where $Z = X + iY$. Set the X Scale to run from -0.5 to 1.4, and set the Y Scale to run from -1.25 to 1. See Figure 8-8.

Exercise 8-7. Plot a basin boundary straddle trajectory on the common boundary of the basin of infinity and the basin of a stable period 2 orbit of the Henon map (H)

$$(X,Y) \rightarrow (0.5 - X*X + 0.9*Y, X)$$

See Figure 8-9.

Exercise 8-8. Plot a saddle straddle trajectory and an accessible saddle straddle trajectory of the Henon map (H)

$$(X,Y) \rightarrow (1.5 - X*X + *Y, X)$$

See Figure 8-10.

Exercise 8-9. The time-2π map of the forced damped Pendulum equation

$$X''(t) + 0.2*X'(t) + \sin X(t) = 2*\cos(t)$$

has two stable fixed points. Set the Y Scale to run from -5 to 5.

a. Plot a basin boundary straddle trajectory on the common boundary of the basin of a stable fixed point and the basin of a second stable fixed point. See Figure 8-11, upper left window.

b. Plot a saddle straddle trajectory in the basin of each of the two stable fixed points. See Figure 8-11, upper right and lower left window.

Exercise 8-10. Plot a basin boundary straddle trajectory on the common boundary of the basin of infinity and the basin of a stable period 2 orbit of the Henon map (H)

$$(X,Y) \rightarrow (1.4 - X*X - 0.3*Y, X)$$

Set the X Scale to run from -3 to 3, and set the Y Scale to run from -3 to 11. See Figure 9-4, lower right window.

Exercise 8-11. Plot a basin boundary straddle trajectory and an accessible basin boundary straddle trajectory on the common boundary of the basin of a stable period 3 orbit and the basin of infinity of the complex map (C)

$$Z \rightarrow Z*Z - 0.11 + i0.6557$$

where $Z = X + iY$. Set the X Scale to run from -1.5 to 1.5 and set the Y Scale to run from -1.5 to 1.5. See Figure 8-3, upper right window.

Exercise 8-12. Plot a basin boundary straddle trajectory on the common boundary of the basin of a stable fixed point and the basin of attractor infinity of the complex Cubic map

$$Z \rightarrow 0.8*Z*Z*Z - Z*Z + 1$$

where $Z = X + iY$. Set the X Scale to run from -1.5 to 2 and set the Y Scale to run from -1.1 to 1.1. See Figure 8-3, lower left window.

8.5 REFERENCES related to DYNAMICS

This chapter in particular presents tools that were developed as the program was written. As the methods as for the plotting of various kinds of straddle orbits were developed, they were incorporated into the program for testing. Several of these papers display figures that were made with this program. Those papers below marked with (*) are of a more general nature and are appropriate for general readings.

This paper gives a detailed study of an example in which basin straddle trajectories between various basins are computed.
P. Battelino, C. Grebogi, E. Ott, J. A. Yorke, and E.D. Yorke (1988), Multiple coexisting attractors, basin boundaries and basic sets, Physica D 32 (1988), 296-305

The following paper describes the points on the boundary of a basin for maps in the plane. In typical cases of dissipative systems, all the points accessible from the basin are on the stable manifolds of periodic saddles of a single period. When the boundary is fractal there are infinitely many inaccessible periodic orbits in the boundary with infinitely many different periods. This theoretical paper was initiated by discoveries made with an early version of the Accessible Boundary Straddle Trajectory method (ABST).
K.T. Alligood and J.A. Yorke, Accessible saddles on fractal basin boundaries. Ergodic Theory and Dynamical Systems 12 (1992), 377-400

The next paper presents the procedure for computing Accessible Basin boundary Straddle Trajectories (ABST). It presents many examples and it also includes rigorous statements and proofs of the foundations of the method.
* H.E. Nusse and J.A. Yorke, A numerical procedure for finding accessible trajectories on basin boundaries, Nonlinearity 4 (1991), 1183-1212.

The next two papers present the theory of Saddle Straddle Trajectories (SST and ASST). The second includes statements and proofs of the foundations of the methods.
* H.E. Nusse and J.A. Yorke, A procedure for finding numerical trajectories on chaotic saddles, Physica D 36 (1989), 137-156
H.E. Nusse and J.A. Yorke, Analysis of a procedure for finding numerical trajectories close to chaotic saddle hyperbolic sets. Ergodic Theory and Dynamical Systems 11 (1991), 189-208

CHAPTER 9

UNSTABLE AND STABLE MANIFOLDS

9.1 INTRODUCTION and the METHOD

Poincaré called the periodic orbits of maps the soul of the dynamics. The study of the stable and unstable manifolds of these orbits does indeed reveal a great deal about the dynamics. The **unstable manifold** of a fixed point p of a map F may be defined as the set of points q_0 that have a backward orbit coming from p, that is, a sequence of points q_i with $i = -1$, $-2, \ldots,$ so that $F(q_{i-1}) = q_i$ for which $q_i \to p$ as $i \to -\infty$. The **stable manifold** of a periodic orbit may be defined for invertible processes as the unstable manifold for the inverted system. If a point p is a saddle fixed point of a map in the plane, then the stable and unstable manifolds are both curves that pass through p. The routines described below are not restricted to planar systems, but the manifold being computed must be a curve.

Assume now that the stable or unstable manifold to be computed is one dimensional. Hence the unstable manifold is a curve passing through that point. Trajectories that start on this curve move along this curve away from the periodic trajectory. The unstable manifold of the fixed point has two components or sides, and the point is called a **flip fixed point** if each component is mapped to the other. Such cases make computation slightly more difficult so internally the program looks at the second iterate of a fixed point or $2*PR$th iterate of a periodic orbit of period PR. For the $2*PR$th iterate, each of the two components of the manifold maps to itself, that is, it is invariant and is computed as follows. The point ya, that has been chosen close to the periodic point at y1 and very close to the unstable manifold, maps to a point yb (using the $2*PR$th iterate), which will almost always be even closer to the manifold but will be further from y1. The manifold curves slowly near y1 and the segment from ya to yb closely approximates a piece of the manifold. See Figure 9-1. Careful computation of the iterates of the segment yield a detailed picture of that component of the manifold.

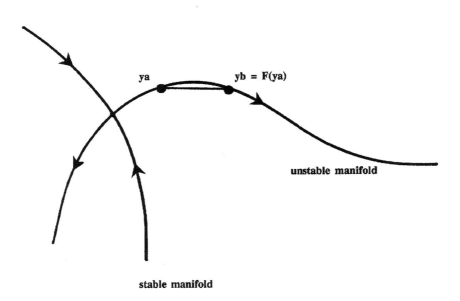

Figure 9-1: Stable and unstable manifold
*This schematic picture shows the stable and unstable manifold of a fixed point p of the map F, where F is the 2*PR*th *iterate of the process.*

The positions of the vectors ya and yb = F(ya) are indicated on the unstable manifold.

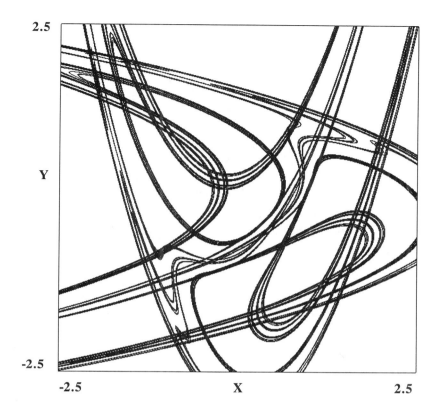

2.5

Y

-2.5

-2.5   X   2.5

Figure 9-2: Stable and unstable manifolds
This figure shows the stable and unstable manifolds of the accessible fixed point, approximately (-1.22,-1.22), on the chaotic saddle of the Henon map
$$(X, Y) \rightarrow (1.5 - X*X + Y, X)$$
Figure 8-10 treats the same map.

Topic of discussion. *Find the inverse of this map and explain how its form is reflected in the symmetry of the stable and unstable manifolds in this figure.*

The method for computation of the manifolds

The key to an algorithm for plotting unstable manifolds is an algorithm for computing the kth iterate of a line segment under a map F. Here F will denote the 2*PRth iterate of the process. See Figure 9-1. In plotting the kth iterate of a line segment from ya to yb under F, we choose a succession of points $\{p_i\}$ along this segment with p_0 = ya. Denote G = F^k If $G(p_i)$ has been plotted on the screen (or is within SD screen diameters of the screen), p_{i+1} is chosen so that $G(p_{i+1})$ is no further than a fixed distance from $G(p_i)$ on the screen, and then the line segment between them is plotted (see PIX). If the distance between them is too small, the routines try to choose a slightly larger next step, from p_{i+1} to p_{i+2}.

Technical Note. The manifold routine reports the "frac of ya, yb" which means the current position of p(i) along the [ya,yb] segment as a fraction of the length of the segment. It also reports the "change in frac" which means the distance between p_i and p_{i+1} as a fraction of the length of the segment. The phrase "minimum change in frac" is the smallest change thus far encountered.

In many cases $G(p_i)$ will not be in the extended screen region and might even be 10^{1000} units from the screen; see Chapter 7 for the definition of extended screen region. Such a point cannot even be computed without causing an overflow and is not computed. There may be other parts of the segment closer to yb whose iterates under G are in the screen area. These iterates must still be plotted. Our solution to this difficulty is as follows. The goal is to plot $G(p_i)$ for those p_i for which $F^j(p_i)$ is in the extended screen region for all j ≤ k. In many cases for SD sufficiently large, if $G(p_i)$ is in the screen region, then all the earlier iterates of p_i must also have been in the extended screen region. The above value of SD is valid for all k. Denote H = F^m. If $H(p_i)$ is in the extended screen region but F∘H(p_i) is not, where m < k, then choose p_{i+1} so that $H(p_i)$ and $H(p_{i+1})$ are not further apart than the specified distance PIX. Higher iterates like F∘F∘H(p_i) are not computed. For each p_i this number m is recomputed.

One might erroneously expect that plots of unstable manifolds would be grossly inaccurate due to the sensitivity of initial data near a saddle periodic orbit, but plots are usually very accurate, provided no warnings are printed on the screen. In computing the kth iterate of a point on a segment, there may be a considerable error, but that error is in the direction tangent to the iterate of the segment. The iterate of the segment is stretched in the same direction due to the sensitivity. Hence the error in plotting G(p(i)) results in a point which is shifted along the unstable manifold, but is nonetheless quite close to the manifold. See You, Kostelich and Yorke (1991) for detailed justification of this algorithm.

Another, but much more restrictive algorithm to compute stable and unstable manifolds of the fixed point of the Henon map, was introduced by Franceschini and Russo (1981). This latter algorithm is based on a parametric representation of the manifolds, and their error estimates hold locally near the fixed point.

Requirements of the procedures

The routines SL and SR compute the stable manifold if it is one dimensional and the inverse is available to the program. The map or differential equation does not need to be 2-dimensional. The inverse is available for all the differential equations and several of the maps implemented in *Dynamics*. The routine also needs to be able to use the Quasi-Newton method to find the fixed point or periodic orbit, so it must have the partial derivatives of the process available. That is equivalent to saying that it must be set up to compute Lyapunov exponents. If the partial derivatives are not available, you can instead choose ya and yb yourself and then use command ABI, choosing ya very close to the periodic point and choosing yb to be either the PR_{th} iterate of ya or the $2*PR_{th}$ iterate of ya. Alternatively ya and yb can be chosen on opposite sides of the periodic point, with the periodic point approximately halfway between ya and yb. Then the routine ABI will compute the first MI iterates of the segment.

Summary for plotting manifolds

- Set the period PR and a good value for Screen Diameters (commands PR and SD).
- (**Optional**) Choose y1 near a periodic point of period PR (e.g. using Quasi-Newton method or OI). The manifold routines automatically look for a periodic orbit of period PR but it might not find the periodic point you want. You can use the routine RP to look for all periodic saddles and attractors and repellors of period PR.)
- Plot Unstable (or Stable) Left manifold (commands UL, SL).
- Plot Unstable (or Stable) Right manifold (commands UR, SR).

9.2 UNSTABLE AND STABLE MANIFOLD MENU UM

There are two separate commands to compute the two pieces of the unstable manifold meeting at a periodic point, namely UL (Unstable Left) and UR (Unstable Right). Similarly, there are two separate commands to compute the two pieces of the stable manifold meeting at a periodic point, namely SL (Stable Left) and SR (Stable Right).

Enter the command UM, and the menu appears on the screen in approximately the following format.

```
UM  -  UNSTABLE AND STABLE MANIFOLD MENU

AB   -  connect ya to yb (to yc to yd to ya)
ABI  -  do AB and draw its MI  (= 1) Iterates

SL   -  plot Left side of Stable manifold at y1
SR   -  plot Right side of Stable manifold at y1
UL   -  plot Left side of Unstable manifold at y1
UR   -  plot Right side of Unstable manifold at y1

        Related commands in other menus:
PR   -  PeRiod:  1
SD   -  Screen Diameters (large is slow and cautious):  1.00

The STABLE MANIFOLD commands SL & SR require inverse,
   available for most maps and Differential Equations.
Routines SL, SR, UL, UR first look for a PR periodic
   point near y1, so choose y1 near a periodic point.
```

UM

The "Unstable and stable manifold Menu" command UM calls the Unstable and stable manifold Menu. The stable and unstable manifold routines (UL, UR, SL, SR) first attempt to find a periodic orbit of period PR. These routines start looking using y1 as the initial point from which it applies the Quasi-Newton method several times. It helps if the user has selected y1 close to a periodic orbit of period PR. The unstable manifold of a periodic orbit can be computed using these routines **if** the periodic orbit is found with considerable precision. Therefore, the small cross can be moved near the suspected position of the periodic orbit. The Quasi-Newton method puts the small cross (that is y1) at the position of the point of orbit. The default period is PR = 1, so it looks for a fixed point. The manifold routines set ya and yb very close to the unstable manifold and then in effect applies the routine ABI.

The role of the number of screen diameters SD

If you are computing an unstable manifold that lies entirely within the region of the screen, then SD is unimportant. On the other hand, in many cases especially with the stable manifold, the manifold leaves the area of the screen and then returns, repeating this process over and over. The manifold may go an immense distance before returning to the screen. The algorithm aims at eliminating the computation of parts far from the screen and yet aims at getting an accurate picture of parts in the screen. In particular, if one piece of the manifold just dips into the screen and leaves again, we want that piece to be drawn, and we want to plot all of it that is within the screen area. In order to handle such situations, the program computes the manifold when it is within the extended screen region (that is, within a larger box 2*SD+1 times as wide as the screen and 2*SD+1 times as high as the screen, with the screen in the center of this big box). The extended screen region introduces a buffer zone. The bigger SD is, the slower the computation will be, but it may also be more reliable, missing less. The theory indicates that for many cases, if SD is big enough, then the plot will be extremely accurate, with no pieces skipped. If a branch of the stable manifold suddenly appears in the middle of the screen, then something is being skipped and that means that SD is too small and a warning is printed. For stable manifolds of the fixed points of the Henon map, using the default parameters, SD = 1 is satisfactory, but for the stable manifold of the period two points, SD must be chosen much larger. In this case SD = 10 appears satisfactory. The choice of SD will also depend on the coordinates of the screen so that if a detailed blow-up is plotted, SD must be correspondingly larger.

Commands for computing manifolds

AB

This command requires ya and yb being set. The behavior of this command depends on whether yc is set and if it is, whether yd is set. If the vectors ya and yb are set but not yc and both are in the screen area, then after entering this command, a line segment [ya,yb] is drawn between them. The default values of ya and yb are the lower-left and upper-right corner of the screen. If all three are set (but not yd) then in addition to [ya,yb] the line segment [yb,yc] (joining yb with yc) is drawn. If all four are set, the quadrilateral is drawn using the four as vertices.

If yc (or yd) has been set and if you do not want it set, then set yc (or yd) equal to a vector that is NOT SET. For example, if y7 is not set, type

$$RV < Enter > \quad C < Enter > \quad 7 < Enter >$$

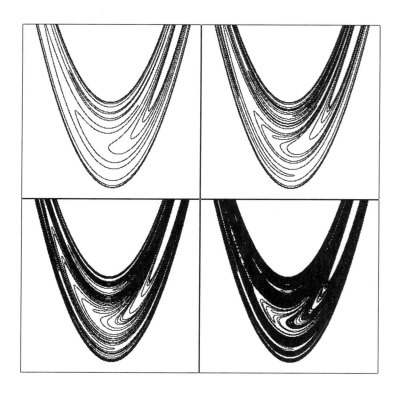

Figure 9-3a: Stable manifold

This figure shows longer and longer pieces of the stable manifold of the fixed point p ≈ (0.89,0.89) for the Henon map (H)

$$(X,Y) \rightarrow (1.42 - X*X + 0.3*Y, X)$$

where the variable X is plotted horizontally, and the variable Y is plotted vertically.

The upper left window displays an initial part of the stable manifold of p, plotted using command SL for computing the Left side of the Stable manifold. The picture in the upper right window shows a longer piece of the stable manifold. Similarly, the pictures in the lower left window and lower right window show still longer pieces of the stable manifold. The pictures in all four windows were plotted using command SL. The stable manifold seems to be filling up the basin of the chaotic attractor, though not uniformly. In all four windows, the X Scale is from -3 (left) to 3 (right) and the Y Scale is from -12 (bottom) to 12 (top).

Topic of discussion. *Explain the dramatic difference if SR is used.*

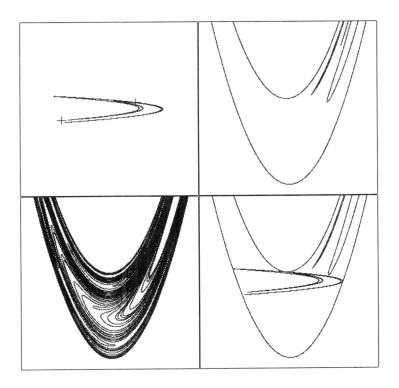

Figure 9-3b: Unstable and stable manifolds of the two fixed points

The fixed points are p ≈ (0.89,0.89) and q ≈ (-1.59,-1.59), marked by crosses. For computing these pictures, the same map and scale are used as in Figure 9-3a.

In the upper left window, the unstable manifold of p is plotted. The picture in the upper right window is a part of the stable manifold of the fixed point q. Commands SL and SR were both required to obtain this picture. The picture in the lower left window is a part of the stable manifold of p, and is similar to the lower left window of Figure 9-2a. The picture in the lower right window is a part of the stable manifold of q superimposed with a trajectory on the chaotic attractor. The attractor here looks just like the unstable manifold shown in the upper left window.

Topic of discussion. *Explain the difference in the way these stable manifolds relate to the basin of infinity.*

ABI

The "AB Iterates" command ABI draws the segment from ya to yb (and possibly, further segments) and its MI Iterates. This assumes that ABT < 2; see command ABT for the case when ABT ≥ 2. If the number of Maximum Iterates MI is 0, then the command ABI is equivalent to the command AB. Assume MI > 0. Let S be the segment that AB would plot. ABI plots S and the PRth iterate of S and (if MI > 1) the successive PRth iterate up to and including the MI*PRth iterate of S. If for example MI is 2 and PR is 3, then S and its third and sixth iterate will be drawn. Example 2-5 illustrates the use of drawing the first iterate of a quadrilateral.

The basic tool is the ability to compute the curvilinear iterate of a line segment. This is provided in the routine ABI which in its simplest form (that is when yc has not been set and PR = 1) draws a straight line between ya and yb and the MI iterates of it under the current map. The routine plots each pixel between ya and yb. It then iterates the map PR times and plots the resulting points. It repeats this (if MI > 1) up to a total of MI iterates. The routine chooses a sequence of points on the straight line from ya to yb, choosing the spacing in an adaptive method, attempting to choose them so that the successive iterates move a fixed amount, usually about 6 one thousandths of the screen size (see PIX).

The program iterates each point PR times between plotting. When yb is chosen to be the PRth iterate of ya, then the successive iterates link together into a single curve.

SL

The "Stable manifold Left" command SL is like UL except the program first automatically substitutes the inverse of the system (see also command V) if it is available. If it is not available, the routine will not work. The stable manifold is the unstable manifold of the inverse. When the routine is terminated, the inverse is again invoked, now yielding the original system.

Enter command SL, and the Actions, Hints and Options menu for plotting manifolds menu appears and includes the following:

ACTIONS, HINTS and OPTIONS for PLOTTING MANIFOLDS
ABSTEP – desired distance of yb from y1: 0.1
PIX – distance in pixels between usual manifold plots: 4 (set to 1 for greatest precision)
PR – iterates per plot: 1
SD – Screen Diameters (large is slow and cautious): 1.00

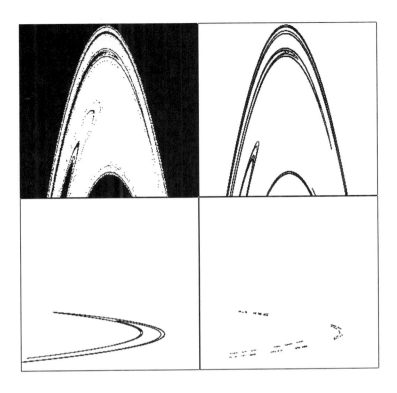

Figure 9-4: Manifolds, basin, and straddle trajectory
The four views here are for the Henon map (H)

$$(X, Y) \rightarrow (1.4 - X*X - 0.3*Y, X)$$

where the variable X is plotted horizontally, and the variable Y is plotted vertically. There are two attractors, an attracting period 2 orbit and the point at infinity.

In the upper left window, the basin of infinity is plotted. In the upper right window, a part of the stable manifold of p = (-2,-2) is plotted. The picture in the lower left window is a part of the unstable manifold of p. The lower right window shows a Basin boundary Straddle Trajectory, which is on the common boundary of the basin of the attracting period 2 orbit and the basin of attractor infinity. In all four windows, the X Scale is from -3 (left) to 3 (right) and the Y Scale is from -3 (bottom) to 11 (top).

Topic of discussion. *In view of the fact that the basin of infinity is actually connected, explain the isolated dots in the upper left window.*

SR

The "Stable manifold Right" command SR is like SL except that it plots the half of the stable manifold on the right side of the periodic orbit.

UL

The "Unstable manifold Left" command UL is for plotting the left side of the manifold at y1. The routine first applies the Quasi-Newton routine one or more times to the point y1 to guarantee that the point y1 is in fact sufficiently close to a fixed point or periodic point. Next the routine chooses ya to be the point very close to y1 and it chooses yb to be the 2*PRth iterate of ya. Then it draws the 2*PRth iterate of the segment [ya,yb] connecting ya with yb, and then draws the 4*PRth iterate of the segment [ya,yb], and so on.

UR

The "Unstable manifold Right" command UR is like UL except that it plots the right side of the manifold at y1.

Note. If you want to compute the unstable manifold of all the points of a periodic orbit where PR > 1, then start by entering commands UL or UR. That sets the vectors ya and yb. Next set PR = 1 and set MI to be a large number, say 100. Then enter command ABI. If the unstable eigenvalue is negative, both sides of the unstable manifold will be plotted. In any case, something is plotted for all points of the periodic orbit.

ABSTEP

Each of the manifold routines tries to locate a fixed point or periodic point of period PR near y1 and it changes y1 to be the point it finds. Next it tries to position vectors ya and yb close to y1 with ya closer and with yb at a distance of ABSTEP from y1. Also, ya maps to yb either in PR or 2*PR iterates of the map (or its inverse).

The default value of ABSTEP is 0.1. For precise computations, you may wish to set PIX = 1 and ABSTEP = 0.001. See also command PIX.

MANI

The "MANifold Iterate" command MANI is for setting the number of iterates of the line segment from ya to yb. The default value is 40. See also command ABSTEP. This command is **not listed in the menus**.

PIX

The PIXel number is for commands ABI, SL, SR, UL, and UR. The stable and unstable manifold routines in effect use the routine ABI for drawing iterates of a segment. When ABI plots the kth iterate of a line segment under a map F (write G for the kth iterate of F), it selects a succession of points p_i along the segment, varying the distance from p_i to p_{i+1} so that $G(p_i)$ and $G(p_{i+1})$ are a fixed distance PIX apart on the computer screen, and their distance is measured in "pixels", that is, picture elements, or plottable points on the screen. Recall for example that a VGA screen is 640 pixels wide by 480 pixels high. The routine then connects $G(p_i)$ and $G(p_{i+1})$ with a line segment. The default value of PIX = 4. Using this command, this number can be decreased for example to 1 so that all points plotted are computed accurately but then the program will run about four times slower.

Note. The fixed point must be found quite accurately for these stable and unstable manifold routines to work, and when this point is a periodic orbit of a differential equation, say a period 2π solution, that is the fixed point of a time-2π Poincaré return map, then the inverse is available by just making the time step negative, when solving the differential equation. (The routines SL and SR do this automatically.) However, for these differential equations, the inverse is not as accurate as for the maps, and if the differential equation step size is moderately big, the fixed points of a Poincaré return map may differ significantly from those obtained when time is reversed. This effect may be reduced by cutting the step size used by the differential equation solver. (See "Steps Per Cycle" command SPC in Section 4.3.)

Hint: Drawing manifolds SLOWLY

Sometimes it is desirable to plot manifolds slowly, if, for example, you wish to plot a small piece of the manifold. First, set PIX to 1. Second, enter the command K so that a cross will be drawn at each point. (Drawing crosses is a time consuming procedure.) When the curve that is plotted approaches the point where you want to stop the plotting, hit the $< \cdot >$ key repeatedly. Each hit gets the plotting process one point further. You can also enter the pause command P before beginning to plot the manifolds. Then use $< \cdot >$ repeatedly to single step through to the position you wish to stop.

Technical note. To study Wada basins (Figure 9-10a) and basin cells (Figure 9-10c), it is necesary to accurately compute segments of stable and unstable manifolds. During the computation of a segment of a manifold, the current initial point on the segment [ya,yb] being used, is stored in y9 while its image is stored in ye.

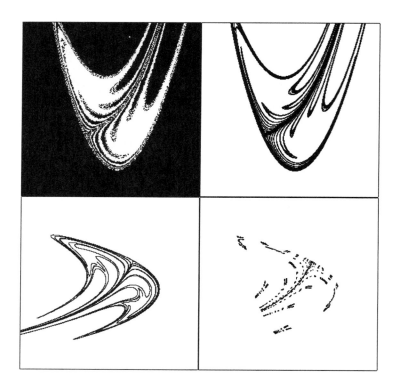

Figure 9-5a: Manifolds, basin, and straddle trajectory
This figure is for the Henon map (H)

$$(X, Y) \;\longrightarrow\; (0.5 - X*X + 0.9*Y, \; X)$$

where the variable X is plotted horizontally, and the variable Y is plotted vertically. There are two attractors, an attracting period 2 orbit and the point at infinity.

The upper left window shows the basin of infinity. In the upper right window, a part of the stable manifold of the fixed point p ≈ (-0.76,-0.76) is plotted. The lower left window shows the unstable manifold of p. The lower right window shows a Basin boundary Straddle Trajectory, which is on the common boundary of the basin of the attracting period 2 orbit and the basin of infinity. In all four windows, the X Scale is from -2.5 (left) to 2.5 (right) and the Y Scale is from -2.5 (bottom) to 2.5 (top).

Topic of discussion. *Find which features map to which features in the upper right window.*

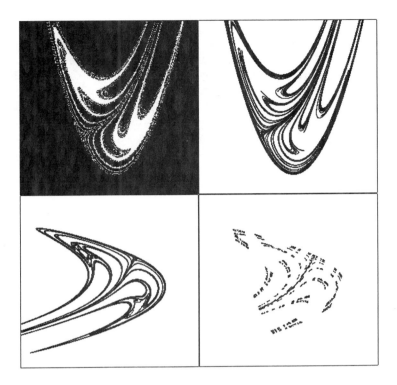

Figure 9-5b: Manifolds, basin, and straddle trajectory

This figure illustrates how the previous figure changes as RHO is changed from 0.5 to 0.6; that is, we now study the Henon map (H)

$$(X, Y) \ \rightarrow \ (0.6 - X*X + 0.9*Y, \ X)$$

The fixed point used is now (0.73,0.73). This figure is otherwise like Figure 9-5a.

Topic of discussion. *The lower right window, there appears to be a thick curve plotted. The two fixed points are at the end of this apparent curve. Plot the UR manifold of the lower one and the SL manifold of the upper one and see how they intersect.*

9.3 EXAMPLES

In Chapter 2, we have already presented a few examples of plotting manifolds. In Example 2-19 the unstable and stable manifolds of a fixed point for the Henon map are plotted. See also Figure 2-18. In the examples below, we assume that you are familiar with

- setting parameters to the desired values (commands RHO, C1, etc.)
- setting the X Scale and Y Scale (commands XS, YS)
- setting vectors (command SV) and replacing vectors (command RV)

All these features are discussed in detail in Example 7-1 in Section 7.3, and you may benefit from going through that example.

Example 9-1: Plotting an unstable manifold of a fixed point

To illustrate the plotting of an unstable manifold of a fixed point, we use the Ikeda map

$$Z \; \rightarrow \; 1 + 0.9*Z*\exp\{i[0.4 - 6/(1 + |Z*Z|)]\}$$

where $Z = X + iY$, the variable X is plotted horizontally, and the variable Y is plotted vertically. The X Scale is from -0.5 to 1.9 and the Y Scale is from -2.6 to 1.2.

When plotting a Basin boundary Straddle Trajectory, this straddle trajectory converges quickly to a fixed point p, which is approximately (1.11,-2.29). The objective of this example is to plot the left side of the unstable manifold of p. The steps needed to plot one side of a manifold are the following. We use the default values of RHO and C1.

1. Set the X Scale to run from -0.5 to 1.9 and set the Y Scale to run from -2.6 to 1.2.

2. Search for the fixed point p (appr. (1.11,-2.29)).

3. Plot the Unstable Left manifold. Enter the "plot Left side of the Unstable manifold at y1" command UL, which should cause the left side of the unstable manifold to be plotted. The resulting picture is given in Figure 9-6.

Figure 9-6: A trajectory with an unstable manifold
The upper left window shows some points on a chaotic trajectory of the Ikeda map (I)

$$Z \rightarrow 1 + 0.9*Z*exp\{i[0.4 - 6/(1 + |Z*Z|)]\}$$

where $Z = X + iY$; the variable X is plotted horizontally, and the variable Y is plotted vertically. The remaining windows show how the trajectory points lie very close to the unstable manifold of the fixed point $p \approx (1.11, -2.29)$.

In the upper right window, the lower left window, and the lower right window longer and longer pieces of the Unstable manifold (Left side) are added to the points shown in the upper left window. The point p is the lowest plotted point in the upper right window. It is on the basin boundary. In all four windows, the X Scale is from -0.5 (left) to 1.9 (right) and the Y Scale is from -2.6 (bottom) to 1.2 (top).

Topic of discussion. *It is unknown whether the unstable manifold comes arbitrarily close to every point of the attractor.*

Example 9-2: Plotting a stable manifold of a fixed point
In this example, the plotting of a stable manifold of a fixed point is illustrated by using the forced damped Pendulum

$$X''(t) + 0.2*X'(t) + \sin X(t) = 2.5*\cos(t)$$

where the variable X is plotted horizontally, and the variable $Y = X'$ is plotted vertically. The X Scale is from -3.14159 to 3.14159, and the Y Scale is from -2 to 4. Use the default values for the parameters, the X Scale, and the Y Scale.

The steps needed to plot such a straddle trajectory are the following.

1. Search for a fixed point p of the time-2π map and store the gained information in y1. We continue with the fixed point p which is approximately (-0.99,-0.33).

2. Plot the Stable manifold of the fixed point p of the time-2π map. First, enter the "plot Left side of the Stable manifold at y1" command SL, which should cause the left side of the stable manifold to be plotted. After it runs a while (perhaps overnight or over a weekend) enter the "plot Right side of the Stable manifold at y1" command SR, which should cause the right side of the stable manifold to be plotted. The resulting picture is given in Figure 9-7. If you enter IN (INcrement color number) you will find that the stable manifold occasionally retraces its path on the screen.

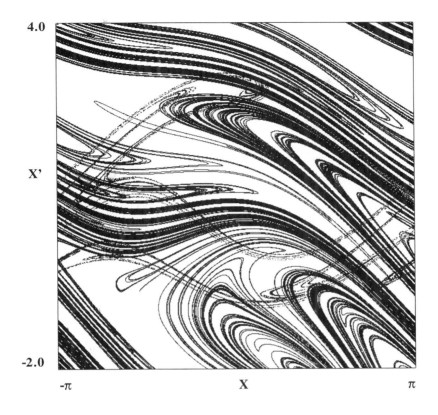

4.0

X'

-2.0

-π   X   π

Figure 9-7: Stable manifold

This figure shows the stable manifold of a fixed point p of the time-2π map
of the forced damped Pendulum

$$X''(t) + 0.2*X'(t) + \sin X(t) = 2.5*\cos(t)$$

The fixed point p is approximately (-0.99,-0.33). The dots of a chaotic
trajectory have been superimposed. It appears plausible that the trajectory
of every (X,X') tends to the chaotic attractor.

Topic of discussion. *It is reasonable to conjecture that the stable*
manifold almost always comes arbitrarily close to all points of the basin
of the attractor. In the above figure the stable manifold has a strong
preference for traversing certain places. Why is that? Is it true of the
stable manifolds of other fixed points of this process?

9.4 EXERCISES

In some exercises below, it is assumed that you are familiar with Accessible Straddle Trajectories. If you have made a file using the "Dump Data" command DD for a related exercise in Chapter 7 or Chapter 8, you can use this file for plotting manifolds. Read this file in, and make the appropriate settings. When an exercise does not specify the X Scale or Y Scale, just use the default values. The statement "Plot the manifolds" means "Plot significant amounts of the stable and unstable manifold". Differential equations will require considerable computation time.

Exercise 9-1. Plot the manifolds of the accessible fixed point (appr. (-1.23,-1.23)) on the chaotic saddle of the Henon map (H) (see Figure 9-2)
$$(X,Y) \rightarrow (1.5 - X*X + Y, X)$$

Exercise 9-2. Plot the manifolds of the 2 fixed points (appr. (-1.59, -1.59) and (0.89,0.89)) of the Henon map (H)
$$(X,Y) \rightarrow (1.42 - X*X + 0.3*Y, X)$$
Describe the similarities and differences. See Figure 9-3.

Exercise 9-3. Plot the manifolds of the fixed point (-2,-2) of the Henon map (H)
$$(X,Y) \rightarrow (1.4 - X*X - 0.3*Y, X)$$
Set the X Scale from -3 to 3, and set the Y Scale from -3 to 11. Verify that this fixed point is on the common boundary of two basins. Is this point "accessible" (see Chapter 8) from a basin? If so, which one? If a branch of a stable or unstable manifold is an "access arc", how does this manifold behave? See Figure 9-4.

Exercise 9-4. Plot the manifolds of the fixed point p of the Henon map
$$(X,Y) \rightarrow (RHO - X*X + 0.9*Y, X).$$
for the following values of RHO and approximations of p (see Figure 9-5):
(a) RHO = 0.5 and p ≈ (-0.76,-0.76); (b) RHO = 0.6 and p ≈ (0.73,0.73).

Exercise 9-5. Plot the manifolds of the fixed point p of the Henon map
$$(X,Y) \rightarrow (RHO - X*X + 0.9*Y, X).$$
for the following values of RHO and approximations of p:
(a) RHO = 1.2 and p ≈ (1.05, 1.05); (b) RHO = 1.5 and p ≈ (-1.28,-1.28);
(c) RHO = 2.0 and p ≈ (-1.47,-1.47). See Figure 9-9.

Exercise 9-6. Find the fixed point p on the boundary of the basin of the chaotic attractor and plot the position of this fixed point (command KK) for the Ikeda map (I)

$$Z \rightarrow 1 + 0.9*Z*\exp\{i[0.4 - 6/(1 + |Z*Z|)]\}$$

where Z = X + iY. Set the X Scale from -5 to 7, and set the Y Scale from -5 to 7. Verify that p is on the common boundary of two basins, and that it is accessible from both basins. Next plot the manifolds of p. See Figure 9-6 for the "Unstable Left" manifold (blow-up).

Exercise 9-7. There is a period 3 point of the time-2π map of the forced damped Pendulum equation (P)

$$X''(t) + 0.2*X'(t) + \sin X(t) = 2*\cos(t)$$

at appr. (2.22,-0.22). It is accessible from the basin of the fixed point B, where B $\approx$ (-.48,-0.61). Set the Y Scale from -5 to 5. Plot the stable manifolds of all 3 points of the orbit. Hint: see command ABI. See Figure 7-7 for the corresponding basins.

Exercise 9-8. Plot the manifolds of the fixed point (appr. (-1.296, -1.296) of the Henon map (H)

$$(X,Y) \rightarrow (1 - X*X + 0.475*Y,X)$$

There are 3 attractors (including infinity). Is this fixed point accessible from any of these basins? See Figure 7-2 and Figure 7-6 for the corresponding basins.

Exercise 9-9. Plot the manifolds of the fixed point at appr. (-2.02, -2.02) of the Henon map (H)

$$(X,Y) \rightarrow (2.66 - X*X + 0.3*Y, X)$$

Set the X Scale from -3 to 3, and set the Y Scale from -3 to 3. There is a period 3 attractor. How is this fixed point positioned relative to both basins? Is it accessible from a basin? See Figure 9-8.

Exercise 9-10. Plot the stable manifold of the fixed point ($\approx$ (-0.99, -0.33) of the time-2π map of the forced damped Pendulum equation (P)

$$X''(t) + 0.2*X'(t) + \sin X(t) = 2.5*\cos(t)$$

See Figure 9-7.

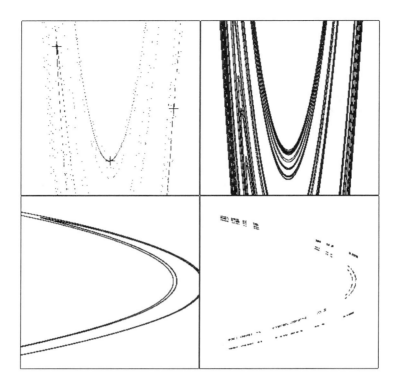

Figure 9-8: Stable and unstable manifold, basin, and straddle trajectory
The figure shows various aspects of the Henon map (H)

$$(X, Y) \rightarrow (2.66 - X*X + 0.3*Y, X)$$

There are two attractors: an attracting period 3 orbit and the point at infinity. The upper left window shows the basin of the period 3 orbit. The positions of the period 3 points are marked by crosses. The upper right window displays a part of the stable manifold of the fixed point p, where p ≈ (-2.02,-2.02). The picture in the lower left window is a part of the unstable manifold of p.

The lower right window shows a Basin boundary Straddle Trajectory, which is on the common boundary of the basin of the attracting period 3 orbit and the basin of infinity.

In all four windows, the variable X is plotted horizontally and the variable Y is plotted vertically, and the X Scale is from -3 (left) to 3 (right) and the Y Scale is from -3 (bottom) to 3 (top).

Topic of discussion. *Is there a period 3 orbit for RHO = 1.4?*

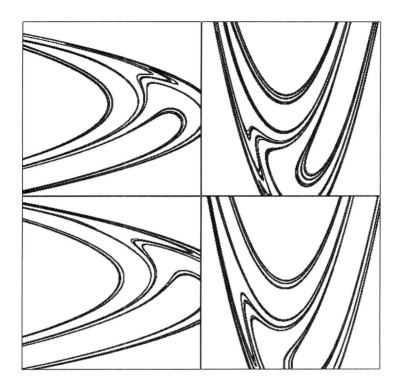

Figure 9-9: Stable and unstable manifolds
*This figure shows the stable and unstable manifolds of the fixed point p
for the Henon map (H)*
$$(X, Y) \rightarrow (RHO - X*X + 0.9*Y, \ X)$$
There is one attractor, namely the point at infinity.

*In the upper two windows, RHO = 1.5. The upper left window shows the
unstable manifold of the fixed point p $\approx$ (-1.28, -1.28), and the upper
right window shows the stable manifold of this point p.*

*In the two lower windows, RHO = 2.0 and p $\approx$ (-1.47,-1.47). The lower
left window shows the unstable manifold of p $\approx$ (-1.47,-1.47), and the lower
right window shows the stable manifold of p.*

*In all four windows, the variable X is plotted horizontally and the
variable Y is plotted vertically, the X Scale is from -2.5 (left) to 2.5
(right) and the Y Scale is from -2.5 (bottom) to 2.5 (top).*

Topic of discussion. *For the inVerse of these maps (command V), what
do the stable and unstable manifolds look like?*

9.5 REFERENCES related to DYNAMICS

The following paper describes the algorithms used in the routines of this chapter and examines in detail the accuracy of the plots. It analyzes the question of whether any pieces of the manifolds are missed and whether any points are plotted in the wrong location. It gives alternative algorithms in which the accuracy can be rigorously guaranteed. That alternative algorithm is much slower than the one in *Dynamics* and it seems to produce exactly the same pictures.

Z. You, E. Kostelich, and J.A. Yorke, Calculating stable and unstable manifolds. International J. of Bifurcation and Chaos 1 (1991), 605-624

E.J. Kostelich, J.A. Yorke, and Z. You, Plotting stable manifolds: error estimates and noninvertible maps, Physica D 93 (1996), 210-222

The following papers introduce the basic notion of "basin cell". A **cell** is a region with a piecewise smooth boundary, whose edges are alternately pieces of stable and unstable manifolds of a periodic orbit. A **basin cell** is a cell that is a trapping region. Of course, the periodic orbit that generates the cell must be well chosen in order that the cell being a basin cell. Whether a given cell is a basin cell, is in some sense easy to verify numerically by using *Dynamics*. However, the error estimates have still to be carried out.

* H.E. Nusse and J.A. Yorke, Wada basin boundaries and basin cells, Physica D 90 (1996), 242-261

H.E. Nusse and J.A. Yorke, The structure of basins of attraction and their trapping regions, Ergodic Th. & Dynamical Systems **17** (1997), 463-481

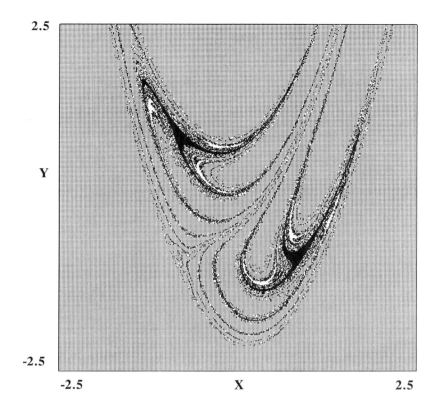

Figure 9-10a: Three basins of attraction (Wada basins)

This figure shows three basins of attraction of the Henon map

$$(X,Y) \;\rightarrow\; (0.71 - X*X + 0.9*Y, \; X)$$

The routine BA has plotted the basin of infinity and the basins of a period two attractor and of a period six attractor of the Henon map. The black area is the basin of a period two attractor. The white area is the basin of attraction the period six attractor. We used command E5 (with PAT = 0) to Erase the plotted region using color 5. The grey area is the basin of infinity, and is obtained by shading the color of the basin of infinity using the Erase commands E1 with PATtern PAT = 6.

Each of the basins is a Wada basin, that is, each point that is on the boundary of one of them, is on the boundary of all three. For references, see Section 7-5.

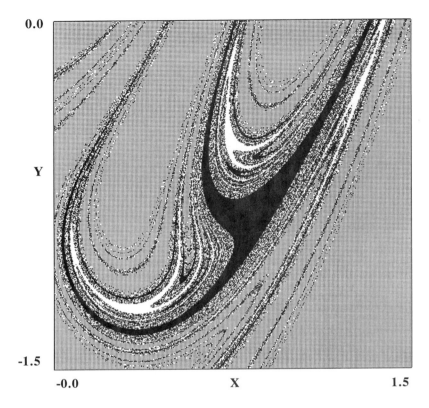

0.0

Y

-1.5

-0.0   X   1.5

Figure 9-10b: Blow-up of the three basins of attraction

This figure shows a blow-up of the three basins of attraction of the the previous figure. Since the routine BA cannot be used, for example, the fixed point attractor in not in the selected region, the routine BAS was utilized to create this picture.

First, we set storage vector y3 to be one of the two points of the period two attractor, (plot a trajectory of a point in the basin of color 3 (the black basin in Figure 9-10a), hit <Esc> and Replace y3 by y0). Next, set storage point y5 to be a point of the period six attractor in the basin of color 5 (the white basin in Figure 9-10a).

Set the Radius of Attraction appropriately (for example, RA = 0.01). To have each of the basins uni-colored, set the basin coloring scheme and enter BN. Then enter command BAS and the three basins are plotted. After the picture has been finished, erase the colors as described in the caption of Figure 7-9a.

Topic of discussion. *What kind of pictures do you expect if more blow-ups are being made?*

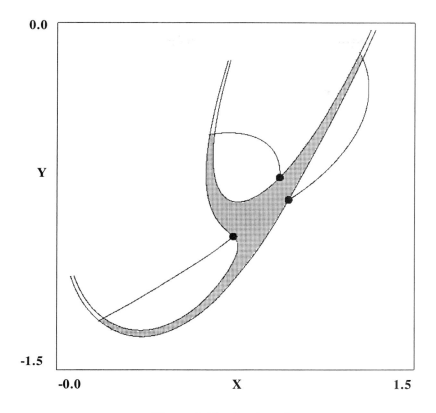

Figure 9-10c: Basin cell

The three arrows indicate the three periodic points of period 6 for the Henon map with parameters as in the two previous figures. These three points constitute a periodic orbit of period 3 for the second iterate of the map. The black region, bounded by parts of the stable and unstable manifolds of these three periodic points, is a cell. The image of this cell when the map is applied once, it outside the cell, but the image of the cell when the map is applied twice, is contained in the original cell. Therefore, the cell is a trapping region for the second iterate of the map. The image of the cell when the map is applied once, is also a basin cell for the second iterate of the map. These basin cells corresponds to the black basin in Figure 9-10b (and 9-10a).

Basin cells characterize the structure of the corresponding basin. The black basin in Figure 9-10a can be viewed as the two basin cell plus the three channels that connect to each of them. These channels are infinitely long and wind in a very complicated pattern without ever crossing each other. For references, see Section 9-5.

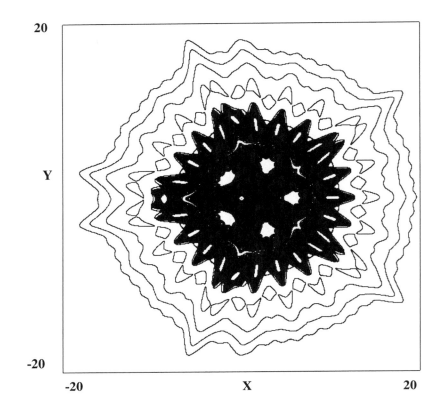

Figure 9-11: Trajectories of the Gumowski/Mira map

The figure shows seven trajectories of the initial condition (x_0, y_0) of the Gumowski/Mira map (GM) for C1 = 1, C2 = 0 and RHO = 0.25

$$(X, Y) \;\longrightarrow\; (Y + F(X),\; -X + F(Xnew)$$

where $F(u) = 0.25*u + 1.5*u^2/(1+u^2)$.

The initial conditions of the trajectories are: (1,1), (6,6), (10,8), (12,10), (12,12) and (13,13).

CHAPTER 10

FINDING PERIODIC ORBITS

10.1 INTRODUCTION and the METHODS

An intensively used method for finding periodic orbits is Newton's method and variants thereof. We describe Newton's method and the Quasi-Newton method later in this section. Newton's method uses the initialization point y1, marked by the small cross, as its initial point. These methods converge to a periodic point with the desired period if the initial condition is reasonably close to it, provided +1 is not an eigenvalue of the Jacobian matrix of the nth iterate of the process, where n is the period (see below).

Interactively finding periodic orbits

To find a periodic orbit using Newton's method, it is often necessary to have a good starting point. The program has an interactive tool, II, for finding a good starting point. For example, to find a periodic point of period 2, first set "PeRiod" PR to 2:

$$\textbf{PR} < Enter > \quad 2 < Enter >$$

Then enter "Initialize and Iterate" command II:

$$\textbf{II} < Enter >$$

Move the mouse arrow to the desired position and click the left button briefly. That point will be marked with a cross and the program will show

---

where the second iterate of the cross is. Alternatively, use the arrow keys. Then as you move the small cross using the arrow keys, the program shows where the second iterate of that point is and denotes it by the big cross. Move the small cross again and the big cross moves to the second iterate of the new position of the small cross. Thus the problem of finding a good starting point becomes one of moving the small cross so that the two crosses come together. Of course they might come together because you are near a fixed point. In this case, since the first iterate is also plotted briefly, you would see all three points moving together, though this intermediate point can be hard to observe. Hit $<Esc>$ to terminate II and type $Q1 <Enter>$ to take 1 Quasi-Newton step. The Q1 routine tells how close the small cross is to being a periodic point of period PR = 2 and how close it is after one Quasi-Newton step. More precisely, the program reports the "error", that is, the distance between the initial point and its PRth iterate. This procedure can be used to locate a variety of fixed points and period two points. The time-2π map of the Pendulum (P) with the default parameter values has at least 8 fixed points and 20 period two orbits; see Figure 2-12.

Randomly finding periodic orbits

The program has a routine RP that Randomly seeks Periodic points of period PR. This routine has a basic step in which it chooses a point in the screen area at random and then applies up to 50 Quasi-Newton steps. If the point is in the basin of infinity and is diverging or if the Quasi-Newton method applied to the initial point converges to a point which is not periodic, then nothing is plotted. The routine plots when it has found a point which is within a distance of 10^{-11} from a periodic point of period PR. The routine RP keeps repeating this basic step, each time choosing a new random seed for starting the process.

The routine RP reports how many different orbits it has found. To obtain periodic orbits of minimum period PR, use the command RPX.

The Quasi-Newton method

The Quasi-Newton method is a variant of the Newton method. This variant guarantees that after the Quasi-Newton step the new point is not farther from being a periodic orbit than was the initial point, that is, the "error" described above does not increase. It starts out by trying to take a Newton step. The Newton method calculates the step in phase space that would be needed to make some quantity zero if the map was linear + constant. Since the map is nonlinear, this step may make matters worse. If that makes the error bigger, it instead takes only half the recommended step and tests the error. It continues cutting until it does yield an improvement. Improvement can be gotten by taking a sufficiently small step in the recommended direction.

Quasi-Newton takes small steps when necessary. Iterating the Quasi-Newton method will result in convergence to a **local** minimum of the error. If that's not 0, too bad. Try a different starting point for the method. Using the Quasi-Newton method is essentially always preferable to using the plain Newton method. If a smaller step is used, it will say so on the computer screen, so the user will know if the program deviates from the Newton step.

To apply the Quasi-Newton method 5 times, just type Q5<Enter>.

Newton's method: how it is done

A fixed point x satisfies

$$F(x,p) = x$$

where F is the map and p is a parameter. For a point u, define

$$\text{error}(u) = \|F(u,p) - u\|$$

where $\| \cdot \|$ denotes the Euclidean norm. Starting from a point u which is not a fixed point, the objective is to estimate a step du so that u + du is fixed for the linear approximation of F, linearizing about u. Write A for the term

$$D_u F(u,p),$$

the matrix of partial derivatives of the map, taken with respect to the phase space variables. Then solve for the linearized equation for du:

$$F(u,p) + A\, du = u + du$$

so

$$du = (A - I)^{-1}(u - F(u,p))$$

where I is the identity matrix. This assumes that the inverse of A-I exists, which will be the case unless +1 is an eigenvalue of A. The "error" at a point u, which is reported when the program is run, is the norm of F(u,p)-u. Newton's method does not guarantee that error(u+du) is smaller than error(u) but if it is not smaller, at least it decreases initially on the line segment between u and u+du, for points near u. The **Quasi-Newton** method takes advantage of this by checking to see if the error at u+du is smaller than at u, and if not, still taking a step in the specified direction du but a shorter distance, cutting the step in half repeatedly until the error actually decreases, that is, until

$$\text{error}(u + 2^{-n} du) < \text{error}(u) \text{ for some } n.$$

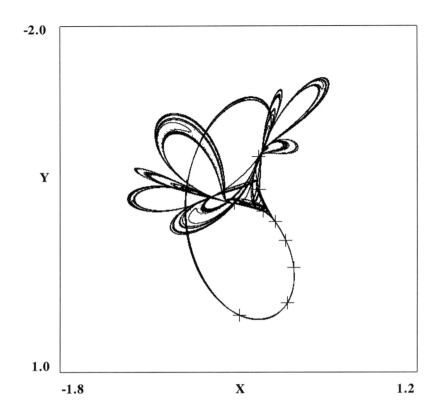

-2.0

Y

1.0

-1.8   X   1.2

Figure 10-1a: A period 10 orbit on an attractor
This figure shows a period 10 orbit of the Tinkerbell map (T)
(X, Y) → (X*X - Y*Y + 0.9*X - 0.6013*Y, 2*X*Y + 2.0*X + 0.5*Y)
The location of this periodic orbit is shown by crosses, (command KKK).

After the period was set to 10 (command PR) and the "take 5 Quasi-Newton steps" command Q5 was invoked, the vector y1 was within a distance of 0.000000000001 of a period 10 point. To mark all the points of the periodic orbit, (1) the permanent cross command KKK was invoked, (2) the program was paused, (3) the trajectory command T was entered, and then < · > was hit ten times to plot the trajectory in a single step mode.

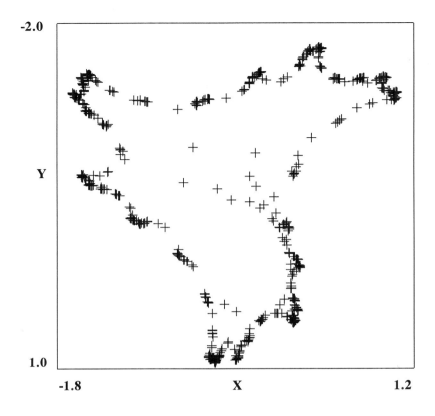

Figure 10-1b: Periodic points with period 10

This figure presents period 10 orbits of the Tinkerbell map (T) for the same parameters as in Figure 10-1a.

Command KKK has been entered to have a permanent cross at each periodic point with period 10. The routine RPX found 64 distinct orbits of exactly period 10. These periodic orbits do not include fixed points and periodic orbits with period 2 or 5.

*Comparison with the chaotic attractor reveals that many of these periodic points are **not** on the attractor.*

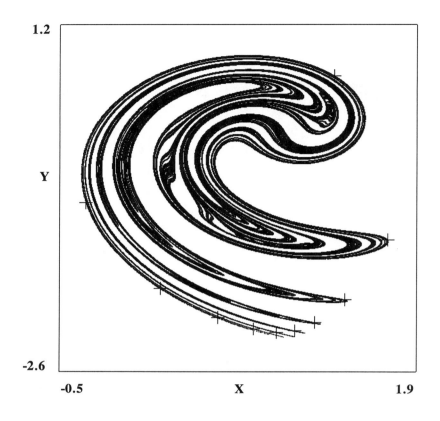

Figure 10-2a: A period 10 orbit on an attractor

This figure shows a period 10 orbit of the Ikeda map (I)

$$Z \rightarrow 1 + 0.9 * Z * exp\{i[0.4 - 6/(1 + |Z*Z|)]\}$$

where $Z = X + iY$. The location of this periodic orbit is shown by crosses.

With PR = 10, command II was invoked. Then the small cross was moved along the edge of the attractor using the arrow keys until a point was found where the big cross was near the small cross.

Thereafter, the Quasi-Newton method Q9 was used several times to locate y1 at one of the periodic points. To mark all the points of the periodic orbit, the command KKK was invoked. Then the "Pause" command P and the "Trajectory" command T were entered, and the trajectory was plotted by hitting $<.>$ ten times.

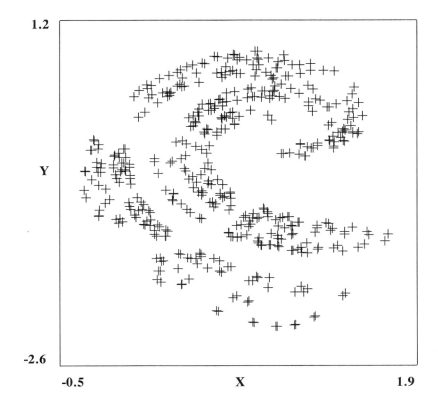

Figure 10-2b: Periodic points with period 10

*This figure presents period 10 orbits of the Ikeda map (I) for the same
parameters as in Figure 10-2a.*

*Command KKK has been entered to have a permanent cross at each
periodic point with period 10. The routine RPX found 46 distinct orbits of
exactly period 10. These periodic orbits do not include fixed points and
periodic orbits with period 2 or 5.*

*Comparison with the chaotic attractor reveals that these periodic
points are on the attractor.*

The procedure when x is a periodic orbit of period PR is the same as the above, when we consider the above function to be the PRth iterate of the map. For estimates of the errors, see also Verner (1978).

When first looking for a periodic point, the following orbit routines FO and FOB take a very conservative approach, taking only small steps in the direction recommended by Newton's method. These orbit routines are discussed in Chapter 11.

Note. Commands in this chapter are for finding periodic orbits, and the eigenvalues of their Jacobian matrix when the phase space is two dimensional.

Summary for finding periodic orbits

Method 1 (crude but educational method)
● Set the period (command PR) of the periodic orbit to be found.
● Initialize and Iterate (command II) and use the mouse to locate periodic points of period PR.

Method 2
● Set the period (command PR) of the periodic orbit to be found.
● Set vector y1 appropriately (by moving small cross).
  Method 1 can be used for finding y1 close to a periodic orbit.
● Apply Quasi-Newton method repeatedly (for example, command Q9).

Method 3 (most effective)
● Set the period (command PR) of the periodic orbit to be found.
● Search for periodic orbits of period PR (commands RP, RPK, RPX).
  If Method 3 is used in the paused mode by tapping the period key < · > for each new initial condition used, then the coordinates are written on the screen (and stored in the vector y1).

10.2 PERIODIC ORBIT MENU POM

The Newton method looks for a periodic orbit starting from the current value of storage vector y1. The new value is put back in y1. When any cursor key is hit, the cross appears at y1.

Enter command POM, and the Periodic Orbit Menu appears on the screen in approximately the following format.

```
┌──────────────────────────────────────────────────────────┐
│  POM  – –       PERIODIC ORBIT MENU                        │
│                                                            │
│  PR    –  PeRiod of the orbit sought:  1                   │
│                                                            │
│  Q1    –  take one Quasi-Newton step                       │
│  QK    –  take K Quasi-Newton steps, where K  =  1, ..., 999│
│  RP    –  Randomly seek Periodic orbits                    │
│  RPK   –  RP plotted with crosses                          │
│  NMS   –  Newton Maximum Step size if set:   NOT SET       │
│                                                            │
│           MENU                                             │
│  FOM   –  Follow Orbit Menu                                │
└──────────────────────────────────────────────────────────┘
```

POM

The command POM calls the Periodic Orbit Menu to the screen. The Quasi-Newton method can be used for finding periodic orbits of period PR. For finding a period PR orbit, the Quasi-Newton method is set up to decrease the "error", that is, the difference between the initial condition y1 and the PRth iterate of that point. The program puts the better estimate of a periodic orbit in both y and y1. Recall y1 can be set using the arrow keys. Using any arrow key makes the cross appear at y1. To use the routine, put y1 (marked by the small cross on the screen) at the point where the Newton method should begin.

Periodic orbit commands

PR

The "PeRiod" PR is used by the Quasi-Newton commands, the Follow Orbit commands, as well as the "Initialize and Iterate" command II, the "circle and its Iterates" command OI, and the stable and unstable manifold commands. To find an orbit with period 3, then set PR to 3. With the default value of 1 (that is, PR = 1) the Newton method seeks period 1 points, that is, fixed points. The Quasi-Newton Method (when applied successively several times) may then find an orbit of period 3 or of period 1 (which counts as an orbit of period 3) or it may fail to converge. The period really means the number of plotted dots; thus if the user is iterating a differential equation and is plotting every Runge-Kutta step, and the forcing period is say equivalent to the time it takes for 30 Runge-Kutta steps (i.e., "steps per cycle" SPC = 30), then the period should be 30 or 60 or 90 to find orbits whose period is 1 or 2 or 3 cycles, respectively; if you are plotting only once every cycle as is often the case with the pendulum (P), then period 3 would mean 3 cycles.

Warning: the Quasi-Newton Method may fail (the program may crash) for large choices of PR. The method requires computation of a matrix of partial derivatives and this matrix often has entries on the order of 10^{PR}. Hence, save your data and picture (use DD or TD commands) frequently, so you can recover them if the program crashes. On the other hand, the program may work fine when PR is 10 or more.

Q1

The command Q1 is to take one Quasi-Newton step.

Eigenvalues and eigenvectors

When you apply Q1, the quasi-Newton method computes for two-dimensional processes the eigenvectors and eigenvalues of the Jacobian matrix of the PRth iterate of the map evaluated at the starting point y1, and then a Newton step is taken. The command Q0 reports these eigenvalues and eigenvectors.

QK

The "take K Quasi-Newton steps" command QK is for applying the Quasi-Newton method K times, looking for a point of period PR. The Quasi-Newton method is applied consecutively, where K can be between 1 and 999. For example, entering command Q5 results in that the Quasi-Newton method will be applied 5 times.

RP, RPK

The command RP is for seeking Randomly Periodic points of period PR. The routine RP uses the Quasi-Newton method from randomly selected initial points. See also command RPK.

The command RPK is for seeking Randomly Periodic points of period PR. The routine RPK differs from RP in that a cross is drawn at each period PR point, while RP merely plots a point.

NMS

The optional setting for "Newton Max Step" NMS makes Newton Method restrict itself to small steps. If NMS is set, it specifies the largest step allowed. If Newton's method recommends a larger step, it is reduced in size but the direction of the step is unchanged.

When NMS is used with BAN, or RP, or RPK, it is useful to increase MC, the Maximum number of steps used, since if small steps are taken, many steps may be required to reach the periodic orbit.

Enter command RP or a variant thereof, and the Actions, Hints and Options menu for RP commands menu appears and includes the following:

ACTIONS, HINTS and OPTIONS for RP commands
RPX   – like RP except only plots orbits of eXact period
RPA   – like RPX except only plots Attractors and repellers
RPC   – RP etc. will Color orbits with color PR = 1; now   OFF
QUASI – use quasi-Newton if ON, else Newton; now   ON

RPX

The command RPX is for seeking Randomly Periodic points of eXactly period PR, that is, the minimum period of that orbit is PR. If for example the PeRiod PR is 10, routine RP will find periodic orbits of periods 1, 2, 5, and 10 because they all are also period 10 orbits. RPX will only find orbits whose minimum period is 10.

RPA

The command RPA is for seeking Randomly Periodic points of eXactly period PR which are either Attractors or repellers. That is, command RPA is like RPX only it plots Attractors and repellers. See command RPX.

RPC

The command RPC is a toggle. If it is ON, it plots the periodic orbits with color PR. For example, if PR = 5, the orbits are colored with color 5.

QUASI

See Chapter 7 for its description.

10.3 EXAMPLES

In Example 2-19b we applied the Quasi-Newton method 5 times (command Q5) to find a fixed point on the chaotic attractor of the Henon map. The following examples illustrate the use of applying the Quasi-Newton method.

Example 10-1: Finding a period 7 orbit on the Henon attractor
To illustrate finding a periodic orbit, we use the Henon map (H)

$$(X,Y) \rightarrow (1.4 - X*X + 0.3*Y, X)$$

where the variable X is plotted horizontally, and the variable Y is plotted vertically. The X Scale is from -2.5 to 2.5, and the Y-Scale is from -2.5 to 2.5, which are the default values.

The attractor for this map has been discussed in Example 2-4. The objective of this example is to find a periodic orbit with period 7 on the chaotic attractor.

1. Set the parameter C1 to be 0.3, and set RHO to be 1.4:

C1 <*Enter*>   **0.3** <*Enter*>

RHO <*Enter*>   **1.4** <*Enter*>

2. Since the trajectory of the default value (0,2) of y1 is diverging (see Example 2-4), Set Vector y1 to be (0,1):

SV <*Enter*>

SV1 <*Enter*>

SV11 <*Enter*>   **1** <*Enter*>

Initialize and plot the Trajectory:

I <*Enter*>   **T** <*Enter*> <*Enter*>

resulting in the chaotic attractor. Hit <Esc>.

3. Set the period to be 7:

PR <*Enter*>   **7** <*Enter*>

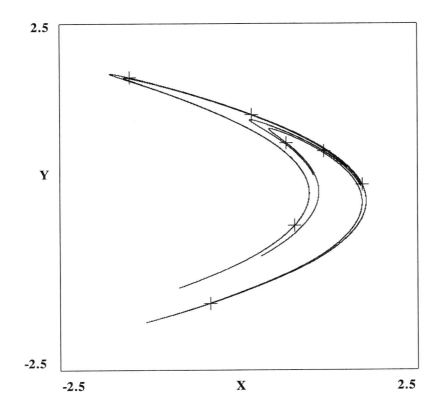

Figure 10-3: A periodic orbit on the Henon attractor

This figure shows a period 7 orbit on the attractor of the Henon map (H)

$$(X, Y) \;\rightarrow\; (1.4 - X*X + 0.3*Y, \; X)$$

The location of this periodic orbit is shown by crosses.

After the period PR was set to 7 and the "take 8 Quasi-Newton steps" command Q8 was invoked, the vector y1 was within a distance of 10^{-14} of a period 7 point. To mark all the points of the periodic orbit, (1) the permanent cross command KKK was invoked, (2) the program was paused, (3) the trajectory was first Initialized (I) and the Trajectory routine T called, and then $< \cdot >$ was hit seven times to plot the trajectory in the single step mode.

4. Take 8 Quasi-Newton steps:

$$Q8 <Enter>$$

which yields a vector which is approximately (0.17982,1.19162). Of course, if you set the vector y1 differently, another point may come out.

5. To show that the periodic orbit found is indeed a period 7 orbit on the attractor, pause the program:

$$P<Enter>$$

Now plot the Trajectory:

$$T<Enter> <Enter>$$

6. To make the orbit visible plot permanent crosses at each position by single stepping through the orbit:

$$KKK<Enter>$$

and then hit the "period" key

$$< \cdot >$$

for one more dot (see Chapter 2). Now hit

$$< \cdot >$$

six times. The final point should be indistinguishable from the starting point.

The resulting picture consisting of the chaotic attractor and the 7 crosses representing the periodic orbit sought, is given in Figure 10-3.

Example 10-2: Finding a period 10 orbit on a chaotic attractor

To illustrate finding a periodic orbit, we now use the Ikeda map (I)

$$Z \rightarrow 1 + 0.9*Z*\exp\{i[0.4 - 6/(1 + |Z*Z|)]\}$$

where $Z = X + iY$, the variable X is plotted horizontally, and the variable Y is plotted vertically.

The unstable manifold of the fixed point on the chaotic attractor has been discussed in Example 9-1. The purpose of this example is to find a periodic orbit with period 10 on the chaotic attractor. The following steps may be carried out to find such a periodic orbit.

1. Note that the parameters do not need to be set since they are the default values. Set the period to be 10:

PR *< Enter >* **10** *< Enter >*

2. Plot the Trajectory:

T *< Enter >* *< Enter >*

resulting in the chaotic attractor.

3. Pause the program:

P *< Enter >*

4. Move the small cross to the left edge of the attractor using the arrow keys.

5. Take 9 Quasi-Newton steps:

Q9 *< Enter >*

6. Repeat step 5 as often as necessary to obtain a point which is within a desired distance, say 10^{-14}, of a period 10 point. It may be necessary to use the arrow keys to move to a new initial point using II.

7. To show that the periodic orbit found is indeed a period 10 orbit (and not period 1, 2 or 5) on the attractor, plot the trajectory by hitting

< space bar >

since the program is in Pause mode (step 3). Otherwise, enter command T.

8. To make the orbit visible, hit

$$<Esc>$$

and plot permanent crosses on each position by single stepping through the orbit:

$$\textbf{KKK} <Enter>$$

Pause the program, initialize and plot the trajectory:

$$\textbf{P}<Enter>$$

$$\textbf{I}<Enter> \quad \textbf{T}<Enter> <Enter>$$

and then hit

$$<\cdot>$$

ten times. The final point should be indistinguishable from the starting point.

The resulting picture consisting of the chaotic attractor and the 10 crosses representing the periodic orbit sought, is given in Figure 10-2.

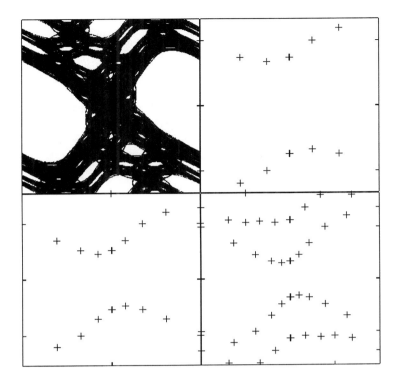

Figure 10-4: Periodic orbits

This figure shows some periodic orbits of the Rotor map (R)

$$(X, Y) \rightarrow ((X + Y) \bmod(2\pi), Y + 1.08*sin(X+Y))$$

where the variable X is plotted horizontally, and the variable Y is plotted vertically. The X Scale is from -3.14159 (left) to 3.14159 (right) and the Y Scale is from -3.14159 (bottom) to 3.14159 (top).

The upper left window shows a chaotic trajectory, which is similar to Figure 5-6a. In the upper right window, the crosses indicate period 3 points. The routine RPX found 5 distinct orbits of exactly period 3.

In the lower left window and lower right window, the crosses indicate the positions of period 4 and period 5 points, respectively. The routine RPX found 5 distinct orbits of exactly period 4 and 8 distinct orbits of exactly period 5.

This map has some closed curves, all of whose points have period PR (e.g. PR = 7). The Quasi-Newton method does not find these points because the Jacobian matrix of the PRth iterate of the map has +1 as an eigenvalue. The routine RP finds (apparently) all the saddles of period PR. The routine would also find attractors and repellors if there were any. One elliptic fixed point is also found on the edge of the picture.

Example 10-3: Searching for all period 10 orbits

To illustrate searching for periodic points with period 10, we now use the Tinkerbell map (T)

$$(X,Y) \rightarrow (X*X - Y*Y + 0.9*X - 0.6013*Y, \; 2*X*Y + 2.0*X + 0.5*Y)$$

where the variable X is plotted horizontally, and the variable Y is plotted vertically. The X Scale is from -1.8 to 1.2 and the Y Scale is from 1 to -2. The parameters are the default values, but not the X Scale and Y Scale. The purpose of this example is to search for periodic points of period 10.

1. Set the X Scale and the Y Scale:

$$\textbf{XS} < Enter > \quad \textbf{-1.8 1.2} < Enter >$$

$$\textbf{YS} < Enter > \quad \textbf{1 -2} < Enter >$$

2. Set the PeRiod to be 10:

$$\textbf{PR} < Enter > \quad \textbf{10} < Enter >$$

3. Search at Random for the Periodic points with period 10:

$$\textbf{RPK} < Enter > \; < Enter >$$

A resulting picture is given in Figure 10-1b.

Note. If the period PR is 10, then RP finds orbits of period 10 including orbits of period 1, 2, and 5. See example 2-11. Command RPX is like RP except that it only finds periodic orbits whose **minimum** period is eXactly 10.

The routine RP works by randomly choosing a seed, that is, an initial point, and then applies the Quasi-Newton method described in this chapter (Section 10.1). It rapidly repeats this process over and over. In some cases, such as when the period PR $\geq$ 10, some periodic orbits will be extremely hard to detect. Do not assume it finds all the periodic orbits.

In order to have a permanent cross at each dot, invoke the command KKK before you enter the command RPX.

Note. If RPK is used and you hit $< \cdot >$, then the program pauses. Each time you hit $< \cdot >$, the routine tries one new initial point. During this mode of operation, each time a periodic point is found, y1 is moved to that point.

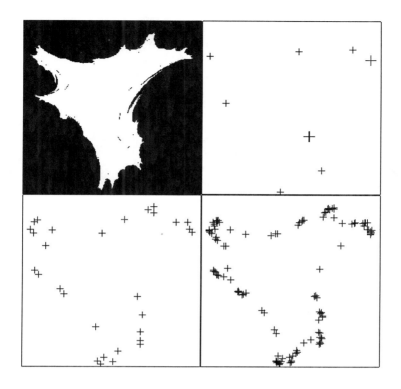

Figure 10-5: Periodic orbits

This figure shows periodic orbits of the Tinkerbell map (T)

$(X, Y) \rightarrow (X*X - Y*Y + 0.9*X - 0.6013*Y, 2*X*Y + 2.0*X + 0.5*Y)$

where the variable X is plotted horizontally, and the variable Y is plotted vertically. The X Scale is from -1.8 (left) to 1.2 (right) and the Y Scale is from 1 (bottom) to -2 (top). The location of the periodic points is shown by crosses.

In the upper left window, the black area is the basin of infinity. This picture is similar to Figure 7-4. In the upper right window, the larger crosses indicate fixed points, and the smaller ones indicate period 3 points. The routine RPX found 2 fixed points and 2 distinct orbits of period 3.

In the lower left window and lower right window, the crosses indicate the positions of period 5 and period 7 points, respectively. The routine RPX found 6 distinct orbits of exactly period 5 and 18 distinct orbits of exactly period 7.

10.4 EXERCISES

Exercise 10-1. Find a period 10 orbit of the Tinkerbell map (T)
$(X,Y) \rightarrow (X*X - Y*Y + 0.9*X - 0.6013*Y, 2*X*Y + 2.0*X + 0.5*Y)$
Search for all period 10 orbits. How many periodic orbits of period 10 exist? See Figure 10-1.

Exercise 10-2. Find a period k orbit for the Henon map (H)
$(X,Y) \rightarrow (1.4 - X*X + 0.3*Y, X)$.
where k = 7, 8, 9. Is there a period 3 orbit? See Figure 10-3 for a period 7 orbit.

Exercise 10-3. Find a period k orbit for the Henon map (H)
$(X,Y) \rightarrow (2.12 - X*X - 0.3*Y, X)$
where k = 3, 4, 5. How many distinct period 5 orbits exist?

Exercise 10-4. Find a period 10 orbit of the Ikeda map (I)
$Z \rightarrow 1 + 0.9*Z*exp\{i[0.4 - 6/(1 + |Z*Z|)]\}$
where $Z = X + iY$. Search for all period 10 orbits. How many periodic orbits of period 10 exist? See Figure 10-2.

Exercise 10-5. Find at least eight fixed points of the time-2π map of the forced damped Pendulum equation (P)
$X''(t) + 0.2*X'(t) + \sin X(t) = 2.5*\cos(t)$
How many periodic orbits of period 2 exist? See Figure 2-12.

Exercise 10-6. Find period 5 and period 7 orbits for the Tinkerbell map (T)
$(X,Y) \rightarrow (X*X - Y*Y + 0.9*X - 0.6013*Y, 2*X*Y + 2.0*X + 0.5*Y)$
How many periodic orbits of period 7 exist? See Figure 10-5.

Exercise 10-7. Find a period 3, a period 4, and a period 5 orbit for the Rotor map (R)
$(X,Y) \rightarrow ((X+Y) \bmod(2\pi) - \pi, Y + 1.08*\sin(X+Y) \bmod(2\pi))$
How many periodic orbits of period 5 at least exist? See Figure 10-4.
* Find a closed curve of period 7 points using ABI or II or OI with MI = 7.

CHAPTER 11

FOLLOWING PERIODIC ORBITS

11.1 INTRODUCTION and the METHOD

As a parameter of a map is varied, periodic orbits are often seen to appear and shift in position and perhaps period double or blink out of existence. The orbit-following capability is for tracing their behavior as a parameter is varied. The orbit following routine runs for two dimensional maps, including for example the time-2π maps of periodically forced differential equations like the forced damped pendulum (P) and the Duffing equation (D). The pendulum for example is plotted as a time-2π map by taking 30 Runge-Kutta steps before plotting.

The follow orbit commands repeatedly vary the parameter PRM (which by default is RHO), and the position y in phase space so that it is a periodic orbit with period PR. The resulting path may not be monotonic in the parameter. White dots indicate test points investigated by the routine though they are not periodic points. They are often seen as the program initially seeks out a periodic point. If the initial position of y is not a periodic orbit, the program will automatically try to locate one; Chapter 10 of this handbook tells how to locate a periodic point with period PR.

For example, start *Dynamics*, select the Ikeda map, set the period PR to 5 and follow the periodic orbit:

I < *Enter* >

PR < *Enter* >   5 < *Enter* >

FO < *Enter* >

Reduce the printing with

T2 < *Enter* >

C < *Enter* >

After the family of period 5 orbits is plotted in the phase space while RHO is varied, you can change X Coordinate to be RHO, set the X Scale to run from 0.9 to 1.9:

XCO < *Enter* >   **RHO** < *Enter* >

XS < *Enter* >   **0.9 1.9** < *Enter* >

C < *Enter* >

to get another view. In this case, the coordinate y[1] of y is being plotted vertically against RHO horizontally. See also Figure 2-13.

In both cases, the periodic orbit changes as RHO varies. The plotting is slow due to the large amount of information being plotted out, including the Jacobian matrix. Enter "Text Level 2" command T2 (equivalent to TL followed by T2) and then enter "Refresh" command R to refresh the screen, and the plotting will proceed faster. After Text Level 1 (T1, followed by the "Refresh" command R) plotting will proceed even faster with less text printed.

When the curve leaves the screen, hit < Esc > and return to the Main Menu. Enter command I (Initialize), thereby reinitializing y and also resetting the parameter RHO to the RHO value of the small cross. Enter "Follow Orbit Backwards" command FOB and the screen will refresh, and the path of orbits will start decreasing, that is, extending the existing curve to the left.

In some cases, to obtain reliable results it may be necessary to decrease the "Follow periodic Orbit Step size" FOS; see below for details. The default value of FOS is 0.005.

Colors and orbit index

The orbit index is a number associated with a periodic orbit, and this number is useful in understanding the patterns of bifurcations the orbit undergoes. For typical orbits, the number is -1, 0, or +1, and the color the program uses to plot reflects the orbit index. The orbit index is a conserved quantity in the sense that if one adds the orbit indexes of the periodic orbits of all periods that exist just before a bifurcation, then that equals the corresponding sum just after that bifurcation. Suppose a periodic orbit has (minimum) period PR. Let A be the Jacobian matrix of the

PRth iterate of the map, at one of the periodic points. The orbit index depends on the eigenvalues of A and we assume that +1 and -1 are not eigenvalues.

Let m be the number of real eigenvalues < -1, and let n be the number of real eigenvalues > +1. The **orbit index** is defined as follows.
The orbit index is 0 if m is odd; (such orbits are plotted in **red** by the orbit follower and are called **flip saddles**); OTHERWISE,
The orbit index is -1 if n is odd; (such orbits are plotted in **green** by the orbit follower and are called **regular saddles**).
The orbit index is +1 in the remaining cases; (such orbits are plotted in **blue** by the orbit follower and include attractors and centers, and repellors).

If PR is an even multiple of the period of the orbit and it is not an attractor, then it is plotted in **maroon** and its orbit index cannot be determined without reducing PR.
If the Jacobian determinant has absolute value greater than 1, then the point is printed in brighter color, that is, the color number is increased by 8. The different blues, reds and greens can be seen using "Color Table" command CT.

Saddle-node and period-doubling bifurcation and the orbit index
In a saddle-node bifurcation, two orbits exist for parameter values on one side of the bifurcation parameter value, and these are an attracting fixed point (orbit index +1) and a regular saddle (orbit index -1) for a total index of 0. On the other side of the critical parameter value, there are no periodic orbits, yielding a total of 0. Notice that we ignore periodic orbits that are not involved in this local bifurcation.
In a saddle-node bifurcation, there are a number of possible configurations. Assume for example, that before the critical parameter value there is an attracting periodic orbit with period PR (orbit index +1) which becomes a flip saddle of period PR after the bifurcation (orbit index 0). In order for the sum to be constant, there must be another periodic orbit, and that is the period 2*PR orbit. If it coexists with a period PR orbit before the bifurcation, it must have orbit index -1 and be a regular saddle, while if it instead exists after the bifurcation, it must have index +1 and be an attractor. Notice that the orbit index is applied to the orbit as a whole, not to individual periodic points, so that an attracting orbit of period PR and one of period 2*PR both have index +1 (and not PR or 2*PR).

Numerical method for following periodic orbits

When first looking for a periodic point, the "Follow periodic Orbit" routines FO and FOB take a very conservative approach, taking only small steps in the direction recommended by Newton's method.

If $(u(s), p(s))$ is a curve of fixed points of $F(u,p)$, where the parameter s can be interpreted as the arc length along the curve, then $F(u(s),p(s)) - u(s) \equiv 0$ and differentiating yields

$$A \frac{du}{ds} + b \frac{dp}{ds} - \frac{du}{ds} = 0$$

where $A = A(u,p)$ is the matrix of partial derivatives of F with respect to u and the vector $b = b(u,p)$ is the partial derivative of F with respect to p. The vector

$$(\frac{du}{ds}, \frac{dp}{ds})$$

is tangent to the curve of fixed points. This fact suggest a way to follow the curve of fixed points. Denote a unit vector tangent to the curve of fixed points (in (u,p) space) by

$$(v_u, v_p)$$

It satisfies

$$A(u,p)v_u + b(u,p)v_p - v_u = 0 \qquad (1)$$

When the matrix $(A - I)$ is nonsingular (I denotes the identity matrix),

$$v_u = (I - A)^{-1}bv_p$$

so $((I - A)^{-1}b, 1)$ is a vector tangent to the curve, which can be rescaled to become a unit vector. At first sight, it appears the curve cannot be followed past a point where $A - I$ is singular, but this appearance is sometimes an illusion caused by our choice of coordinates. If

$$S(0) = (u_0, p_0)$$

is a point along the curve $(u(s),p(s))$ at which $p(s)$ has a local minimum or maximum, then v_p is 0 and the unit vector v_u satisfies

$$(A - I)v_u = 0$$

so that $(v_u, 0)$ is tangent to the curve. If there is a line of solutions, then there is no problem in choosing u.

The simplest way to follow a curve of solutions of $F(u,p) = u$, once a particular solution $S(0) = (u_0, p_0)$ has been found, requires defining a unit vector $v = v(u,p)$ with coordinates $(v_u(u,p), v_p(u,p))$ in (u,p)-space in some small neighborhood of (u_0, p_0) by

$$A(u,p)v_u + b(u,p)v_p = v_u \qquad (1)$$

$$< (u_0, p_0), (v_u, v_p) > \; > 0 \qquad (2)$$

where $< \cdot, \cdot >$ denotes inner product. The latter condition inequality (2) allows to distinguish between the two unit vector solutions of the equation (1). Typically Eqs. (1,2) will have a unique solution v for (u,p) near $S(0)$, thereby providing us with a vector field. Furthermore v is tangent to the curve of fixed points, so the curve can be found by finding the solution of differential equation

$$\frac{du}{ds} = v_u \; , \quad \frac{dp}{ds} = v_p \qquad (3)$$

through the point $S(0)$. The independent variable s may be interpreted as arc length along the path since v is a unit vector. *Dynamics* uses a fourth order Runge-Kutta method for solving (3), or rather for taking one Runge-Kutta step along the curve. The step size FOS has a default value of 0.02. Starting at $S(0)$ and taking one Runge-Kutta step yields a point (u_1^*, p_1).

Since these curves of fixed points can turn rather suddenly, we add a constraint that
the angle between $v(u_0, p_0)$ and $v(u_1^*, p_1)$
must be less than 18 degrees or more precisely

$$< v(u_0, p_0), v(u_1^*, p_1) > \; > 0.95 \qquad (4)$$

If it is not, cut the step size, return to $S(0)$ and repeat the process with smaller step size until (4) is satisfied.

Unlike the ordinary differential equation solvers, we can evaluate how good our point (u_1^*, p_1) is by comparing $F(u_1^*, p_1)$ with u_1^*; if the component v_p is not too small, that is, $|v_p| > 0.03$ at (u_1^*, p_1), the routine uses the Quasi-Newton method to improve u_1^*, yielding a new value u_1 that is much closer to the curve. Hence, the total error can be kept small.

The above description tells how to take a step along a curve of fixed points. This step is repeated, producing a plot of the curve. The step is only taken if

$$\| F(u_0, p_0) - u_0 \|$$

is small, that is, 0.01 times the length of the diagonal of the screen.

At saddle-node bifurcations, there typically is a unique (line) solution of

$$A \ du + b \ dp = du \qquad (5)$$

but (A-I) is not invertible, so our method of solution fails. The probability of landing exactly on such a point is low. When it does occur, the routine simply solves (5) at a nearby point and uses that solution. At period-halving bifurcation points (5) has a plane of solutions, and the curve follower may switch to the low period branch or it may continue on the period PR branch.

Summary for following periodic orbits
- Search for the periodic point using Quasi-Newton method (optional).
- Set the period PR of the periodic orbit to be followed.
- Set X COordinate (XCO), and set the X Scale (XS).
- Set the "Follow periodic Orbit Step size" FOS.
- Follow periodic Orbit (command FO) and/or
  Follow periodic Orbit Backwards (command FOB).

11.2 FOLLOW ORBIT MENU FOM

Enter "Follow Orbit Menu" command FOM,

$$\text{FOM} < Enter >$$

and the following menu appears on the screen.

```
FOM  – –     FOLLOW ORBIT MENU

FO    –  Follow Orbit as parameter is varied
FOB   –  Follow Orbit Backwards
FOS   –  Follow Orbit Step size:  0.005
FOV   –  reVerse direction for FO

         Related commands:
PR    –  PeRiod of the orbit sought:  1
PRM   –  PaRaMeter to be changed:  rho
         Allowed settings for PRM:  C1  RHO
QK    –  take K Quasi-Newton steps, where K = 1, ..., 999
```

Commands for following periodic orbits

FO

The "Follow periodic Orbit" command FO makes the program repeatedly vary the parameter (which by default is rho) and change y so that it is a periodic orbit with period PR. The resulting path may not be monotonic in the parameter. Initially the parameter increases in small steps when FO is called, but this direction of change will sometimes automatically reverse. The small cross is not changed, so command I can be used to reinitialize the starting point y with the starting parameter values. Command I also reinitializes the parameter which will have changed when FO is used.

Following a family of orbits may lead to a saddle-node bifurcation where a saddle meets a stable orbit. In this case, when y reaches the saddle node parameter value, the program continues onto the other branch, reversing the direction of parameter change. A similar event occurs when the family encounters another family with half the period. That is, there is a period halving bifurcation. The family may switch to the lower period family, though PR is not changed, or it may stay on the higher period family.

FOB

The "Follow periodic Orbit Backwards" command FOB makes the program repeatedly vary the parameter (which by default is rho) and change y so that it is a periodic orbit with period PR. This command is like FO except that initially the parameter **decreases** in small steps when FOB is called.

FOS

The "Follow periodic Orbit Step size" command FOS is for setting the maximum step size FOS, which is by default 0.02. As the program follows a curve of periodic points in (u,p)-space where u is in phase space and p is the parameter being varied, it tries to take steps of constant arc length in (u,p) space. Occasionally, the step size is decreased, especially when the curve is changing directions rapidly. The user may find it necessary to decrease the step size FOS if the program is having trouble following the path of orbits.

FOV

As the orbit path is followed, command FOV instructs the routine to reverse direction. If the family of curves has left the screen, this command tells it to backtrack, thereby returning to the screen.

11.3 EXAMPLES

In Chapter 2, we have presented one simple example of following a periodic orbit. In Example 2-13 a family of period 5 orbits of the Ikeda map is followed while the parameter RHO is varied. In the examples below we consider this map anew. In the first example, a family of period 3 periodic orbits is followed. The second example deals with following a family of period n periodic orbits, where n = 1, 2, 4.

In the examples below, we assume that you are familiar with the following features of the program.
- Setting parameters to the desired values (command RHO, C1, etc.).
- Setting the X Scale and Y Scale (command XS, YS).
- Setting the X COordinate (command XCO).
- Setting the period (command PR).
- Finding a periodic orbit with period PR (command Qk).

Example 11-1: Following periodic orbits of period three

To illustrate the following of periodic orbits, we use the Ikeda map

$$Z \rightarrow RHO + 0.9*Z*\exp\{i[0.4 - 6/(1 + |Z*Z|)]\}$$

where $Z = X + iY$. The variable RHO is plotted horizontally, and variable Y is plotted vertically. The X Scale is from 0 to 2, and the Y Scale is from -3 to 1.5. The objective of this example is to follow a family of periodic orbits with period PR, where PR = 3.

The values for the parameters C1, C2, and C3 are the default values. The steps needed to follow a family of period 3 periodic orbits are the following.

1. Set the period PR to be 3:

$$PR < Enter > \quad 3 < Enter >$$

2. Using the default value of RHO, search for a period 3 periodic orbit; use command Q9 or II. One period 3 point is approximately (0.778, 0.767), and its periodic orbit will be followed.

3. Set the X COordinate to be RHO:

$$XCO < Enter > \quad RHO < Enter >$$

4. Set the X Scale to run from 0 to 2; the Y Scale need not be set, since it is already set by its default values:

$$\mathbf{XS} < Enter > \quad \mathbf{0} \ \mathbf{2} < Enter >$$

5. Follow the period 3 periodic Orbit:

$$\mathbf{FO} < Enter >$$

The resulting picture is given in Figure 11-1.

Note. As soon as the trajectory leaves the screen, hit $<Esc>$ and follow the periodic orbit in reverse direction be entering $\mathbf{FOV} < Enter >$

Note. Since the curve is closed, invoking the "Follow Orbit Backwards" command FOB instead of the command FO would yield a similar picture.

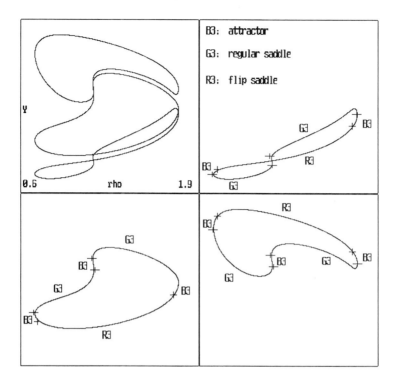

Figure 11-1: Following a family of periodic orbits

The curves in this figure are the result of following a family of period 3 periodic orbits of the Ikeda map (I)

$$Z \rightarrow RHO + 0.9*Z*exp\{i[0.4 - 6/(1 + |Z*Z|)]\}$$

where $Z = X + iY$. First, the period PR was set to be 3, and a period 3 periodic point was found.

On a color screen, different segments of the path are colored Red (flip saddle), Blue (attractors), and Green (regular saddles). These segments are labeled by R3, B3, or G3, since the period is 3.

The following transitions for the period 3 periodic orbit can be observed: flip saddle (R3) $\rightarrow$ attractor (B3) $\rightarrow$ regular saddle (G3) $\rightarrow$ attractor (B3) $\rightarrow$ regular saddle (G3) $\rightarrow$ attractor (B3) $\rightarrow$ flip saddle (R3).

After the picture was completed, the text and labeling of the picture was typed in using the feature TYPE, while the picture was on the screen.

Example 11-2: Following periodic orbits of period two

To illustrate following periodic orbits, we use the Ikeda map (I)

$$Z \rightarrow RHO + 0.9*Z*exp\{i[0.4 - 6/(1 + |Z*Z|)]\}$$

where $Z = X + iY$. The variable RHO is plotted horizontally, and the variable Y is plotted vertically. The X Scale is from 0 to 2, and the Y Scale is from -3 to 1.5. The purpose of this example is to follow periodic orbits with period PR, where $PR = 1, 2$.

The values for the parameters C1, C2, and C3 are the default values. First, follow a family of period 2 periodic orbits.

1. Set the period PR to be 2:

$$PR < Enter > \quad 2 < Enter >$$

2. Using the default value of RHO, search for a period 2 periodic orbit; use command Q9 or II or RP.

3. Set the X COordinate to be RHO:

$$XCO < Enter > \quad RHO < Enter >$$

4. Set the X Scale to run from 0 to 2; the Y Scale need not to be set, since it is already set by its default values:

$$XS < Enter > \quad 0 \ 2 < Enter >$$

5. Follow the period 2 periodic Orbit:

$$FO < Enter >$$

6. Observe that there are two period doubling (or halving) bifurcations. Turn the toggle K on, and when the two crosses are close to such a bifurcation point, for example the rightmost point, pause the program by hitting $< . >$. Set the period to be 1, and follow the family of fixed points:

$$PR < Enter > \quad 1 < Enter >$$

$$FO < Enter >$$

The resulting picture is given in Figure 11-2.

Note. As soon as the trajectory leaves the screen, hit $< Esc >$ and follow the periodic orbit in reverse direction by entering **FOV** $< Enter >$

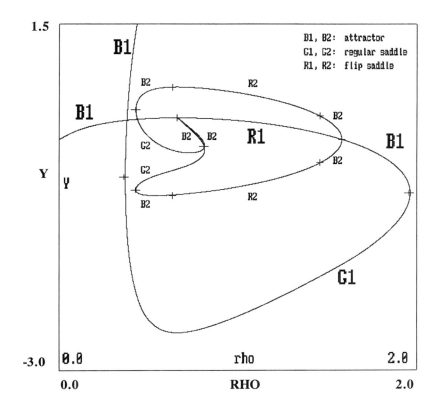

Figure 11-2: Following a family of periodic orbits

The curves in this figure are the result of following a family of period 2 periodic orbits of the Ikeda map (I)

$$Z \rightarrow RHO + 0.9*Z*exp\{i[0.4 - 6/(1 + |Z*Z|)]\}$$

where $Z = X + iY$. The parameter RHO is plotted horizontally, and the variable Y is plotted vertically.

First, the period PR was set to be 2, and a period 2 periodic point was found. The X COordinate was set to be RHO, and the "Follow Orbit" command FO was invoked. Thereafter a family of fixed points was followed.

In the figure B2 indicates period 2 attractor, G2 indicates period 2 regular saddle, R2 indicates period 2 flip saddle, B1 indicates fixed point attractor, G1 indicates fixed point regular saddle, R1 indicates fixed point flip saddle. Observe the period-doubling and saddle-node bifurcations.

Topic of discussion. *The family of period 2 orbits appear to come to a sharp point. Explain what is really happening there.*

11.4 EXERCISES

In the Exercises below, "follow periodic orbit" means Follow the Orbit (forwards) and Follow the Orbit Backwards. When in an Exercise the X Scale or Y Scale is not specified, use the default values.

Exercise 11-1. Follow in the phase space a period 3 and a period 5 periodic orbit of the Ikeda map (I)
$$Z \rightarrow RHO + 0.9*Z*exp\{i[0.4 - 6/(1 + |Z*Z|)]\}$$
where Z = X + iY. See Figure 2-13 for a family of period 5 orbits.

Exercise 11-2. Follow a period 3 and period 5 periodic orbit of the Ikeda map (I)
$$Z \rightarrow RHO + 0.9*Z*exp\{i[0.4 - 6/(1 + |Z*Z|)]\}$$
where Z = X + iY. Set the X COordinate to be RHO, and set the X Scale to run from 0 to 2. (Hint: for RHO = 1, period 3 and 5 points are approximately (0.777969, 0.767169) and (0.305904,-0.770871), respectively.) See Figure 11-1 and Figure 2-13.

Exercise 11-3. Follow in the phase space a period 1, a period 2, a period 4, and a period 8 periodic orbit of the Ikeda map (I)
$$Z \rightarrow RHO + 0.9*Z*exp\{i[0.4 - 6/(1 + |Z*Z|)]\}$$
where Z = X + iY.

Exercise 11-4. Follow a period 1, a period 2, a period 4, and a period 8 periodic orbit of the Ikeda map (I)
$$Z \rightarrow RHO + 0.9*Z*exp\{i[0.4 - 6/(1 + |Z*Z|)]\}$$
where Z = X + iY. Set the X COordinate to be RHO, and set the X Scale to run from 0 to 2. See Figure 11-2.

Exercise 11-5. Follow a period 7, period 8, and period 9 periodic orbit for the Henon map (H)
$$(X,Y) \rightarrow (RHO - X*X + 0.3*Y, X)$$
Set the X COordinate to be RHO, and set the X Scale to run from 1 to 2. (Hint: use command RP to locate the orbits; for RHO = 1.4, period 7, 8, and 9 points are approximately (0.179821,1.191622), (0.739075,0.946088), and (0.525075,1.052868), respectively.)

Exercise 11-6. Follow a period 3, a period 4, and a period 5 periodic orbit for the Henon map (H)

$$(X,Y) \rightarrow (RHO - X*X - 0.3*Y, X)$$

Set the X COordinate to be RHO, and set the X Scale to run from 1.5 to 3. (Hint: for RHO = 2.12, period 3, 4, and 5 periodic points are approximately (-0.779592,-1.519592), (0.339065,1.389970), and (-0.751156,1.669012), respectively.)

Exercise 11-7. Follow a fixed point and a period 2 periodic orbit for the Henon map (H)

$$(X,Y) \rightarrow (RHO - X*X + 0.9*Y, X)$$

Set the X COordinate to be RHO, and set the X Scale to run from -0.2 to 1. (Hint: for RHO = 0.6, a fixed point and a period 2 point are approximately (-0.826209,-0.826209) and (-0.719754,0.819754), respectively.)

11.5 REFERENCES related to DYNAMICS

This program was written in part to make accessible the tools we have developed and used in our papers. These papers are interwoven with this program in that as the methods were developed, such as following periodic orbits, they were incorporated into the program and were then developed further in the program to be of greater use in illustrating the concepts for which we were aiming. Those papers below marked with (*) are of a more general nature and are appropriate for general readings.

The following paper shows that in some sense, the vector field v is "almost always" defined in some neighborhood of a curve of fixed points. It argues that a wide class of problems can be solved using such curve followers. This paper formed the basis of our ideas for tracking periodic orbits.
  * S.N. Chow, J. Mallet-Paret, and J.A. Yorke, Finding zeros of maps: Homotopy methods that are constructive with probability one, Math. Comp. 32 (1978), 887-899

The following paper attacks the problem of following a curve in practice, for example a curve of fixed points, and the approach in *Dynamics* uses the Runge-Kutta solver with its 18 degree rule mentioned above.
  T.-Y. Li and J.A. Yorke, A Simple Reliable Numerical Algorithm for Following Homotopy Paths, In: "Analysis and computation of fixed points", 73-91, Academic Press, 1980.

Individuals only interested in following curves of fixed points or periodic points, for example for periodic orbits with a high period, might also wish to investigate using a method tailored solely to this problem; see for example Doedel and Kernévez (1986).

The orbit index was developed and used in a series of papers. Using it one can see why infinite cascades of period-doubling bifurcations occur. A basic idea in the paper below is to follow a family of orbits and its branches, never following the flip saddle (red) branches. This gives a unique path for a computer to follow.
  * K.T. Alligood, E.D. Yorke, and J.A. Yorke, Why period-doubling cascades occur: periodic orbit creation followed by stability shedding, Physica D 28 (1987), 197-205

CHAPTER 12

DIMENSION

12.1 INTRODUCTION and the METHODS

There are several notions of dimension in the theory of dynamical systems. A variety of methods have been proposed for computing the dimension of complicated sets. One of these is the box-counting dimension It does not need to be an integer. For sets that appear on your screen, the box-counting dimension is between 0 and 2.0. Example 2-25 gives an estimate of 1.667 ± 0.059 for the box-counting dimension for a set (which happens to be a chaotic attractor) of the time-2π map of the forced damped Pendulum differential equation. In Figure 5-6a, the chaotic set seems to have nonzero area. When this area is indeed positive, the box-counting dimension is 2. The box-counting dimension is bounded from above by n when the phase space is n-dimensional. The box-counting dimension is frequently used in all the sciences. The reason for this lies in the easy and automatic computability by machine. It is straightforward to count boxes and to maintain statistics allowing dimension calculation. We now describe the definition of this dimension.

Let S be a set in a given region M. For every $\varepsilon > 0$, let $N(\varepsilon,S)$ denotes the minimum of n-dimensional cubes in a grid of length ε needed to cover the set S in the specified region M. If $N(\varepsilon,S) \approx \varepsilon^d$ for small ε, we say that the box-counting dimension of the set S is d. More precisely, the **box-counting dimension** (or **capacity dimension**) of the set S in the region M is defined as

$$D_{bd}(S) = \lim_{\varepsilon \to 0} \frac{\ln N(\varepsilon,S)}{-\ln \varepsilon}$$

We assume here that the limit exists; if it does not exist, then we say that the box-counting dimension of S does not exist.

The box-counting dimension is an example of a **fractal dimension**. There are several other definitions of fractal dimensions. They may yield different answers. Notice that the box-counting dimension can be computed for all kind of sets, while the Lyapunov dimension is only for attractors of dynamical systems; see Chapter 5 for a discussion of the Lyapunov dimension. For a long time, "fractal dimension" in the literature was equivalent to "box-counting dimension". But things have changed a bit.

In 1983 Grassberger and Procaccia, and Takens introduced independently another notion of dimension, namely the correlation dimension. This fact was pointed out in Grassberger (1990), Takens (1985) and Theiler (1987). The correlation dimension is a notion that is considered nowadayas to be more important than the box-counting dimension. The correlation dimension was obtained by considering correlations between "random" points on a chaotic attractor, but the same notion will hold for other invariant sets as well such as chaotic saddles. Consider the set $S_N = \{X_i: i = 1, ...,N\}$ of points on an invariant set, obtained, for example, from a time series, that is, $X_i \equiv X(t+i\tau)$ with a fixed time increment τ between successive measurements. When there is exponential divergence (this is the case for chaotic attractors or chaotic saddles), most pairs (X_i,X_j) with $i \neq j$ will be dynamically uncorrelated pairs of essentially random points. We now describe the definition of this dimension.

Let $S_N = \{X_i: i = 0, ...,N-1\}$ denote a set of N points on an invariant set, obtained from a trajectory, and let S denote the union of the S_N's, that is, $S = \bigcup_{N \geq 1} S_N$. For example, for a map F, if $\{x_n\}_{n \geq 0}$ with $x_{n+1} = F(x_n)$ is a trajectory, then for some fixed positive integer τ, let $X_n = x_{n\tau}$. Note that $X_{n+1} = F^\tau(X_n)$. For every $\varepsilon > 0$, let $C(\varepsilon,S_N)$ denote the proportion of pairs of different points in S_N that are less than ε apart, that is,

$$C(\varepsilon,S_N) = \#\{ \{X_i,X_j\} \in S_N \times S_N : \|X_i-X_j\| < \varepsilon \text{ and } i \neq j\}/N(N-1)$$

If the limit exists, we define

$$C(\varepsilon,S) = \lim_{N \to \infty} C(\varepsilon,S_N).$$

If $C(\varepsilon,S) \approx \varepsilon^d$ for small ε, we say that the correlation dimension of the set S is d. More precisely, the **correlation dimension** of the set S is defined as

$$D_{cd}(S) = \lim_{\varepsilon \to 0} \frac{\ln C(\varepsilon,S)}{- \ln \varepsilon}$$

The computer allows only a finite number of iterates. As in virtually all of our calculations, a certain imprecision remains. We now describe the methods that have been used to compute the box-counting and the correlation dimensions.

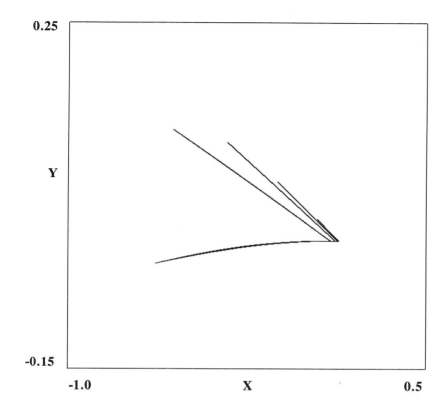

Figure 12-1: Chaotic attractor

Figure 12-1 shows a chaotic attractor for the Nordmark map (N)

$$(X, Y) \rightarrow (X + Y + 0.1, -0.2*X) \qquad \text{if } X \leq 0$$
$$(X, Y) \rightarrow (-2*sqrt(X) + Y + 0.1, -0.2*X) \text{ if } X > 0$$

The bifurcation diagram in Figure 6-17 shows that for many values of the parameter RHO, there exists a chaotic attractor. The chaotic attractor in this figure is the attractor for RHO = 0.1 and was plotted INcrementally, that is, IN was ON so that the routine CD could use the resulting densities.

Computations of the dimensions were made with the following settings. The number of color planes was set to 16 and the grid size was set to 1024 × 1024. This was done by setting SIZE = 16 1024 1024.

The estimate for the box-counting dimension of the attractor is 1.09 ± 0.034 and the estimate for the correlation dimension is 1.197 ± 0.2 .

Numerical method for the computation of the box-counting dimension

The routine BD examines a box in the core memory representing the central area of your screen. The box is normally 256 pixels by 256 pixels (or 512 by 512 or even 1024 by 1024 if the core includes such a box). Parts of your picture outside this box do not influence the computation. The program checks to see if there are points of your picture in this box. Then it divides the box into smaller boxes, first boxes containing a single pixel, next boxes that are 2 pixels by 2 pixels, then 4 by 4, then 16 by 16, etc. Hence, at stage k, the routine BD divides the large box into a grid boxes of size 2^k by 2^k pixels. It counts the number $N(k)$ of these boxes which contain at least one point of your picture. Hence, $N(0)$ is the number of occupied 1 pixel boxes. Of course, this number $N(k)$ will depend on the size of the boxes. The **box-counting dimension** is defined as follows: if $N(k) \approx 2^{kD}$ for large k, then D is the box-counting dimension.

To estimate the box-counting dimension, one may do the following. Plot log $N(k)$ on the vertical axis, and (k log 2) on the horizontal axis. Then try to fit a straight line to the plotted points of the diagram and measure its slope D, yielding an estimate of the box-counting dimension.

Recall that the routine BD considers a sequence of grids of boxes where the grid box size is reduced by a factor of 2 from one grid of boxes to the next. In this approach each box from a grid is subdivided into four boxes, each half the size in the next grid. The slope of the line from the data at stage k to the next at stage k+1 in the corresponding log/log diagram is $D(k) = \{\log N(k) - \log N(k+1)\}/\log 2$. The slope is an estimate for the box-counting dimension of the set. In other words, if the number of boxes counted increases by a factor of 2^D when the box size is halved, then the box-counting dimension is equal to D.

Routine BD uses $\{\log N(2) - \log N(7)\}/\log 32$ to estimate the box-counting dimension of the set under investigation. The error estimates are based on intermediate values of $D(k)$, where $2 \leq k \leq 5$. Use BD0 and BD1 to get extra information.

In principle, a better estimate would be obtained if we had used $\{\log N(0) - \log N(7)\}/\log 2^7$, but then it becomes particularly important to be sure that the program has plotted the trajectory long enough that all or most of the boxes that intersect the attractor have had points plotted in them. If BD is computed with too few points plotted, $N(0)$ and hence CD will be too small. This estimate can be obtained by BD0.

Summary for computation of box-counting dimension

● Set the core size (command SIZE)
● Plot trajectory (command T, or SST, or BST etc.).
● Use the "Box-counting Dimension" command BD to get an estimate of the box-counting dimension.

Numerical method for the computation of the correlation dimension

An algorithm for the computation of the Correlation Dimension from data, was given independently by Grassberger and Procaccia (1983) and Takens (1983). For improved methods and error estimates, see, for example, Takens (1985), Theiler (1987), (1990) Grassberger (1990), Molteno (1993) and, Frank, Keller and Sporer (1996). The algorithm we use is relatively crude because it is based only on what has been plotted on the screen (and in the core).

The routine CD examines a box in the core memory representing the central area of your screen, the same box as used by the Box-counting Dimension BD. The box is normally 512 pixels by 512 pixels (or 1024 by 1024 if the core includes such a box). Parts of your picture outside this box do not influence the computation. The program checks to see if there are points of your picture in this box. Then it divides the box into smaller boxes, of size 2^k by 2^k pixels as with command BD. Assume the boxes are numbered from 1 through M. For box number m, let N(m;k) be the number of trajectory points that have hit that box m, where $1 \leq m \leq M$. For every integer $k \geq 0$, define

$$B_k = \sum_{m=1}^{M} [N(m;k)]^2$$

In particular, B_2 is this sum for boxes 4 pixels by 4 pixels, and B_7 is this sum for boxes 128 pixels by 128 pixels.

If the grid size of the picture is 256×256, the correlation dimension estimate "CorDim(2,7)" is given by

CorDim(2,7) = log(B_2/B_7)/((7-2) log 2)

More generally, if the grid size of the picture is $2^K \times 2^K$, the correlation dimension estimate "CorDim(2,K-1)" is given by

CorDim(2,K-1) = log(B_2/B_{K-1})/((K-1-2) log 2)

Error estimates are based on other CorDim(i,j) values. For the computation of Correlation Dimension it is best to use as many color planes as possible (command COLP or command SIZE) and to plot the trajectory (commands T, SST, BST) and we recommend that the toggle IN is ON when the trajectory (commands T, BST, SST). To compute the Correlation Dimension of a picture that has been created by *Dynamics*, enter the "Correlation Dimension" command CD (or CD1 to get more information) and after some computing, the result appears on the screen.

Summary for computation of correlation dimension

● Set the number of color planes and the grid size (command SIZE)
● **optional**: Select a time lag (command IPP)
● Plot trajectory (command T, or SST, or BST etc.) preferably using the INcremental option, that is, with the toggle IN turned ON.
● Use the "Correlation Dimension" command CD to get an estimate of the correlation dimension.

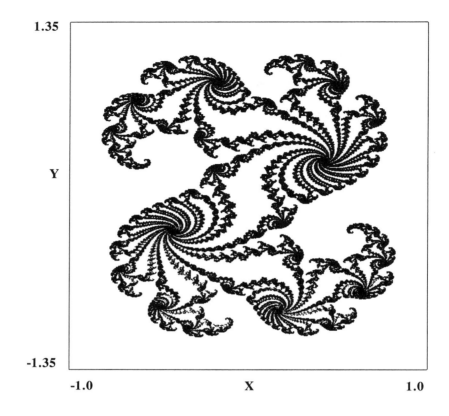

1.35

Y

-1.35

-1.0   X   1.0

Figure 12-2: Box-counting and correlation dimension
for a chaotic trajectory on basin boundary

This figure shows a chaotic trajectory of the complex quadratic map (which is the Cubic map (C) with RHO = 0)

$$Z \rightarrow Z*Z + 0.32 + i0.043$$

where $Z = X + iY$. The figure shows the basin boundary straddle trajectory (BST) on the common boundary of the basin of a stable period 11 orbit and the basin of the attractor at infinity.

For this picture, the routine BD gives 1.687 ± 0.037 as an estimate of the box-counting dimension and the routine CD gives 1.506 ± 0.239 as an estimate for the correlation dimension.

12.2 DIMENSION MENU DIM

Enter the "DImension Menu" command DIM

$$DIM < Enter >$$

and the menu appears on the screen in approximately the following format.

```
DIM  - -    MENU FOR ESTIMATING DIMENSION

The following 4 commands are for estimating the dimension
of a picture on your screen. If most of the screen has
been plotted, forget it, because the dimension is 2.
BD    -  Box counting Dimension (also called capacity)
BD1   -  like BD but more information is provided
CD    -  Correlation Dimension
CD1   -  like CD but more information is provided
CDM   -  Correlation Dimension Method description

To compute the Lyapunov dimension of an attractor, you
must set L is the largest possible value (2),
and then plot a trajectory on the attractor.
The dimension will be printed on the screen. Or enter LL.
L     -  number of Lyapunov exponents:  0
LL    -  List Lyapunov numbers and Lyapunov dimension
T     -  Trajectory

Warning: You may assume  CD < = Lyap dim < = BD
but the program's crude estimates may not reflect this.
```

Commands for computing dimensions

BD, BD1

The "Box-counting Dimension" command BD estimates the box-counting dimension of any picture in a large box in the center of your screen. If you invoke the command SQ5 before creating the picture or if you set RB to 1024×1024, then the BD computation is made of the entire screen picture. See commands SQ5, RB and SIZE.

The command BD1 computes the box-counting dimension just as BD does, but it provides more information.

CD, CD1

The "Correlation Dimension" command CD produces a crude estimate based on the picture you have created in core memory. Like BD, this routine is for pictures of trajectories where most of the screen is blank. Preferably trajectories should be plotted with the option IN, so that the color of a pixel indicates how many times the pixel has been hit, up to the maximum color number. Use of 256 colors is preferable. CD examines only the central region that is 256 x 256 pixels or 512 by 512 pixels, or whatever is the largest power of 2 possible, and it draws a box around this region on the screen.

The command CD1 computes the correlation dimension just as CD does, but it provides more information.

CDM

The command CDM provides a description of the method for estimating the correlation dimension.

12.3 EXAMPLES

In Chapter 2, we have already presented an example of estimating box-counting dimension. In Example 2-25 the computation of the estimate of the box-counting dimension of an attractot for the forced damped Pendulum equation map was discussed. In this section, we summarize for a variety of pictures, the numerical results of the box-counting and correlation dimension. We just present the numerical results on the box-counting and correlations dimensions of pictures of which the majority of them are from previous chapters. These results are summarized in the table below. You may wish to consult previous examples that correspond to the figures, how to create the pictures. One relationship is always true:

Box-counting Dimension $\geq$ Correlation Dimension,

but with the limited precision of our estimates, it is possible for the estimates to violate this relationship.

Results dimension computations				
Process	Figure	routine	Box-counting	Correlation
Henon map	2-1	T	1.227 ± 0.122	1.214 ± 0.356
Henon map	2-4	T	1.245 ± 0.093	1.251 ± 0.056
Henon map	2-19	SST	0.804 ± 0.108	0.762 ± 0.048
Tinkerbell map	2-22	T	1.485 ± 0.164	1.166 ± 0.067
Pendulum eq.	2-25	T	1.626 ± 0.043	1.311 ± 0.046
Ikeda map	3-1	T	1.724 ± 0.109	1.697 ± 0.025
Quasiper. map	5-1b	T	1.650 ± 0.288	1.396 ± 0.074
Quasiper. map	5-1c	T	1.930 ± 0.088	1.887 ± 0.130
Tinkerbell map	8-2	BST	1.234 ± 0.094	0.904 ± 0.087
PLinear map	8-7	SST	1.306 ± 0.082	1.034 ± 0.224
Ikeda map	8-8	SST	1.416 ± 0.043	1.357 ± 0.068
Henon map	8-10	SST	1.163 ± 0.048	0.995 ± 0.144
Nordmark map	12-1	T	1.109 ± 0.034	1.197 ± 0.200
Cubic map	12-2	BST	1.687 ± 0.037	1.506 ± 0.239
Pendulum eq.	12-3	SST	1.535 ± 0.055	1.483 ± 0.044
any process*	12-4		1.851 ± 0.013	–
Goodwin eq.	13-2	T	1.640 ± 0.015	1.533 ± 0.017

Pictures were created with the following settings:
SIZE: 16 1024 1024
IN was set to be ON
(*) box-counting dimen estimate is based on a 720×720 picture

12.4 EXERCISES

Exercise 12-1. Pick any map or differential equation from the table in section 12.3 and create the pictures in a grid size of 512 × 512. Compute the box-counting and correlation dimensions of the resulting picture. Compare the results with the results of the table. If there is a difference, can you explain why?

Exercise 12-2. Compute the box-counting and correlation dimensions of the picture that you get when plotting any trajectory of the LOGistic map with RHO = 3.5699456. Do you think that the resulting numbers are approximately correct? It may be helpful to compute the Lyapunov exponent too when you are going to create the picture.

Exercise 12-3. Do some experimenting for the Henon map with RHO = 1.4 and C1 = 0.3. Compute the box-counting dimension of the resulting pictures of the trajectory in the following cases:
(a) DOTS = 1000; (b) DOTS = 10,000; (c) DOTS = 100,000; (d) DOTS = 1000,000
What do you observe?

Exercise 12-4. Do some experimenting for the Henon map with RHO = 1.4 and C1 = 0.3. Compute the correlation dimension of the resulting pictures of the trajectory in the following cases:
(a) IPP = 1; (b) IPP = 10; (c) IPP = 100; (d) IPP = 1000; (e) IPP = 10,000.
What do you observe?

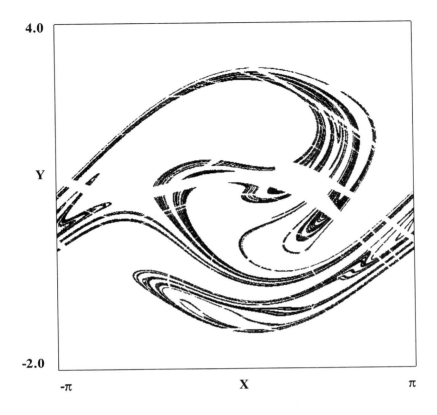

4.0

Y

-2.0

-π   X   π

Figure 12-3: Box-counting and correlation dimension of a chaotic saddle
This figure shows a Saddle Straddle Trajectory for the time-2π map of the forced damped Pendulum differential equation (P)

$$X''(t) + 0.1*X'(t) + \sin X(t) = 0.16 + 1.6*\cos(t))$$

The Saddle Straddle Trajectory is chaotic, since one of the Lyapunov exponents of the straddle trajectory is positive. Use command L to compute the Lyapunov exponents while plotting a straddle trajectory. For this picture, the routine BD gives 1.535 ± 0.055 as an estimate of the box-counting dimension and the routine CD gives 1.483 ± 0.044 as an estimate for the correlation dimension. See also Example 2-25 and Figure 2-25, in which all three the Lyapunov, box-counting and correlation dimension are computed for the attractor that occurs when RHO = 1.5.

Topic of discussion. *Can you describe a phenomenon that may occur when RHO is varied from 1.5 to 1.6?*

12.5 REFERENCES related to DYNAMICS

Computation and theory develop together. This program was written in part to make accessible the tools we have developed and used in the papers of the Maryland Chaos Group. These papers were interwoven with this program in that as the methods were developed, such as the computation of the Lyapunov exponents of trajectories and the Lyapunov dimension of attractors, they were incorporated into the program. Those papers below marked with (*) are of a more general nature and are appropriate for general readings.

An expository paper on dimension, making considerable use of the Generalized Baker map (GB) to provide easily computable examples. The map had been introduced in an earlier expository paper (see E. Ott, Strange attractors and chaotic motion of dynamical systems, Rev. Modern Phys. 53 (1981), 655-671).
    * J.D. Farmer, E. Ott and J.A. Yorke, The dimension of chaotic attractors, Physica 7D (1983), 153-180

The following paper introduced the idea of fractal basin boundary of physically meaningful systems and shows how the dimension of a basin boundary is intimately related to the problem of knowing in which basin initial points are.
    * C. Grebogi, S.W. McDonald, E. Ott and J.A. Yorke, Final state sensitivity: An obstruction of predictability, Phys. Letters 99A (1983), 415-418

The following paper describes why boundaries are sometimes fractal. It introduces the uncertainty dimension as a quantitative measure for final state sensitivity in a system.
    * S.W. McDonald, C. Grebogi, E. Ott, and J.A. Yorke, Fractal basin boundaries, Physica D 17 (1985), 125-153

The next paper looks at the role of "basic sets" that lie on basin boundaries. These are compact invariant sets that determine the fractal dimension of those boundaries. The paper addresses the question of why different regions of a boundary might be expected to have the same dimension.
    C. Grebogi, H.E. Nusse, E. Ott, and J.A. Yorke. Basic sets: sets that determine the dimension of basin boundaries. In: Dynamical Systems, Proc. of the Special Year at the University of Maryland, Lecture Notes in Math. 1342, 220-250, Springer-Verlag, Berlin etc. 1988

The next paper shows that the box-counting dimension, the uncertainty dimension, and the Hausdorff dimension are all equal for one and two dimensional systems which are uniformly hyperbolic on their basin boundary.

H.E. Nusse and J.A. Yorke, The equality of fractal dimension and uncertainty dimension for certain dynamical systems, Commun. Math. Phys. 150 (1992), 1-21

Computers can represent only a finite number of points in Euclidean space. When iterating a map like the Henon map, each computer trajectory if iterated long enough will eventually repeat and thereafter that trajectory is locked into a periodic orbit. The following paper shows that how long it takes to fall into the periodic orbit and indeed the period of the periodic orbit can be estimated from the (1) the number of digits in the computer's numbers together with (2) correlation dimension of the chaotic attractor through which the numerical trajectory travels.

C. Grebogi, E. Ott, and J.A. Yorke, Roundoff-induced periodicity and the correlation dimension of chaotic attractors, Phys. Rev. A 38 (1988), 3688-3692

The estimated correlation dimension of a reconstructed attractor of the reconstructed dynamics in an m-dimensional space, typically increases with m and reaches a plateau (on which the dimension estimate is relatively constant) for a range of large enough m values. The plateaued dimension value is then assumed to be an estimate of the correlation dimension for the attractor in the original full phase space. The next two papers address and discuss this topic extensively.

M. Ding, C. Grebogi, E. Ott, T. Sauer and J.A. Yorke, Plateau onset for correlation dimension: when does it occur?, Phys. Rev. Lett. 70 (1993), 3872-3875; Estimating correlation dimension from a chaotic time series: when does plateau onset occur?, Physica D 69 (1993), 404-424

Many experimenters use dimension as one tool for analyzing data. The following book addresses this and other data related questions.

E. Ott, T. Sauer and J.A. Yorke (Eds.), *Coping with Chaos*, John Wiley & Sons, New York, 1994

One of the main tools of experimentalists when dealing with data from chaotic experiments is to attempt to "embed" the data in a high dimensional space. The next paper analyzes the dimension question of what dimensional space the data should be embedded in to fully represent a chaotic attractor.

T. Sauer, J.A. Yorke and M. Casdagli, Embedology, J. Stat. Phys. 65 (1991), 579-616

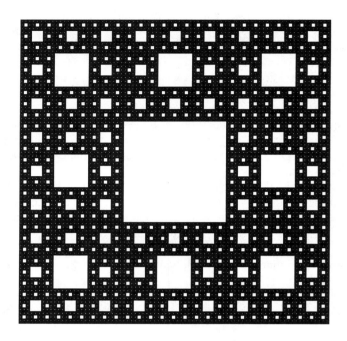

Figure 12-4: Computer art: Fractals and Dynamics
THis picture has been created by using the commands AB and AFD5 only. See also the caption of Figure 4-3a. After a number of steps, the resulting picture resembles a fractal known as Sierpinski's gasket.

CHAPTER 13

ADDING YOUR OWN PROCESS TO DYNAMICS

13.1 INTRODUCTION

Although there is a variety of maps and differential equations in the menus of *Dynamics*, you may want to study a process that is not in the menus. You can add your own processes to the program as follows.

We first describe how to add a map. Start the program as usual. Type

$$\textbf{dynamics} < Enter >$$

and the first menu that appears will have the line:

OWN	– enter your OWN process

When you select this entry, four windows will appear on your screen, presenting an example of how to add a new process. Simply change what is in these four windows to define your process. When adding adding a process, the following keys are important for carrying out this task:

- hit < Esc > to cancel the option of adding a new process to *Dynamics*;
- hit < Tab > to go to the next window;
- hit < F1 > to compile after you have inserted the new process;
- hit < F3 > for adding a map to *Dynamics*;
- hit < F4 > for adding a differential equation to *Dynamics*;
- use < BkSp > for erasing text in any of the windows;
- use < Ctrl-K > for erasing a line of text;
- use < Ctrl-W > for erasing all text in a window.

13.2 ADDING A NEW MAP

When you select the entry "OWN" from the map menu, four windows will appear on your screen, in approximately the following format, presenting an example of how to add a map to *Dynamics*. First some general comments.

- In each window, the '!' means that the rest of the line is a comment. Delete superfluous text using <BkSp>, <Ctrl-K> or <Ctrl-W>. *Note.* For deleting text, <Del> does not work.
- The equations are not case sensitive. Use either upper or lower case letters.
- The computer screen shows only 4-6 lines of each window. You may use more than the number of lines shown. However, if you use more lines, then a part of the information will be hidden whenever you switch to another window, but can be accessed with the arrow keys.
- In the windows you can have more than one equation per line or more than one line per equation. Temporary variables need not be initialized.

Documentation window
Henon map   (x,y) – –>   (rho - x*x + c1*y, x)
Map window
x := rho - x*x + c1*y   y := x
Initialization window
x := 0   y := 2   c1 := -0.3   rho := 2.12   x_lower := -2.5   x_upper :=   2.5   y_lower := -2.5   y_upper :=   2.5
Inverse map window
x := y   y := (x - rho + y*y) / c1   ! Warning: do not use c1 = 0
Esc=Cancel Tab=Next F4=Diff Eqs F1=Compile Ctrl-K=EraseLine

Simply change what is in these four windows to define your map. Of course, there are some rules which are described later. We first describe the contents of the windows followed by some detailed examples.

Documentation window

The Documentation window provides text that will appear whenever you call either the Main Menu or the Parameter Menu. It will also be part of printed copies of pictures, when you print them using Text Level 2 or 3.

- Document your work!
Generally it is best to use four lines of text or less.

- All the text in the Documentation window will appear on the screen when the Main Menu or the Parameter Menu is called. It will also be part of printed copies of pictures, when you print them using Text Level 2 or 3.

Map window

The map that you want to add must be defined in the Map window.
- Allowed **process parameters**: $c1$, ..., $c9$, phi, rho, sigma, beta.

- Allowed **main variables**: r, s, t, ..., z. The value of the main variables that are used are inserted into $y[0]$, $y[1]$, etc. You can find their values using the command YV. The program makes a list in **alphabetical order** of the r, s, t, ..., z variables you use and assigns them to $y[0]$, $y[1]$, $y[2]$, ... If you use r, x, and y (listing them alphabetically), their values will be inserted into $y[0]$, $y[1]$, and $y[2]$, respectively.

- Allowed **phase space variables**: u, ..., z. These variables have a special role. When the quasi-Newton method is used, for example, numerical partial derivatives are computed with respect to these variables, and the Lyapunov exponents indicate how fast these variables move apart. In practice, these six variables will be the only ones used for maps, unless you use r, s, and t for plotting. You might, for example, define r and s to be variables that give a better plot.

- Each main variable used must appear on the left side of an equation exactly once in the Map window.

Initialization window

The initial values that are listed in the Initialization window alert the program to the elements used in your process. Notice that you can have more than one equation per line.

- All the main variables and parameters that appear in the Map window are initialized with value 0 unless you initialize them here.

- C1,..., C9, phi, rho, sigma, beta are the only parameters permitted in equations. Their values will appear in the Parameter Menu PM.

- As described above, the program makes a list in alphabetical order of the variables you use and assigns to them y[0], y[1], y[2], ... If you use two or more phase space variables, then the first two phase space variables (when listed alphabetically) are assigned to XCO and YCO.

    If you use only w and z, then w corresponds to y[0] and z to y[1]. If you wish to plot w horizontally and z vertically, then the Initialization window would include

    XCO := 0   ! w is variable # 0
    YCO := 1   ! z is variable # 1

    Actually the above assignment is automatic and does not have to made explicitly.

    If you use x, y and z, then x corresponds to y[0], y to [y1] and z to y[2]. If you wish to plot y horizontally and z vertically, then the Initialization window would include

    XCO := 1   ! y is variable # 1
    YCO := 2   ! z is variable # 2

- The default scale both horizontally and vertically is 0 to 1. If you wish to change one of these, then as in the example above define the variables

    x_lower   x_upper   and/or   y_lower   y_upper

    Their values will appear in the Parameter Menu PM.

- You may even wish to define Z_lower Z_upper and ZCO if you plan use three dimensional plots.

- Anything defined in this window can be changed by using standard commands of the program. For example, if you choose the defaults, then you can redefine the scale using the X Scale and Y Scale commands XS and YS, which are in the Parameter Menu PM.

Inverse map window

Using the Inverse map window is optional. The inverse is used in calculating stable manifolds and in going backwards in time for DE's (see the inVerse command V). If you want to make use of the inverse (provided it exists) then you must define the inverse in the Inverse map window.

- The program does not check to verify that the map you provide is in fact the inverse. See command V (inVerse map).

No text in the Initialization window for maps

If you erase all text in the Initialization window of the Henon map, then after you have compiled this process, the Parameter Menu is of approximately the following format

```
PM      – –      PARAMETER MENU

OK      –  parameters are fine as set
XCO     –  X COordinate to be plotted:  y[0]
XS      –  X Scale: from  0  to  1
YCO     –  Y COordinate to be plotted:  y[1]
YS      –  Y Scale: from  0  to  1
C1      –  C1 = 0.00000000
RHO     –  rho = 0.00000000
SD      –  Screen Diameters:  1.00


           MENUS
VM      –  Vector Menu for initializations, etc.
WWM     –  When and What to plot Menu
```

The Y Vectors are set as follows:

```
              YV – –   Y VECTORS

Vectors -- y0 = y  = current state,  y1 = cursor position
   state vec y          storage vec y1          storage vec y2
   y[0] = 0;            y1[0] = 0;              y2[0] = NOT SET
   y[1] = 0;            y1[1] = 0;              y2[1] = NOT SET


storage vec ya          storage vec yb          storage vec ye
ya[0] = 0;              yb[0] = 1;              ye[0] = NOT SET
ya[1] = 0;              yb[1] = 1;              ye[1] = NOT SET
```

For the case of the Henon map and other maps, you can set parameters like C1 and RHO, coordinates XCO and YCO, and scales XS and YS as you wish. You also may set the vectors. Hence, you may leave the Initialization window empty.

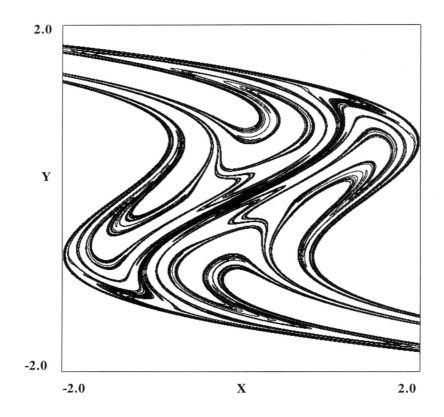

Figure 13-1a: Unstable manifold of a fixed point

This figure shows the unstable manifold of the fixed point (0,0) on the chaotic saddle of the Holmes map (HO)

$$(X, Y) \rightarrow (1.5*X - X*X*X + 0.95*Y, X)$$

You may plot also the stable manifold of the fixed point (0,0). If you do so, the intersection of the unstable manifold and the stable manifold is a chaotic saddle and is shown in Figure 13-1b.

Example 13-1. **Adding the Holmes map to** *Dynamics*

The objective of this example is to show how to add the Holmes map, a cubic map in the plane,

$$(X,Y) \rightarrow (RHO*X - X*X*X + C1*Y, X)$$

to *Dynamics* using the OWN-feature. For our convenience, we use "Own-Holmes map" in the Documentation window. See also Section 13.4, entitled "Storing data and pictures of an OWN-process". Start *Dynamics* and select **OWN**, and add the equations as indicated below.

Documentation window
Own-Holmes map (x,y) – –>   (rho*x - x*x*x + c1*y, x)
Map window
x := rho*x - x*x*x + c1*y y := x
Initialization window
c1 := 0.95 rho := 1.5 x_lower := -2.0   x_upper :=   2.0 y_lower := -2.0   y_upper :=   2.0
Inverse map window
x := y y := (x - rho*y + y*y*y) / c1   ! Warning: do not use c1 = 0
Esc=Cancel Tab=Next F4=Diff Eqs F1=Compile Ctrl–K=EraseLine

After completing the typing, hit <F1> to compile.

Plot the unstable manifold of the fixed point (0,0) (command UM). The resulting picture is shown in Figure 13-1a. To store the picture on disk, we recommend to have the file easily recognized as being a picture file created by a process that has been added to *Dynamics* by the OWN feature. For example, store this picture under the file name O-Ho-um.pic.

Note. As discussed on page 495, you may leave the Initialization window empty. Then C1 = rho = 0 = y1[0] = y1[1] and scales are from 0 to 1.

Note. You can create a picture of a chaotic saddle by entering command SST; see Figure 13-1b.

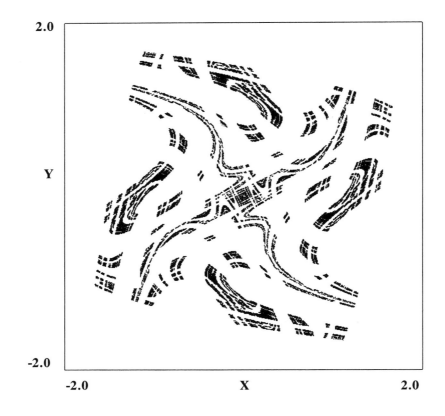

Figure 13-1b: Saddle straddle trajectory
This figure shows a Saddle Straddle Trajectory on a chaotic saddle of the Holmes map (HO)

$$(X, Y) \rightarrow (1.5*X - X*X*X + 0.95*Y, X)$$

The chaotic saddle is the intersection of the stable and unstable manifolds of the fixed point (0,0). It is an invariant set.

Example 13-2. Adding a one-dimensional map to *Dynamics*

The purpose of this example is showing how to add a one-dimensional real cubic map

$$X \rightarrow X*X*X + C1*X + RHO$$

to *Dynamics* using the OWN-feature. For our convenience, we use "Own-RealCubic map" in the Documentation window. See also Section 13.4, entitled "Storing data and pictures of an OWN-process". Start *Dynamics* and select **OWN**, and add the equations as indicated below.

Documentation window
Own-RealCubic map
x - -> x*x*x + c1*x + rho
Plot x at time n (horizontally) against x at time n+1 (vertically)
Map window
x := x*x*x + c1*x + rho
Initialization window
x := 0.5
c1 := -0.5
rho := 1.0
x_lower := -1.0   x_upper :=   1.0
Inverse map window
Esc=Cancel Tab=Next F4=Diff Eqs F1=Compile Ctrl–K=EraseLine

After completing the typing, hit <F1> to compile.

Note. Since there is one phase space variable, *Dynamics* will plot x vertically and horizontally it will plot x at the previous time, that is, at time n it plots (x_{n-1}, x_n) where x_n denotes the value of x at time n.

Note. As discussed on page 495, you may leave the Initialization window empty. Then C1 = rho = 0 = y1 = y0 and the scales are from 0 to 1.

Example 13-3. Adding a Piecewise Linear map to *Dynamics*

In this example we want to show how to add the Piecewise Linear map

$$(X,Y) \rightarrow (C1*X + Y + RHO, C2*X) \quad \text{if } X \leq 0$$
$$(X,Y) \rightarrow (C3*X + Y + RHO, C4*X) \quad \text{if } X > 0$$

to *Dynamics* using the OWN-feature. Although the Piecewise Linear map is a process of *Dynamics*, we carry out this example, because it is of different type than, for example, the Henon map. For our convenience, we use "Own-Piecewise Linear map" in the Documentation window. See also Section 13.4, entitled "Storing data and pictures of an OWN-process". Start *Dynamics* and select **OWN**, and add the equations as indicated below.

You will find out that on PC's, this "OWN" version runs slightly slower than PL.

Documentation window
Own-Piecewise Linear map (x,y) - -> (c1*x + y + rho, c2*x)   if x < 0 or x = 0 (x,y) - -> (c3*x + y + rho, c4*x)   if x > 0
Map window
x := if (x < = 0) then (C1*x + y + rho) else (C3*x + y + rho) y := if (x < = 0) then (C2*x) else (C4*x)
Initialization window
c1 := -1.25   c2 := -0.1   c3 := -2   c4 := -2 rho := 0.4 x_lower := -1.0 x_upper := 1.0 y_lower := -1.0 y_upper := 1.0
Inverse map window
Esc = Cancel Tab = Next F4 = Diff Eqs F1 = Compile Ctrl–K = EraseLine

After completing the typing, hit <F1> to compile.

Note. You may leave the Initialization window empty; see p. 495. Then C1 = C2 = C3 = C4 = rho = 0, and the scales are from 0 to 1.

Example 13-4. **Adding a one-dimensional piecewise linear map to** *Dynamics*
This example exhibits how to add the one-dimensional 3PL map

$$X \rightarrow C1*X + \text{rho if } X \leq \text{beta}$$
$$X \rightarrow C2*X + (C1\text{-}C2)*\text{beta} + \text{rho if beta} < X \leq \text{sigma}$$
$$X \rightarrow C3*X + (C2\text{-}C3)*\text{sigma} + (C1\text{-}C2)*\text{beta} + \text{rho if } X > \text{sigma}$$

to *Dynamics* using the OWN-feature. For our convenience, we use "Own-3PL map" in the Documentation window. See also Section 13.4, entitled "Storing data and pictures of an OWN-process". Start *Dynamics* and select **OWN**, and add the equations as indicated below.

Documentation window
Own-3PL map X → C1*X + rho if X ≤ beta X → C2*X + (C1-C2)*beta + rho if beta < X ≤ sigma X → C3*X + (C2-C3)*sigma + (C1-C2)*beta + rho if X > sigma Plot x at time n (horizontally) against x at time n+1 (vertically)
Map window

x := if x <= sigma then
    (if x <= beta then C1*x + rho else C2*x + (C1-C2)*beta + rho)
    else C3*X + (C2-C3)*sigma + (C1-C2)*beta + rho

Initialization window
c1 := 0.9   c2 := -7   c3 := 0.5 beta := 0.2   sigma := 0.8 x_lower := -4.0 x_upper := 4.0   ! scale for horizontal plot y_lower := -4.0 y_upper := 4.0   ! scale for vertical plot
Inverse map window
Esc=Cancel Tab=Next F4=Diff Eqs F1=Compile Ctrl–K=EraseLine

After completing the typing, hit <F1> to compile.
    A bifurcation diagram for this map (BIFS) is shown in Figure 13-2.
    Note. As discussed on page 495, you may leave the Initialization window empty. In that case, c1 = c2 = c3 = beta = sigma = 0.

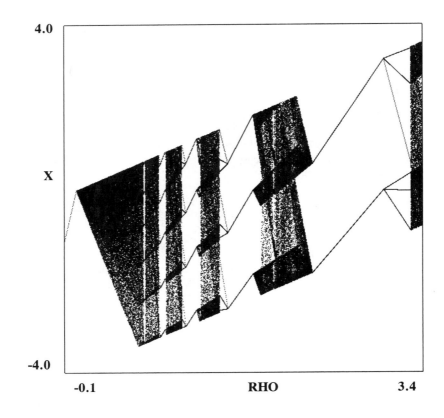

4.0

X

-4.0

-0.1   RHO   3.4

Figure 13-2: Bifurcation diagram

In this bifurcation diagram for the 1D Piecewise Linear map

$$X \;\rightarrow\; C1*X \;+\; rho \text{ if } X \leq beta$$
$$X \;\rightarrow\; C2*X \;+\; (C1-C2)*beta \;+\; rho \text{ if } beta < X \leq sigma$$
$$X \;\rightarrow\; C3*X \;+\; (C2-C3)*sigma \;+\; (C1-C2)*beta \;+\; rho \text{ if } X > sigma$$

rho varies from -0.1 to 3.4, C1 = 0.9, C2 = -7, C3 = 0.5, beta = 0.2 and sigma = 0.8. The diagram exhibits at least 8 **border-collision bifurcations.** *while the diagram for the PL map of Figure 6-16 exhibits exactly one such a bifurcation at rho = 0 if rho is varied, for example, from -0.1 to 0.1.*

Observe the following BC bifurcations (from left to right):

fixed point attractor $\longrightarrow$ *chaotic attractor, period 15* $\longrightarrow$ *period 5, period 12* $\longrightarrow$ *period 4, period 4* $\longrightarrow$ *chaotic attractor, period 9* $\longrightarrow$ *period 3, period 3* $\longrightarrow$ *chaotic attractor, chaotic attractor* $\longrightarrow$ *period 2, and period 2* $\longrightarrow$ *period 6.*

Topic of discussion. *Which border-collision bifurcations of the above can or cannot occur for the map of Figure 6-16? Explain why or why not.*

Example 13-5. Adding a square root map to *Dynamics*

In this example we demonstrate how to add the square root map

$$(X,Y) \; \rightarrow \; (C1*X + Y + RHO, \; C2*X) \qquad \text{if } X \leq 0$$
$$(X,Y) \; \rightarrow \; (C3*sqrt(X) + Y + RHO, \; C4*X) \quad \text{if } X > 0$$

using the OWN-feature. For our convenience, we use "Own-Square root map" in the Documentation window. See also Section 13.4, entitled "Storing data and pictures of an OWN-process". Start *Dynamics* and select **OWN**, and add the equations as indicated below.

Documentation window

Own-Square root map
(x,y) - -> (c1*x + y + rho, c2*x)   if x < 0 or x = 0
(x,y) - -> (c3*sqrt(x) + y + rho, c4*x) if x > 0

Map window

x := if (x <= 0) then (C1*x+y+rho) else (C3*sqrt(x)+y+rho)
y := if (x <= 0) then (C2*x) else (C4*x)

Initialization window

x := 1   y := 1
c1 := 1   c2 := -0.2   c3 := -2   c4 := -0.2
rho := 0.1
x_lower := -1.0 x_upper := 0.5
y_lower := -0.15 y_upper := 0.25

Inverse map window

Esc=Cancel Tab=Next F4=Diff Eqs F1=Compile Ctrl-K=EraseLine

After completing the typing, hit <F1> and compile. Plot the trajectory. The resulting picture is similar to Figure 12-1.

Note. As discussed on page 495, you may leave the Initialization window empty. Then C1 = C2 = C3 = C4 = rho = 0 and scales are from 0 to 1.

---

13. Adding your own process   503

Example 13-6. Using temporary variables:
Adding the Gumowski/Mira map to *Dynamics*
In this example we demonstrate how to add the Gumowski/Mira map

$$(X,Y) \rightarrow (C1*(1+C2*Y*Y)*Y + F(X), -X + F(Xnew))$$
$$F(u) = rho*u + 2*(1-rho)*u*u/(1+u*u)$$

using the OWN-feature. Although the Gumowski/Mira map is a process of *Dynamics*, we carry out this example, because it demonstrates when temporary variables might be used. For our convenience, we use "Own-Gumowski/Mira map" in the Documentation window. See also Section 13.4 for storing data and pictures. Start *Dynamics* and select **OWN**, and add the equations as indicated below. In this example, the temporary variables are F0, F1, and xNew. These do not appear in the Initialization window. See p. 513 for the rules for temporary variables.

Documentation window
Own-Gumowski/Mira map (x,y) - -> (c1*(1+c2*y*y)*y + F(x), -x + F(xNew)) $\qquad$ F(u) = rho*u + 2*(1-rho)*u*u/(1+u*u)
Map window
F0 := rho*x + 2*(1-rho)*x*x/(1+x*x) xNew := c1*(1+c2*y*y)*y + F0 F1 := rho*xNew + 2*(1-rho)*xNew*xNew/(1+xNew*xNew) y := -x + F1 x := xNew
Initialization window
x := 1$\qquad$ y := 1 c1 := 1$\qquad$ c2 := 0$\qquad$ rho := 0.3 x_lower := -20 x_upper := 20 y_lower := -20 y_upper := 20
Inverse map window
Esc=Cancel Tab=Next F4=Diff Eqs F1=Compile Ctrl–K=EraseLine

After completing the typing, hit <F1> to compile.
Note. As discussed on page 495, the Initialization window can be empty.

BASIC RULES FOR ADDING A NEW MAP TO *DYNAMICS*

Simply change what is in these four windows to define your map. Of course, there are some rules. Standard algebraic expressions are allowed.

● Use * to denote multiplication

● Arithmetic operators (* / + -) have their conventional precedence.

● Expressions can use the usual algebraic expressions as well

the number pi (pi = π)

and the following functions:

sin, cos, tan, atan, sqrt, exp, mod,

log (log base e), log10 (log base 10), rand

as well as

absolute value |E| of an expression E and

power E^G, where either E is positive or G is an integer.

● If you are in doubt as to whether you need parentheses in an expression, they do no harm so they can be added.

● First, the values of the main variables (r, s, ..., z) are first inserted into the right hand side of all the equations, and then the equations are evaluated sequentially. Consider the case where x and y both have initial values of 0.0 and the process is defined by

$$x := y + 1$$
$$y := x$$

First, the value 0.0 is inserted for x and y in the right hand side of both equations. The x in the second equation is the initial value. Hence after one iterate, x is 1 and y is 0. The sequential evaluation is only important if temporary variables are used.

● See p. 513 for a discussion of temporary variables. Such variables must be defined in each window in which it is used, so temp := temp+1 cannot be the first line of a window.

random number rand( )

The function **rand( )** selects a pseudo-random number between -1 and 1. For example, temp := rand( ). Notice that rand( ) is a function with no arguments!

mod functions

The function **mod** has two formats, one with two arguments and one with three. The expression **mod(E,G)** (where G > 0) means some integer multiple of G is added to or subtracted from E so that the result lies in the interval [0,G). The standard mathematical way of writing this is E mod(G), but that terminology won't work for *Dynamics*. The expression **mod(E,G,H)** where G < H means some integer multiple of H-G is added to or subtracted from E so that the result lies in the interval [G,H). For example, mod(4,2*pi) = 4 while mod(4, -pi, pi) = 4 - 2*pi.

Note. mod(E,0,H) is equivalent to mod(E,H).

if ( ) then ( ) else ( )

The expression "If (E) then (F) else (G)" denotes a function, since it returns a value. The expression E is a logical expression using

$$=, \ <, \ >, \ <> \ (\text{not equal}), \ <=, \ >= \ .$$

Here are some examples.

Example 1.

The following line declares x to be the maximum of y and z.
x := if (y > z) then y else z   ! notice that x is the maximum of y and z

Example 2.

The following Map window defines the Piecewise Linear map PL.

Map window
x := if (x <= 0) then (C1*x + y + rho) else (C3*x + y + rho)
y := if (x <= 0) then (C2*x) else (C4*x)

The following example uses "if ( ) then ( ) else ( )" and mod. Notice that if ( ) then ( ) else ( ) is a function of three arguments that has a value. In particular the "else" argument cannot be omitted.

Example 3.
The following Map window defines the Rotor map R.

Map window
x := mod(x+y, -pi, pi)
y := if (c1 <> 1) then (y + rho*sin(x+y))
else (mod(y + rho*sin(x+y), -pi, pi))

Note the last equation occupies two lines, but that is not a problem for the compiler. It can be rewritten:

y := if (c1 = 1) then (mod(y + rho*sin(x+y), -pi, pi))
                else (y +rho*sin(x+y))

if( ) and logical decisions
For more complicated logical decisions, "if ( )" uses key words

AND, OR, and NOT.

Hence

(x <= 0) OR (y < 0)

and

$(x^2 + y^2 > 1)$ AND NOT $(x^3 + y^3 < 2)$

are legitimate arguments for "if ( )".

13.3 ADDING A NEW DIFFERENTIAL EQUATION

Equation entry can be in one of two modes, called Map Mode or Differential Equation Mode. Each accepts three windows of inputs in addition to the Documentation window. In Map Mode these are called the map, inverse map, and initialization. In DE Mode these are called vector field, modulo, and initialization.

After selecting OWN, you will be in the Map Mode. To switch to the Differential Equation, press <F4> and four windows will appear on your screen, in approximately the following format, presenting an example of how to add a differential equation to *Dynamics*. In each of the four windows, the '!' means that the rest of the line is a comment.

Documentation window
forced damped Pendulum x'' + c1*x' + c2*sin(x) = rho*(c3 + cos(phi*t))
Vector field window
s' := 1   ! this is time t' := 1   ! this is time mod 2*pi/phi (see Modulo window) x' := y y' := rho*(c3 + cos(phi*t)) - c1*y - c2*sin(x)
Initialization window
t := 0   x := 1   y := 0 x_upper := pi   x_lower := -pi   y_upper := 4   y_lower := -2 c1 := 0.2   c2 := 1.0   c3 := 0.0   rho := 2.5   phi := 1 spc := 30   ipp := 30   ! Take 30 steps per 2*pi/phi and   ! plot once in 30 steps
Modulo window
t := mod (t, 2*pi) x := mod (x, -pi, pi)
Esc=Cancel Tab=Next F4=Diff Eqs F1=Compile Ctrl-K=EraseLine

Using the pendulum process P to create a basins picture is slightly faster on PC's than the corresponding picture made with OWN. For most purposes however, the difference in speed is less important than the convenience of getting the program running. Simply change what is in these four windows to define your map. Of course, there are some rules which are described later. We first describe the contents of the windows followed by some detailed examples.

The four windows follow essentially the same rules as for maps, and use the same variables and functions and parameters. See Section 13.2 for details.

Vector field window

Differential equations to be added are written in the Vector field window.

Initialization window

The step sized used by the differential equation solver is specified in this window.

Modulo window

Using the Modulo window is optional. The equation(s) in this window are applied after each time step of the differential equation solver.

Text in the Initialization window for differential equations

For maps it has been discussed that you are allowed not to enter anything in the Initialization window. All the parameters are set to be zero and the scales run from 0 to 1. The same holds for differential equations like the Lotka/Volterra equations. In addition to these settings of parameters, the step size is set to be 0.01.

For the forced damped pendulum and other periodically forced differential equations, it is required that the

```
spc := 30
```

is included. However, it may be better to include "ipp := 30". Since the forced damped Pendulum equation is in fact acting on a cylinder, you should also include "x_upper := pi   x_lower := -pi"

Therefore, the minimal contents of the Initialization window for the forced damped pendulum is the following:

Initialization window
x_upper := pi   x_lower := -pi
spc := 30   ipp := 30   ! Take 30 steps per 2*pi/phi and
! plot once in 30 steps

Example 13-7. Adding the Lotka/Volterra eqs. to *Dynamics*
In this example we add the Lotka/Volterra eqs.

$$X' = (C1 + C3*X + C5*Y)*X$$
$$Y' = (C2 + C4*Y + C6*X)*Y$$

to *Dynamics* using the OWN-feature. Although the Lotka/Volterra equations are a process of *Dynamics*, we will carry out this example. For our convenience, we use "Own-Lotka/Volterra DE" in the Documentation window. See also Section 13.4 for storing data and pictures. Start *Dynamics* and select **OWN**, hit <F4> and add the equations as indicated.

Documentation window
Own-Lotka/Volterra DE x' = (c1 + c3*x + c5*y)*x y' = (c2 + c4*y + c6*x)*y
Vector field window
t' := 1   ! this is time x' := (c1 + c3*x + c5*y)*x y' := (c2 + c4*y + c6*x)*y
Initialization window
step := 0.01 x_lower := 0   x_upper := 1000 y_lower := 0   y_upper := 1000 x := 10 y := 100 c1 := 0.5   c3 := -0.0005   c5 := -0.00025 c2 := 0.5   c4 := -0.0005   c6 := -0.00025
Modulo window
Esc=Cancel Tab=Next F4=Diff Eqs F1=Compile Ctrl–K=EraseLine

After completing the typing, hit <F1> to compile. The differential equation solver will use the "step" that is specified in the Initialization window. You may change its value, which is listed in the menu DEM.

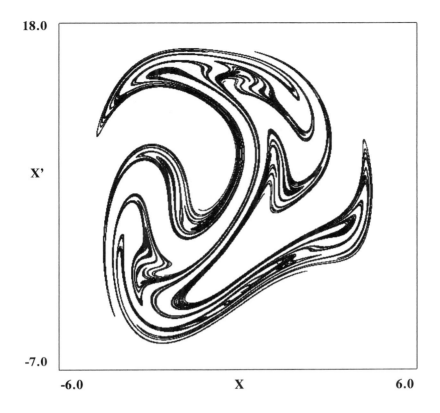

18.0

X'

-7.0

-6.0   X   6.0

Figure 13-3: A chaotic, strange attractor

This figure shows the attractor of the time-2π map of the GoodwiN differential equation (GN)

$$X'' + C1*((X*X-1)/(X*X+1))*X' - C2*X + C3*X*X*X = RHO*sin(phi*t)$$

The parameter values are C1 = 0.1, C2 = C3 = 0.5, phi = 1, and RHO = 37. After 1800000 dots are plotted, the approximate Lyapunov exponents are 0.087 and -0.145; the Lyapunov dimension of the attractor is 1.60.

This attractor is called a "strange attractor" because it appears to have infinitely many layers of curves. Each blow-up of the attractor reveals curves that split into multiple curves. It is called a "chaotic attractor" because it contains a trajectory which has a positive Lyapunov exponent.

Topic of discussion. *There are a number of spikes, both sharp and blunt shown on the outside of the attractor. Find the iterates of these points. Which spikes map to which?*

Example 13-8. Adding the GoodwiN equation to *Dynamics*

In this example we add the GoodwiN equation

$$x'' + c_1*((x^2 -1)/(x^2 +1))*x' - c_2*x + c_3*x^3 = rho*sin(phi*t)$$

to *Dynamics* using the OWN-feature. Although the GoodwiN equation is a process of *Dynamics*, we will carry out this example for comparison. For our convenience, we use "Own-GoodwiN DE" in the Documentation window. See also Section 13.4. Start *Dynamics* and select **OWN**, hit < F4 >, erase the example provided, and add the equations as indicated.

Documentation window
Own-GoodwiN DE $x'' + c_1*((x^2 -1)/(x^2 +1))*x' - c_2*x + c_3*x^3 = rho*sin(phi*t)$
Vector field window
s' := 1   ! this is time t' := 1   ! this is time mod 2*pi/phi (see Modulo window) x' := y y' := -c1*((x*x-1)/(x*x+1))*y + c2*x - c3*x*x*x + rho*sin(phi*t)
Initialization window
t := 0   x := 1   y := 0 x_upper := 6   x_lower := -6   y_upper := 17   y_lower := -7 c1 := 0.1   c2 := 0.5   c3 := 0.5   rho := 37   phi := 1 spc := 80   ipp := 80   ! Take 80 steps per 2*pi/phi and   ! plot once in 80 steps
Modulo window
t := mod (t, 2*pi)
Esc=Cancel Tab=Next F4=Diff Eqs F1=Compile Ctrl–K=EraseLine

After completing the typing, hit < F1 > to compile. Plot the trajectory and the resulting picture is shown in Figure 13-3.

Note. The variable phi is a special variable used in time-periodic differential equations; see below (p. 513). We could use temporary variable x2 replacing the y'-equation with two equations:

x2 := x*x

y' := -c1*((x2-1)/(x2+1))*y + c2*x - c3*x*x2 + rho*sin(phi*t)

BASIC RULES FOR ADDING A DIFFERENTIAL EQ. TO *DYNAMICS*

The rules are essentially the same as for maps; see pages 505-507. In addition:
- For each main variable used, there must be precisely one differential equation.

Temporary variables
In writing equations, you can also use temporary variables. They can have any name beginning with a letter using up to 8 letters and digits, except the names cannot be the names of parameters c1, ..., c9 or main variables r, s, t, ..., z or function names sin, exp, .. or reserved words pi, AND, NOT, ... mentioned earlier in this chapter. For an example, see xNew in Example 13-6. Each temporary variable must be defined in each window in which it is used, so "temp := temp+1" can not be the first line of a window.

Special parameter phi
Adding a differential equation is quite similar to the process for adding a map. Here the parameter phi has a special role and should be used only when there is a time dependent periodic forcing. It appears in arguments of periodic forcing functions. The forced damped Pendulum and the GoodwiN equation are examples of these of these type.

The program knows that a differential equation is going to be used when it finds the differential equation step size has been set, that is the variable "step" (for example, step = .01). The Lotka/Volterra equations is an example of this type. See also "Rules for periodically forced differential equations" below.

Rules for periodically forced differential equations
The pendulum example above has a time periodic term

$$\text{rho} * (C3 + \cos(\text{phi} * T))$$

in its equation

$$X'' + C1 * X' + C2 * \sin(X) = \text{rho} * (C3 + \cos(\text{phi} * T))$$

For such examples, it is common to plot points once every period. The program allows such plotting only if the variable phi is defined. The period is assumed to be 2*pi/phi. Hence even if you have a fixed period in your example, phi should be defined.

In the Initialization window you may also assign

$$SPC \text{ (steps\_per\_cycle).}$$

If phi is defined and SPC is not, it is given the default values 50, and the time step STEP is given the value

$$STEP = 2*PI/(PHI*SPC)$$

for differential equations. The default values are .01 and 50. If you set PHI, it should be greater than 0.

Another variable that can be set in the Initialization window is

$$IPP$$

It has a default value of 1. If you wish the trajectory to plotted once each period, give it the same value you give to PHI. Do NOT write IPP = PHI because PHI does not have a value in the Initialization window.

Example 13-9: Plotting a differential equation in polar coordinates

This example shows how to choose coordinates for plotting which are not usual coordinates. Here we add two coordinates r and s purely for the purpose of plotting. Since they are coordinates, they must be initialized and be given differential equations. Since they are assigned explicit values in the Modulo window, their initializations and differential equations have no influence on their values after each time step. Notice below that in the Modulo window we make the following assignments:

Modulo window
r := (y + 2)*sin(x)
s := (y + 2)*cos(x)

The variables used are r, s, t, x, y in alphabetical order, so r and s correspond to y[0] and y[1]. To plot these, the Initialization window sets:

$$Xco := 0 \ Yco := 1$$

Then trajectories will plot (r,s). Notice that since they are variables, they must be initialized and must have differential equations even though these initializations and equations do not affect the values of r and s when they are plotted.

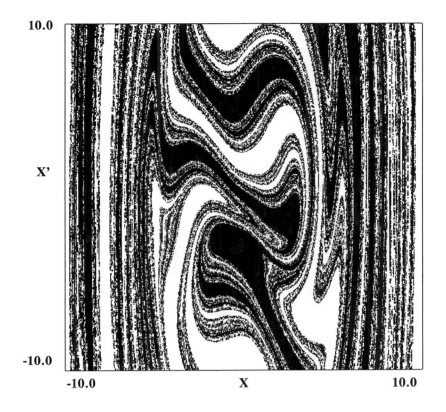

Figure 13-4: Basin of attraction

The black area is the basin of attraction of a stable period 2 orbit of the time-2π map of the GoodwiN differential equation (GN)

$$X'' + C1*((X*X-1)/(X*X+1))*X' - C2*X + C3*X*X*X = RHO*sin(phi*t)$$

There are two stable periodic orbits of period 2, and the white area is the basin of attraction of the other stable period 2 orbit. The parameter values are C1 = 0.75, C2 = C3 = 0.5, phi = 1, and rho = 14.

Start *Dynamics* and select **OWN**, hit < F4 > and add the equations as indicated below.

Warning: In this example, the variables r and s are not the variables of the differential equations and some routines may behave strangely. In the above example changing r and s with arrow keys does not affect x and y.

Documentation window
forced damped Pendulum x'' + c1*x' + c2*sin(x) = rho*(c3 + cos(phi*t)) Polar coordinate plot: x is an angle and y + 2 is the distance to the origin; r = y[0] s = y[1] t = time = y[2] x = y[3] y = y[4]
Vector field window
r' := 0 s' := 0 ! these are dummy equations;   ! they don't affect plots. t' := 1   ! this is time mod 2*pi/phi (see Modulo window) x' := y y' := rho*(c3 + cos(phi*t)) - c1*y - c2*sin(x)
Initialization window

t := 0   x := 1   y := 0
r := 0 s:= 0 ! r and s must be initialized
                    ! even though the values do not affect plots.
x_upper := 6   x_lower := -6   y_upper := 5.5   y_lower := -4.5
c1 := 0.2   c2 := 1.0   c3 := 0.0   rho := 2.5   phi := 1
spc := 30   ipp := 30   ! Take 30 steps per 2*pi/phi and
                                  ! plot once in 30 steps

Modulo window
t := mod (t, 2*pi) x := mod (x, -pi, pi) r := (y + 2)*sin(x) s := (y + 2)*cos(x)
Esc=Cancel Tab=Next F4=Diff Eqs F1=Compile Ctrl–K=EraseLine

13.4 STORING DATA AND PICTURES OF AN OWN-PROCESS

After tapping <F1> and successfully entering the stage for iteration, you can store your equations using the Dump Data command DD or the TO Disk TD (which stores the picture you have created as well as the data).

Here's how this might appear in a DD file. Each window is encapsulated by quotes. The DD file continues on (not shown here) with additional changes you have made, such as changes in parameters or scales, etc. These are not reflected in the map but are contained in the file. See Section 4.4 for the use of the *.dd and *.pic files.

```
own /* process from map menu */
1
"The Henon Map
(x,y) --> (rho - x*x + c1*y,x)
"
"x := rho - x*x + c1*y
y := x"
"x := 0
y := 2
c1 := -0.3
rho := 2.12
x_lower := -2.5 x_upper :=  2.5
y_lower := -2.5 y_upper :=  2.5"
"x := y
y := (x - rho + y*y) / c1"
```

Here's how the pendulum equation might appear in a DD file:

```
own /* process from map menu */
0
"Forced damped pendulum
X'' + C1 * X' + C2 * sin(X) = rho * (C3 + cos(phi * T))
"
"s' := 1  ! this is time
t' := 1 ! this is time mod 2 pi/ phi
x' := y
y' := rho * (C3 + cos(phi * t)) - C1 * y - C2 * sin(x)
"t := 0 x := 1 y := 0
X_upper := pi  X_lower := -pi  Y_upper := 4  Y_lower := -2
C1 := 0.2   C2 := 1.0   C3 := 0.0   rho := 2.5   phi := 1
spc := 30  ipp := 30"
"t := mod(t, 0, 2*pi / phi)
x := mod(x, -pi, pi)"
```

Root name

If you have created your own process using *Dynamics*'s OWN feature, the the default root name is "own". Hence, if you save data using command DD, then the data is saved in file **own.dd**. If you store a picture that has been created by this process, then it is stored in file **own.pic**.

We recommend to change, for example, the root name of the Holmes map to **o-ho**, and the root name of the GoodwiN equation to **o-gn**. Of course, you may want variants for these. As an example, the picture that has been created in Figure 13-1a can be stored in the file "o-ho.pic", but a file name like "o-ho-um.pic" might be better to indicate that the picture is an unstable manifold for the Holmes map that was created by the own-feature of *Dynamics*.

CHAPTER 14

DYNAMICS ON UNIX SYSTEMS

by Brian R. Hunt and Eric J. Kostelich

CONTENTS

For Unix updates and for information on modifying the program, please consult
http://www.ipst.umd.edu/dynamics/

There are a variety of implementations of Unix and the X Window System. As a result, we generally cannot give commands guaranteed to cover all cases. You may need to consult your system manager for assistance in carrying out these instructions.

Getting the source code from the floppy disk

1. The source files are provided on a DOS formatted floppy disk in the file UNIX.TGZ. This file is in Unix "tar" format, compressed with the program "gzip", which is a commonly used file compression program provided by the Free Software Foundation. Your system **must** have "gzip" installed in order to extract the individual source files from this compressed archive. Details of the extraction procedure, and what to do if you do not have "gzip", are described in Step 2 below. Assuming that you do have "gzip", then first you must copy the file UNIX.TGZ from the floppy disk to your Unix workstation. You can do this in one of two ways.

   a. If your Unix workstation has a 3.5" floppy disk drive, and software installed for reading DOS formatted disks, then you can copy UNIX.TGZ directly to your hard disk drive. On many systems, the command you should use is called "mcopy". In that case you would insert the floppy disk into your disk drive and type:

      mcopy   a:unix.tgz   unix.tar.gz

   This will make a copy of the file under the name *unix.tar.gz* in your current directory on the Unix system. If you do not have the program "mcopy" and are uncertain how to copy files from your floppy disk drive, then consult your system manager. Be aware that UNIX.TGZ is **not** a text file and you should be sure to copy it in binary mode, not in text mode. Also, naming the copy of the file *unix.tar.gz* is advised but not strictly necessary. You should now go to Step 2.

   b. If your Unix workstation does not have a floppy disk drive, or if the software to read DOS formatted disks is unavailable, then you must have access to a PC from which you can upload files to your Unix workstation. How exactly you will do the file transfer depends heavily on your particular hardware and software configuration. Consult your system manager if you are unsure how to proceed. With whatever program you use, it is important to make sure that you transfer UNIX.TGZ in binary mode, **not** in text mode. In Step 2 below it is assumed that you rename the file, either during or after the transfer, to be called *unix.tar.gz*, though this is not strictly necessary.

---

2. Assuming the file now to be called *unix.tar.gz* on the hard disk drive of your Unix workstation, you should type the following command:

gzip -cd unix.tar.gz | tar xf −

(If the name of the file is in captital letters, then you must type its name using capital letters in the command above.) If the "gzip" program is available for you to run, then the command above will create a subdirectory called "dydir" in the current directory and put the source files in the subdirectory. Type "cd dydir" and you will be ready to compile the program.

If you get an error message that the command is not found when you try to run "gzip", then either the program is not installed on your system or is not in the directory "path" used to find the programs you commonly run. We suggest replacing "gzip" with "/usr/local/bin/gzip" and repeating the command above. If that does not work, then consult your system manager to see if "gzip" is available. If not, then you or your system manager can check the *Dynamics* web page,

http://www.ipst.umd.edu/dynamics/

for instructions on where to get "gzip".

If you get any other type of error message from the command above, it is likely that *unix.tar.gz* was corrupted by being transfered from the floppy disk in text mode rather than binary mode. Make sure you are using binary mode in your file transfer and try again.

Compiling Dynamics on Unix/X11 workstations

1. You must compile the program on a machine running a version of the Unix operating system, having a C compiler, and supporting version 11 of the X window system (for short, "X11"), release 4 or later (the current release number is 6). You must run the program from a workstation with a bitmapped screen that also supports X11 release 4 or later. If you have an earlier release, see item 2 below. You cannot run the program from a conventional, text-only terminal.

Please note that you **must** have the X11 object code library (usually called libX11.a) and header files (also known as "include files") installed on your workstation in order to compile and run *Dynamics*. There are no extra source code files that we can include to work around this.

For example, many older Sun, Silicon Graphics and Apollo workstations have windowing systems, but they are not built on X windows. In particular, the SunView windowing system (on Sun 3 and older machines) and the 4Sight/NeWS windowing system (on older Silicon Graphics machines) are **not** built on X windows. If you receive error messages from the compiler to the effect that files like X11/X.h cannot be found, then you should consult with your system manager to determine whether X windows software is installed on your machine.

2. Edit "makefile" to set up the appropriate C compiler flags and to reflect the location of the software libraries on your computer. On many systems, no modifications to the makefile are necessary; however, there are still some options you can select by editing the makefile that may improve the performance of the program. Directions for making modifications are included in the makefile; follow them carefully. In particular, read the lines following "# USER MODIFICATIONS". (Note that lines beginning with '#' are comments.)

3. Type

make

to compile and link the program. The executable file is called "dynamics".

Note that the X11 object library (which is a file that usually is called libX11.a), as well as the C language header files, may be located in different directories depending on your implementation of X11. The usual location for libX11.a is in /usr/lib, but on other systems, the object library may be located in /usr/local/lib, or elsewhere. If you have trouble finding it, consult your system manager to determine where it is installed on your workstation. Then include the appropriate directory on the X11DIR line in the makefile.

The C language header files usually are located in /usr/include/X11. If they are located in a different directory, then you must edit the INCLUDEDIR line in the makefile as described below under "Porting the program to other Unix platforms".

Installing the program

You may wish to "install the program" so that others can use it remotely, and the program will know where to look for auxiliary files.

1. Edit the makefile and set INSTALLDIR to the fully qualified path name of the directory which will hold support files. The support files are mostly text files used to construct help menus. In the default version,

---

INSTALLDIR is set to /usr/local/lib/dydir. The directory must be created if it does not already exist. Likewise, the executable program "dynamics" must be placed somewhere where your shell looks for executable programs. This is the directory specified by BINDIR. In the default makefile, BINDIR is set to /usr/local/bin.

If you do not have system privileges on your machine, then set INSTALLDIR to be a subdirectory of your own home directory, and set BINDIR to be a subdirectory of your home directory (e.g., $HOME/bin). INSTALLDIR and BINDIR can be the same directory if you want.

2. Type "make install" to copy the executable program and all support files to the proper places.

Running the program

1. If you have not followed the instructions above to install the program, then set the DYNAMICS environment variable to the full name of the directory that holds the support files. For example, if the files are in the subdirectory dydir of your home directory /homes/myname, then type:

setenv DYNAMICS /homes/myname/dydir

Often, you may refer to your home directory as $HOME in a command like this. It is most convenient to put the command above in one of your shell's startup files (typically $HOME/.cshrc for the C shell or $HOME/.profile for the Bourne and Korn shells).

2. Set the DISPLAY environment variable to reflect the name of the X window display for your graphics. Generally this is the name of the computer at which you are sitting followed by ":0". If you are logged onto the graphics console of the machine from which you will run the program, then DISPLAY should be set by the system to something like "localhost:0" or simply ":0".

3. In some cases, such as on Linux machines and certain older Sun workstations, you must first start the X server manually. The details vary considerably from one installation to the next. However, on most Sparcstations equipped with OpenWindows software, you can start the X server by typing the command "openwin". If your Sun workstation does not have X window software, then you **cannot** run the *Dynamics* program on it. (In particular, the *Dynamics* program will **not** run under SunView.)

4. The operation of the program is otherwise the same as described in previous chapters. Keep in mind though that *Dynamics* runs in a window and that in order to interact with the program, you must direct input to that window using the mouse. Depending on your setup, you may have to click on the title bar of the window, click in the window, or just move the mouse pointer into the window.

5. Any printing that you request will create a file whose name has the form "<root>.prn" in the current directory, where "<root>" is by default the name of the process (differential equation or map) you are running. This file contains the print codes for whatever printer you have selected. A list of supported printers is given in Chapter 1. You must then send the contents of the file to your system's print manager using a Unix command like lpr, lp, or qpr. See "Printing Pictures" below for details on how to print and how to select the name of the print file.

Customizing the program

1. The X11 version of *Dynamics* uses the X resource facility in a primitive way and allows for rudimentary customization of the program. If you want to use this facility, you should first define the XENVIRONMENT environment variable to the full name of the file containing your X window customizations. In most cases, this will be the file ".Xdefaults" in your home directory; then set XENVIRONMENT to $HOME/.Xdefaults in your shell's startup file as described in step 1 of "Running the program" above.

2. The following resources can be specified:
   font
   geometry
   borderWidth

   By default, the font used to display text is "9x15", which is present on most X11 implementations. If it is not, or if you don't like the way it looks, you can specify a different font, for instance "9x15bold", by placing a line of the form

   dynamics*font: 9x15bold

in the file named by XENVIRONMENT.

Note. The program assumes that the requested font is a fixed-size font, i.e., a typewriter-like font like Courier and not a proportionally-spaced one like Times-Roman. If you specify a proportionally-spaced font, then characters will not be placed correctly on the screen.

You can also change the default size of the window by placing a line of the form

dynamics*geometry: 800x800

in the file named by XENVIRONMENT. In this example, the geometry specifies a window that is 800 pixels wide and high. In general, the geometry specification takes the form

dynamics*geometry: WxH+X+Y

where W, H are width and height respectively, and X, Y are optional horizontal and vertical offsets from the upper left corner of the display. All units are in pixels.

Note. Window managers are not required to honor the requested size, so the window you get may not be exactly the specified size.

In order to get the changes in your XENVIRONMENT file to take effect, you may need to run the command

xrdb $XENVIRONMENT

If this command does not exist or if it does not work, then log out and log in again. The *Dynamics* program should start up with the specified font and geometry.

Common problems

1. You get an error message that says "can't get display".

   Suggestion: If you are running the program on the same machine as the graphical display, check that the DISPLAY environment variable is set according to the directions on p. 501 and make sure that the X server is running.

   If you are running the program on a remote machine, verify that you have permission to display graphics requests from the remote machine. You will probably have to run the "xhost" utility to do the appropriate authentication. See manual page for xhost or check with your system manager for details. See also "Running Dynamics from a remote machine", below.

2. You get an error message that says "can't load font".
   Suggestion: Try loading a different font. Use the "xlsfonts" utility to list the available fonts and select a fixed-width one from the list. Place the name of the font in your XENVIRONMENT file as described above, and rerun.

Porting the program to other Unix platforms

1. This program has been compiled and run on computers made by Hewlett Packard, IBM (RS/6000), Sun, Digital Equipment Corporation, Silicon Graphics, as well as PCs running Linux, FreeBSD, and other Unix implementations.

2. If you want to port the program to a different machine, you must be sure that the proper include files and the X11 library file are available and that keyboard keys are mapped properly (see "Keyboard notes", below). Most difficulties can be avoided by adding appropriate options in the makefile. Common difficulties:
   a) Your machine does not have the X11 include files and/or library file in a standard place.
      Suggestion: If the problem is missing include files, then you will get error messages immediately after typing "make" and the problem will be clear from the context of the messages.

      If the library file can't be found, the error will not occur until the very end of the compilation. In either case, determine the directory that contains the missing files and place the appropriate option in the makefile. For example, if the include files are in /usr/local/include/X11, then the option "-I/usr/local/include" on the INCLUDEDIR line should be used. Likewise if the X11 library file libX11.a is in /usr/local/lib, then add the option "-L/usr/local/lib" to the X11DIR line.
   b) Your machine lacks an include file mentioned in "yinclud.h".
      Suggestion: Remove the line that generates the error message from yinclud.h and try to compile again.

3. Changes may also be required to the LIBS line in the makefile to specify the name of the X11 library. Additional loader flags can also be placed there, if necessary.

Keyboard notes

The program assumes that you have 10 function keys, labeled F1 to F10. Some keyboards don't have this many, or they may be labeled differently. If this is the case, then you will have to make changes to the file "ydefines.h" to map the missing keys to something else. You may find it helpful to peruse the X11 include file that defines keyboard codes, in order to see what substitutions can be made. On most systems this file is called "/usr/include/X11/keysymdef.h".

In addition, the program uses the keys <Home>, <PgUp>, <End> and <PgDn>. Many Unix workstations have PC-like keyboards that include these keys. They may not be present on your machine. The upper case '1', '2', '3', and '4' will also work, that is '!' = <Home>, '@' = <PgUp>, '#' = <End>, and '$' = <PgDn>. Or you can edit "ydefines.h" to map them to some other keys.

After you edit "ydefines.h", type "make" to recompile the program.

Running Dynamics from a remote machine

Suppose you are seated at the console of a local machine called "local.cpu" and you can remotely log into a remote machine called "remote.cpu". If both machines have X11 capability, then you can run *Dynamics* on the remote machine (perhaps remote.cpu has faster floating point hardware) and display the graphics on local.cpu.

This requires a few preliminary steps.

1. On local.cpu type the command

xhost remote.cpu

This gives the remote machine permission to open a graphics window on your local console. Now remotely log in ("rlogin remote.cpu") to the computer remote.cpu. You of course will need an account and password on that computer also.

2. On remote.cpu type the command (assuming you're in the C shell)

setenv DISPLAY local.cpu:0

This tells the remote machine to display the graphics on your local console.

3. Execute the *Dynamics* program in the usual way on remote.cpu. When you are done you can "logout" from remote.cpu.

Notes

a. This procedure can fail if the local and remote machines are separated by a busy network. For example, it is theoretically possible to run *Dynamics* on a computer in California and display the results on a machine in Virginia, but this is not likely to work well because of the amount of traffic on the Internet.

b. If you routinely run *Dynamics* between the same pair of machines you can add the commands in 1) and 2) above to your login command files as appropriate.

c. Step 1) above may be unnecessary if you compute on a closely networked set of workstations. It is possible for your system manager to give remote.cpu blanket permission to display graphics on local.cpu.

d. The computer remote.cpu may run *Dynamics* faster when used remotely than if you were to log in directly on remote.cpu because local.cpu will be assisting with the graphics.

Printing pictures

When you are using a PC, pictures can be sent directly to a printer using the Print Picture command PP. On a Unix computer, the command PP sends the picture to a file with a name of the form <root>.prn (where <root> is a name you can select using the Root file Name command RN). Then you must send <root>.prn to your printer using your local printing commands. If you choose the Epson or HP printer type, the print file will contain binary data, so you may need to modify your print command accordingly. If your print command is "lpr", you should type

lpr   -x   root.prn

If your computer system and printer support PostScript printing, we recommend that you use the PS or PSC printer type. Many Hewlett Packard printers now are able to print PostScript files. Consult your system manager if you are unsure how to print.

Bug reports

If you encounter unusual difficulties that are not otherwise covered in this discussion, please send email to dynamics-authors@ipst.umd.edu.

CHAPTER 15

APPENDIX

15.1 THE MAP Q

Most of maps and differential equations are given explicitly in the program, but the map Q was too complicated to do this. The map Q for the study of Quasiperiodicity is defined as follows.

    y[0] = old_y[0] + C1 + RHO*p1/twopi   mod 1;
    y[1] = old_y[1] + C2 + RHO*p2/twopi   mod 1;

where the functions p1 and p2 involve the lowest order sin() terms:

    p1 =   A[1]*sin(twopi*(X+k[1])) + A[2]*sin(twopi*(Y+k[2]))
          + A[3]*sin(twopi*(X+Y+k[3])) + A[4]*sin(twopi*(X-Y+k[4])));
    p2 =   B[1]*sin(twopi*(X+j[1])) + B[2]*sin(twopi*(Y+j[2]))
          + B[3]*sin(twopi*(X+Y+j[3])) + B[4]*sin(twopi*(X-Y+j[4])));

We wanted p1 and p2 to representative maps that were periodic in X = y[0] and Y = y[1] so we imagined p1 and p2 to be expanded in a Fourier series and then we took the lowest order terms and chose their coefficients using a random number generator. Those choices now follow. The constants j[], k[], A[], B[1] were chosen using a random number generator and so are complicated:

A[1] = -.2681366365;	k[1] = .985460843;
A[2] = -.910675594;	k[2] = .5044604561;
A[3] = .3117202638;	k[3] = .9470747252;
A[4] = -.0400397884;	k[4] = .233501055;

B[1] = .0881861167;	j[1] = .9903072286;
B[2] = -.5650288998;	j[2] = .3363069701;
B[3] = .1629954873;	j[3] = .2980492123;
B[4] = -.8039888198;	j[4] = .1550646728;

The following constants which may be chosen arbitrarily are given with their default values.

C1 = 42./100.;
C2 = 3./10. ;
RHO = 6./10.;

15.2 LIST OF MENU COMMANDS

```
AXM   – –      DRAW AXES MENU (2 dimensional)

XAX   –  X AXis
XAX1  –  X Axis with tic marks
XAX2  –  X Axis with double tic marks
XAXI  –  X Axis Intercept (height)
YAX   –  Y AXis
YAX1  –  Y Axis with tic marks
YAX2  –  Y Axis with double tic marks
YAXI  –  Y AXis Intercept (X value)
```

```
AX3M   – –  ROTATE AXES MENU (3 dimensional)

ANGLE  –  ANGLE of rotation
AD     –  Axes Default orientation
AX     –  make the X axis the rotation Axis
AY     –  make the Y axis the rotation Axis
AZ     –  make the Z axis the rotation Axis
D2     –  2 Dimensional mode
D3     –  3 Dimensional mode
ZCO    –  Z COordinate
ZS     –  Z Scale
```

```
BASPM   – –    BAS PLOTTING SCHEME MENU

BT     –  plot Basin with color number = capture_Time mod(16)
BT1    –  plot Basin with color number = capture_Time up to 16
BT2    –  BT with color number = numColors capture_Time / MC
BN     –  plot Basin with color number = Number of ball entered
BMOD   –  plot Basin with color number = 1+capture_Time(modPR)
BP     –  plot periodic attractor period and chaotic attractors
BASA   –  All: plot for All diverging trajectories -- default
BASN   –  Not: plot for No diverging trajectory
BAST   –  Top: plot whenever diverging with Y coord > 0
BASR   –  Right: plot when diverging with X coord > 0
BASL   –  Left: plot when diverging with X coord < 0
BASB   –  Bottom: plot when diverging with Y coord < 0
TV     –  Top+V: plot when inVerse diverges with Y coord > 0
```

```
┌─────────────────────────────────────────┐
│  BCM  – –   Background Color Menu         │
│                                           │
│  BLA  –  BLAck background color           │
│  BLU  –  BLUe background color            │
│  GRE  –  GREy background color            │
│  WHI  –  WHIte background color           │
└─────────────────────────────────────────┘
```

```
┌──────────────────────────────────────────────────────────────────┐
│  BIFM  – –      BIFURCATION DIAGRAM MENU                           │
│                                                                    │
│  BIFS   –  make diagram on the Screen                              │
│  BIF    –  make diagram on the printer                             │
│  BIFD   –  iterates (Dots) for each PRM                            │
│  BIFI   –  re-Initialize y for each PRM                            │
│  BIFP   –  Print parameter values on paper (for BIF only)          │
│  BIFPI  –  PreIterates for each PRM                                │
│  BIFR   –  the Range of parameter PRM                              │
│  BIFV   –  number of Values of PRM (about 720 per page)            │
│  PRM    –  PaRaMeter to be varied                                  │
└──────────────────────────────────────────────────────────────────┘
```

```
┌──────────────────────────────────────────────────────────────────────┐
│  BM   – –      BASIN OF ATTRACTION MENU                                │
│                                                                        │
│  BA   –  Basins and Attractors                                         │
│  BAN  –  BAsin for Newton method for points of period PR               │
│  BAS  –  plot BASin                                                    │
│  MC   –  Maximum Checks per initial point                              │
│  RA   –  Radius of Attraction for storage vectors                      │
│  WW   –  check Where the trajectory of y1 goes and When it gets        │
│            within RA of y2, ..., y7 or diverges within MC iterates     │
└──────────────────────────────────────────────────────────────────────┘
```

```
┌──────────────────────────────────────────────────────────────────────┐
│  BRM     – –      BASIN RESOLUTION MENU                                │
│                                                                        │
│  RB    –  Basin flexible Resolution, now:  100  ×  100  grid           │
│  RL    –  Low Resolution:                  100  ×  100  grid           │
│  RMED  –  MEDium Resolution:               240  ×  200  grid           │
│  RMH   –  Medium-High Resolution:          320  ×  272  grid           │
│  REGA  –  EGA Resolution:                  640  ×  360  grid           │
│  RVGA  –  VGA Resolution:                  640  ×  480  grid           │
│  RH    –  High Resolution (= core grid):   720  ×  720  grid           │
└──────────────────────────────────────────────────────────────────────┘
```

```
┌─────────────────────────────────────────────────┐
│ BXM  – –      BOX MENU                            │
│                                                   │
│ B     – draw Box                                  │
│ B1    – draw Box with tic marks                   │
│ B2    – draw Box with double tic marks            │
│ BB    – set Box equal to entire window            │
│ BLL   – set Box Lower-Left corner at y1           │
│ BUR   – set Box Upper-Right corner at y1          │
└─────────────────────────────────────────────────┘


┌─────────────────────────────────────────────────┐
│ CM   – –      COLOR MENU                          │
│                                                   │
│ CTM  – Color Table Menu                           │
│ EM   – Erase color Menu                           │
│ CI   – get Color Information                      │
│ CNT  – CouNT pixels of each color in picture      │
│ COL  – 256 colors on the screen                   │
│ IN   – INcrement color number when plotting       │
│ SDEF – Screen DEFault mode                        │
│ F7   – decrease color number by 1                 │
│ F8   – increase color number by 1                 │
│ F9   – choose color number                        │
└─────────────────────────────────────────────────┘


┌─────────────────────────────────────────────────┐
│ CTM  – –      COLOR TABLE MENU                    │
│                                                   │
│ CT   – display Color Table                        │
│ CI   – get Color Information                       │
│ SC   – Set Color                                  │
│ SCT  – Set Color Table                            │
│                                                   │
│ BGY  – Blue to Green to Yellow                    │
│ BRI  – BRIght colors                              │
│ BRY  – Blue to Red to Yellow to white             │
│ DEF  – DEFault color table                        │
│ RGB  – Red to Green to Blue to white              │
│ RNB  – RaiNBow: Red Orange Yellow – –             │
│              Green Blue Indigo Violet             │
│ RRR  – dim Red to Red to bright Red               │
│ RYB  – Red to Yellow to Blue                      │
│ RYW  – Red to Yellow to White                     │
│ WYR  – White to Yellow to Red                     │
└─────────────────────────────────────────────────┘
```

```
┌─────────────────────────────────────────────────────────────────────┐
│                    DIFFERENTIAL EQUATIONS MENU                        │
│                                                                       │
│  CC    –  Chua's Circuit                                              │
│  CF    –  Cylinder-Flow                                               │
│  D     –  forced double-well Duffing equation                         │
│  GN    –  GoodwiN equation                                            │
│  HAM   –  HAMiltonian system, h = (p^2 + q^2)/2 + u^2*v^2 = const     │
│  L     –  Lorenz system                                               │
│  LPR   –  Lorenz-Poincare-Return-map                                  │
│  LV    –  Lotka/Volterra equations                                    │
│  P     –  forced damped Pendulum equation                             │
│  PD    –  Parametric Duffing equation                                 │
│  ROS   –  ROeSsler equation                                           │
│  R3    –  Restricted-3-body-problem                                   │
│  SG    –  Samardzija/Geller odd-symmetry lorenz-like system           │
│  VP    –  forced Van der Pol equation                                 │
│  VP2   –  2 forced coupled Van der Pol equations                      │
└─────────────────────────────────────────────────────────────────────┘
```

```
┌─────────────────────────────────────────────────────────────────────┐
│  DEM     – –    DIFFERENTIAL EQUATION MENU                            │
│                                                                       │
│  EULER  –  Euler solver, fixed step size                             │
│  RK4    –  4th order Runge-Kutta solver, fixed step size             │
│  RK5    –  5th order Runge-Kutta solver, variable step size          │
│  RK6    –  6th order Runge-Kutta solver, fixed step size             │
│  E      –  compute single step Error every  5   steps                │
│  LE     –  Local Error for RK5                                        │
│  RET    –  Poincare RETurn map (for process HAM) and for             │
│              periodically forced diff. equations (using phi)         │
│  DF     –  plot the Direction Field                                   │
│  STEP   –  STEP size for differential equation                        │
│  SPC    –  Steps_Per_Cycle                                            │
└─────────────────────────────────────────────────────────────────────┘
```

```
┌─────────────────────────────────────────────────────────────────────┐
│  DIM     – –     MENU FOR ESTIMATING DIMENSION                       │
│                                                                       │
│  BD    –  Box counting Dimension (also called capacity)              │
│  BD1   –  like BD but more information is provided                    │
│  CD    –  Correlation Dimension                                       │
│  CD1   –  like CD but more information is provided                    │
│  CDM   –  Correlation Dimension Method description                    │
│  LL    –  List Lyapunov numbers and Lyapunov dimension                │
└─────────────────────────────────────────────────────────────────────┘
```

```
DM    - -    DISK MENU

RN    -    Root Name (to change root name)
DD    -    Dump Data to disk file h.dd
PCX   -    copy picture to PCX file root.pcx readable by MS Windows
PPP   -    if on, Printer output goes to file root.prn
PSC0  -    copy picture to PostScript Color file with no text header
PSC2  -    copy picture to PostScript Color file with text header
PSP0  -    copy picture to PostScript file with no text header
PSP2  -    copy picture to PostScript file with text header
REC   -    RECord all keystrokes in file root.rec
TD    -    copy picture and parameters To Disk file root.pic
TDD   -    call TD every  TDD  Dots (no call if TDD = 0)
TDM   -    call TD every  TDM  Minutes (no call if TDM = 0)
```

```
EM    - -    ERASE COLOR MENU

EK    -    Erase color K where K < 16
EOD   -    Erase all ODd numbered colors
EEV   -    Erase all EVen numbered colors
PAT   -    erase using shaded PATtern
```

```
FM    - -    FILE MENU

FD    -    retrieve picture (From Disk)
PP    -    Print Picture
TD    -    save picture (To Disk)
DYN   -    quit & start new map or Differential Eqn.
OWN   -    quit & create OWN process
Q     -    Quit Dynamics program
```

```
FOM   - -    FOLLOW ORBIT MENU

FO    -    Follow Orbit as parameter is varied
FOB   -    Follow Orbit Backwards
FOS   -    Follow Orbit Step size
FOV   -    reVerse direction for FO
```

```
┌─────────────────────────────────────────────────────────────┐
│ H          – –   HELP MENU                                    │
│                                                               │
│ INDEX   –  list all the commands whose descriptions           │
│               have a word you choose                          │
│ MOUSE   –  what you can do with a MOUSE                        │
│ TUT     –  for the quick start TUTorial                       │
└─────────────────────────────────────────────────────────────┘
```

```
┌─────────────────────────────────────────────────────────┐
│ KM    – –      KRUIS (CROSS) MENU                         │
│                                                           │
│ I      –  Initialize y using y1                           │
│ II     –  plot y1 & its MI = 1 Iterates                   │
│ K      –  big cross along trajectory                      │
│ KK     –  permanent cross at y                            │
│ KK1    –  permanent cross at y1                           │
│ KKK    –  permanent cross at each point                   │
│              as a trajectory is plotted                   │
│ KKS    –  permanent cross Size                            │
│ O      –  draw circle at y1                               │
│ OI     –  draw circle at y1 & its MI Iterates             │
│ W      –  show coordinates of y1                          │
└─────────────────────────────────────────────────────────┘
```

```
┌───────────────────────────────────────────────────────────────────┐
│ LM    – –    LYAPUNOV MENU                                          │
│                                                                     │
│ LL    –  List Lyapunov exponents (results of L thus far)            │
│ L     –  number of Lyapunov exponents                               │
│                                                                     │
│ If Lyapunov exponents are computed (L > 0), then:                   │
│ PT    –  T Plots Time horizontally and Lyap. exps. vertically if ON │
│ BAS   –  plot chaotic basin or create Chaos plot                    │
│ BIF   –  plot Lyapunov exponents vs. parameter on printer           │
│ BIFS  –  plot Lyapunov exponents vs. parameter on the screen        │
└───────────────────────────────────────────────────────────────────┘
```

```
┌─────────────────────────────────────────────────┐
│                  MAP MENU 1                       │
│                                                   │
│  C      –  complex Cubic map                      │
│  CAT    –  linear map on unit square              │
│  CIRC   –  CIRCle map                             │
│  CW     –  CobWeb map, a 1-dim map                │
│  DR     –  Double Rotor map                       │
│  H      –  Henon map                              │
│  I      –  Ikeda map                              │
│  KY     –  Kaplan/Yorke map                       │
│  LOG    –  LOGistic map                           │
│  Q      –  Quasiperiodicity map                   │
│  R      –  pulsed Rotor & standard map            │
│  T      –  Tinkerbell map                         │
│  TT     –  TenT map                               │
│                                                   │
│  QX     –  Quit and eXit the program              │
└─────────────────────────────────────────────────┘
```

```
┌─────────────────────────────────────────────────┐
│                  MAP MENU 2                       │
│                                                   │
│  CG    –  Chossat/Golubitsky symmetry map         │
│  DDR   –  Degenerate Double Rotor map             │
│  GB    –  Generalized Baker map                   │
│  GM    –  Gumowski/Mira map                       │
│  H2    –  Henon with 5th order polynomial         │
│  HZ    –  Hitzl/Zele map                          │
│  LPR   –  Lorenz system's Poincare Return map     │
│  N     –  Nordmark truncated map                  │
│  PL    –  Piecewise Linear map                    │
│  RR    –  Random Rotate map                       │
│  TRI   –  TRIangle map                            │
│  YAK   –  2 competing populations                 │
│  Z     –  Zaslavskii rotation map                 │
│                                                   │
│  QX    –  Quit and eXit the program               │
└─────────────────────────────────────────────────┘
```

```
╔══════════════════════════════════════════════════════════════╗
║                   MOUSE SET VECTOR MENU                        ║
║                                                                ║
║  MV     –  set y and y1 to current mouse arrow position        ║
║                                                                ║
║  MV0    –  Set Vector y        MV1   –  Set Vector y1          ║
║  MV2    –  Set Vector y2       MV3   –  Set Vector y3          ║
║  MV4    –  Set Vector y4       MV5   –  Set Vector y5          ║
║  MV6    –  Set Vector y6       MV7   –  Set Vector y7          ║
║  MV8    –  Set Vector y8       MV9   –  Set Vector y9          ║
║                                                                ║
║  MVA    –  Set Vector ya       MVB   –  Set Vector yb          ║
║  MVC    –  Set Vector yc       MVD   –  Set Vector yd          ║
║  MVE    –  Set Vector ye                                       ║
║                                                                ║
║  MQ9    –  apply Quasi-Newton 9 times                          ║
║  MWW    –  initialize and WW:                                  ║
║                check When and Where the trajectory goes.       ║
╚══════════════════════════════════════════════════════════════╝
```

```
╔══════════════════════════════════════════════════════════════╗
║  NEM    – – NUMERICAL EXPLORATIONS MENU                        ║
║                                                                ║
║  T      –  plot Trajectory                                     ║
║  DYN    –  quit & start new map or Differential Eqn.           ║
║  OWN    –  quit & create OWN process                           ║
║  P      –  Pause the program                                   ║
║  Q      –  Quit Dynamics program                               ║
║                                                                ║
║            MENUS                                               ║
║  BIFM   –  BIFurcation diagram Menu                            ║
║  BM     –  Basin of attraction Menu                            ║
║  DIM    –  DImension Menu                                      ║
║  FOM    –  Follow (periodic) Orbit Menu                        ║
║  LM     –  Lyapunov (exponents) Menu                           ║
║  POM    –  Periodic Orbit Menu                                 ║
║  STM    –  Straddle Trajectory Menu                            ║
║  UM     –  Unstable and stable manifold Menu                   ║
╚══════════════════════════════════════════════════════════════╝
```

```
PM      - -     PARAMETER MENU

XCO    -  X COordinate to be plotted
XS     -  X Scale
YCO    -  Y COordinate to be plotted
YS     -  Y Scale
BETA   -  set value of beta
Ck     -  set value of Ck, where k = 1, ..., 9
RHO    -  set value of rho
SD     -  Screen Diameters
SIGMA  -  set value of sigma
```

```
PNM     - -     PRINTER MENU

EP    -   EPson or compatible dot matrix printer
HP    -   HP laser printer
PSC   -   PostScript Color printer
PSP   -   PostScript Printer
```

```
PNOM    - -     PRINTER OPTIONS MENU

FF     -  Form Feed, eject printer sheet
PP     -  Print Picture (core copy of picture)
PP0    -  T0 + PP + restore old text level
PP2    -  T2 + PP + restore old text level
PPP    -  Printer output goes to file h.prn
TL     -  Text Level (or commands T0,...,T3)
```

```
POM     - -     PERIODIC ORBIT MENU

PR    -   PeRiod of the orbit sought

Q1    -   take one Quasi-Newton step
QK    -   take K Quasi-Newton steps, where K = 1, ..., 999
RP    -   Randomly seek Periodic orbits
RPK   -   RP plotted with crosses
NMS   -   Newton Maximum Step size if set
```

```
┌─────────────────────────────────────────────────────────────────────┐
│ RDM  - -   READ DISK FILES MENU                                       │
│                                                                       │
│ AFD   - Add picture From Disk file root.pic into core picture         │
│ AFD1  - Add picture into window 1 (reducing size of pic)              │
│ AFD2  - Add picture into window 2                                     │
│ AFD3  - Add picture into window 3                                     │
│ AFD4  - Add picture into window 4                                     │
│ AFD5  - Add picture From Disk in format and position you specify      │
│ AFDR  - AFD Ratio changes size of window for AFD1,...,AFD4            │
│ FD    - read picture From the Disk file root.pic                      │
└─────────────────────────────────────────────────────────────────────┘
```

```
┌───────────────────────────────────────────────────────────┐
│ RM    -     RECOVER MENU                                    │
│                                                             │
│ CENT - move small cross to CENTer of screen                 │
│ I    - reInitalize trajectory to small cross position       │
│ SD   - increase Screen Diameters SD                         │
│ SV1  - Set value of initialization Vector                   │
└───────────────────────────────────────────────────────────┘
```

```
┌───────────────────────────────────────────────────────────┐
│ SCM    - -  SIZE OF CORE MENU                               │
│                                                             │
│ SAME  - set core picture size equal to screen               │
│ SIZE  - set core picture, width, height, no.colors          │
│ SQ5   - core picture 512 pixels wide by 512 high            │
│ SQ7   - core picture 720 pixels wide by 720 high            │
│ SQ9   - core picture 960 pixels wide by 960 high            │
│ HIGH  - core picture is  HIGH  dots high                    │
│ WIDE  - core picture is  WIDE  dots wide                    │
│ COLP  - set number of COLor Planes                          │
└───────────────────────────────────────────────────────────┘
```

```
┌───────────────────────────────────────────────────────────┐
│ SM     - -     SCREEN MENU                                  │
│                                                             │
│ C     - Clear current window                                │
│ CC    - Complete Clear                                      │
│ CS    - Clear Screen (but not core copy)                    │
│ R     - Refresh screen                                      │
│ ROT   - ROTate square picture 90 degrees clockwise          │
│ SHIFT - SHIFT or flip picture                               │
│ TL    - Text Level (or commands T0,...,T4)                  │
│ TS    - Size of Type used with command TYPE                 │
│ TYPE  - TYPE text on screen at small cross                  │
└───────────────────────────────────────────────────────────┘
```

```
┌──────────────────────────────────────────────────────────────────────┐
│ SOM    – –    STRADDLE OPTIONS MENU                                    │
│                                                                        │
│ ABST   –   compute (from ya's basin) Accessible BST trajectory         │
│ ASST   –   compute Accessible SST trajectory                           │
│ BAF    –   tells BST to use the BA picture                             │
│ DIV    –   DIVisions of line ya-yb for ABST, SST, ASST, GAME           │
│ GAME   –   like ASST, ABST, and SST, only                              │
│                you choose the new ya and yb each time                  │
│ SDIS   –   subdivide ya-yb until ya and yb are close                   │
│ SMC    –   Maximum number of Checks for Straddle methods               │
└──────────────────────────────────────────────────────────────────────┘
```

```
┌──────────────────────────────────────────────────────────────────────┐
│ STM    – –    STRADDLE TRAJECTORY MENU                                 │
│                                                                        │
│ BST    –   plot Basin boundary Straddle Trajectory using ya and yb     │
│ SST    –   plot Saddle Straddle Trajectory using ya and yb             │
└──────────────────────────────────────────────────────────────────────┘
```

```
┌──────────────────────────────────────────────────────────────────────┐
│ TL    –    TEXT LEVEL for SCREEN and PRINTER                           │
│                                                                        │
│ T0    –   no text                                                      │
│ T1    –   minimal text                                                 │
│ T2    –   ideal text                                                   │
│ T3    –   lots of text (default)                                       │
│ T4    –   send current y values to printer                             │
│ T5    –   send continuous stream of y values to                        │
│               printer as they are computed                             │
│ T6    –   send all screen text to the printer                          │
└──────────────────────────────────────────────────────────────────────┘
```

```
┌──────────────────────────────────────────────────────────────────────┐
│ UM    –    UNSTABLE AND STABLE MANIFOLD MENU                           │
│                                                                        │
│ AB    –   connect ya to yb (to yc to yd to ya)                         │
│ ABI   –   do AB and draw its  MI  Iterates                             │
│                                                                        │
│ SL    –   plot Left side of Stable manifold at y1                      │
│ SR    –   plot Right side of Stable manifold at y1                     │
│ UL    –   plot Left side of Unstable manifold at y1                    │
│ UR    –   plot Right side of Unstable manifold at y1                   │
└──────────────────────────────────────────────────────────────────────┘
```

```
╔══════════════════════════════════════════════════════════════╗
║ VM    – –     VECTOR MENU                                      ║
║                                                                ║
║ CENT  –  CENTer small cross (y1)                               ║
║ DIAG  –  DIAGonal: set ya = lower left screen corner           ║
║                    and yb = upper right screen corner          ║
║ I     –  Initialize: change y to equal y1                      ║
║ I1    –  Initialize y1: change y1 to equal y                   ║
║ RV    –  Replace one storage Vector by another vector          ║
║ SV    –  Set values of a storage Vector                        ║
║ YK    –  list coordinates of yK for K = 0,1,...,9 or a,b,c,d,e ║
║ YV    –  list coordinates of several storage vectors           ║
╚══════════════════════════════════════════════════════════════╝
```

```
╔══════════════════════════════════╗
║ WM   – –   WINDOW MENU            ║
║                                   ║
║ C    –  Clear current window      ║
║ CC   –  Complete Clear            ║
║ OW   –  Open new Window           ║
╚══════════════════════════════════╝
```

```
╔════════════════════════════════════════════════════════════════════╗
║ WWM   – –      WHEN AND WHAT TO PLOT MENU                            ║
║                                                                      ║
║ ALL   –  plot dots simultaneously in ALL small open windows          ║
║ CON   –  CONnect consecutive dots                                    ║
║ DOTS  –  DOTS to be plotted                                          ║
║ IPP   –  Iterates Per Plot                                           ║
║ MI    –  Maximum Iterates for ABI, II, OI                            ║
║ PI    –  PreIterates before plotting                                 ║
║ PRM   –  PaRaMeter to be varied by  +/-                              ║
║ PS    –  Parameter Step that +/- change PRM                          ║
║ PT    –  T Plots Time horizontally if ON                             ║
║ V     –  inVert process                                              ║
╚════════════════════════════════════════════════════════════════════╝
```

```
╔════════════════════════════════════════════════════════════════════════╗
║ ZOOM   – –    ZOOM MENU                                                  ║
║                                                                          ║
║ ZF   –  Zoom Factor (usually > 1)                                        ║
║ ZB   –  draw the Box that would be used by ZI                            ║
║ ZI   –  Zoom In towards the small cross by a factor ZF                   ║
║ ZO   –  Zoom Out away from the small cross by a factor ZF                ║
║ DW   –  restore Default coordinates and scales in current Window         ║
╚════════════════════════════════════════════════════════════════════════╝
```

15.3 TABLES OF THE FIGURES

CIRCle map (CIRC)			
x $\rightarrow$ rho + x + C1*sin(x) mod 2*pi			
C1	rho	Fig.	Abbreviated title
1.0	PRM	2-23	Lyapunov exponent bifurcation diagram
YCO	XCO	2-24a	Arnol'd tongues
YCO	XCO	2-24b	Chaos plot

CobWeb map (CW)					
x $\rightarrow$ -C3*arctan(C2*x)/C1 + (1-C3)*x + rho*C3/C1					
C1	C2	C3	rho	Fig.	Abbreviated title
0.25	4.0	YCO	XCO	7-10a	Chaos plot
0.25	4.0	0.5	PRM	7-10b	Bifurcation diagram
0.05	10.0	YCO	XCO	7-11	Chaos plot

complex Cubic map (C) with z = x + iy					
z $\rightarrow$ rho*z*z*z + (C1+iC2)*z*z + (C3+iC4)*z + C5+iC6					
C1	C5	C6	rho	Fig.	Abbreviated title
1.0	XCO	YCO	0.0	2-14	Mandelbrot set
1.0	0.32	0.043	0.0	2-15	Basins and Attractors
1.0	XCO	YCO	0.0	2-27	Period-m plots
1.0	0.32	0.043	0.0	3-3b	Blow-ups of a basin
1.0	-0.11	0.6557	0.0	3-4a	Zooming in
1.0	PRM	0.0	-0.5	6-14	Bifurcation diagram
1.0	0.32	0.043	0.0	7-3	Basins of attraction
1.0	-0.11	0.6557	0.0	7-5	Basins of attraction
1.0	-0.11	0.6557	0.0	8-3	Basins of attraction
1.0	-0.11	0.6557	0.0	8-3	BST trajectory
-1.0	1.0	0.0	0.8	8-3	Basins of attraction
-1.0	1.0	0.0	0.8	8-3	BST trajectory
1.0	0.32	0.043	0.0	8-5	Basins of attraction
1.0	0.32	0.043	0.0	8-5	BST trajectory
1.0	0.32	0.043	0.0	8-5	ABST trajectories
1.0	0.32	0.043	0.0	12-2	BST trajectory

For all the figures:
C2 = C3 = C4 = 0 (default values)

Computer art: Fractals and Dynamics			
process	parameters	Fig.	Abbreviated title
Henon map	Fig. 2-4	1-4	Henon & Sierpinski's triangle
any process	any values	4-3a	Sierpinski's triangle
Tinkerbell	default	4-3b	Tinkerbell & Sierpinski's triangle
any process	any values	12-4	Sierpinski's gasket

GoodwiN equation (GN)
$$x'' + C1*((x*x\text{-}1)/(x*x+1))*x' - C2*x + C3*x*x*x = rho*sin(phi*t)$$

C1	C2	C3	rho	phi	Fig.	Abbreviated title
0.1	0.5	0.5	37	1.0	13-3	A chaotic, strange attractor
0.75	0.5	0.5	14	1.0	13-4	Basins of attraction

Gumowski/Mira map (GM)
$$(x,y) \rightarrow (C1*(1 + C2*y*y)*y + F(x), \text{-}x + F(xNew))$$
$$F(u) = rho*u + 2*(1\text{-}rho)*u*u/(1+u*u)$$

C1	C2	rho	Fig.	Abbreviated title
1.0	0.0	0.3	2-1e	One chaotic and 9 nonchaotic trajectories
1.0	0.0	-0.1	4-2	Four different trajectories
1.0	0.0	0.25	9-11	Five different trajectories

Henon map (H)
$(x,y) \rightarrow (rho - x*x + C1*y, x)$

C1	rho	Fig.	Abbreviated title
-0.3	2.12	2-1a	Trajectory, axes and tic marks
-0.3	2.12	2-1b	Various options for tic marks
-0.3	2.12	2-1c	Periodic points of period 5
-0.3	2.1	2-1d	Chaotic trajectory
-0.3	2.0	2-1d	Chaotic trajectory
-0.3	1.9	2-1d	Attracting period 4 orbit
-0.3	1.8	2-1d	Attracting period 2 orbit
-0.3	2.12	2-2	Trajectory versus time
0.3	1.4	2-4	Henon attractor
0.3	1.4	2-5	A quadrilateral and its iterate
0.0	PRM	2-7	Bifurcation diagram
PRM	1.25	2-8	Bubbles in bifurcation diagram
0.475	1.0	2-9a	All basins and attractors
0.475	1.0	2-9b	Shading of basins
-0.3	1.31	2-10	Basin of infinity
-0.3	1.32	2-10	Basin of infinity
-0.3	1.395	2-10	Basin of infinity
-0.3	1.405	2-10	Basin of infinity
-0.3	2.12	2-11	Periodic points of period 10
0.3	1.4	2-17	Unstable manifold of fixed point
-0.3	2.12	2-18	Stable manifold of fixed point
0.3	4.2	2-19	Saddle Straddle Trajectory
0.3	4.2	2-19	Stable and unstable manifolds
-0.3	2.12	2-20	Basin of infinity and chaotic attractor
-0.3	2.12	2-21	Basin boundary Straddle trajectory
0.3	1.4	3-3a	Attractor and some blow-ups
0.3	1.4	5-3	Approximate Lyapunov exponents
PRM	1.25	6-2	Bifurcation diagram with "bubbles"
0.0	PRM	6-4a	Lyapunov exponent bifurcation diagram
0.0	PRM	6-4b	Bifurcation diagram
0.3	PRM	6-6a	Lyapunov exponent bifurcation diagram
0.3	PRM	6-6b	Bifurcation diagram
0.5	PRM	6-8	Discontinuous bifurcation diagram
0.5	PRM	6-9	Bifurcation diagrams (BIFI = 1)
0.5	PRM	6-10	Bifurcation diagrams (BIFI = 8)
PRM	1.25	6-11	Blow-up of bifurcation diagram
0.3	PRM	6-12a	Lyapunov exponent bifurcation diagram
0.3	PRM	6-12b	Bifurcation diagram

Henon map (H)			
$(x,y) \rightarrow (rho -x*x + C1*y, x)$			

C1	rho	Fig.	Abbreviated title
0.475	1.0	7-2	Generalized basin (BAS-method)
0.475	1.0	7-6	3 basins using BAS-method
0.5	0.855	7-8	Basin using BAS-method
0.9	1.2	8-4	Saddle Straddle Trajectory
0.9	3.5	8-6a	Basin of infinity
0.9	3.5	8-6b	Basin boundary Straddle trajectory
0.9	0.5	8-9	Basin boundary Straddle trajectory
1.0	1.5	8-10	Saddle Straddle Trajectory
1.0	1.5	9-2	Manifolds of a fixed point
0.3	1.42	9-3a	Pieces of stable manifold
0.3	1.42	9-3b	Chaotic attractor
0.3	1.42	9-3b	Unstable manifold of fixed point
0.3	1.42	9-3b	Attractor and unstable manifold
-0.3	1.4	9-4	Two basins
-0.3	1.4	9-4	Stable and unstable manifolds
-0.3	1.4	9-4	Basin boundary Straddle trajectory
0.9	0.5	9-5a	Basin of infinity
0.9	0.5	9-5a	Stable and unstable manifolds
0.9	0.5	9-5a	Basin boundary Straddle trajectory
0.9	0.6	9-5b	Basin of infinity
0.9	0.6	9-5b	Stable and unstable manifolds
0.9	0.6	9-5b	Basin boundary Straddle trajectory
0.3	2.66	9-8	Basin of period 3 orbit
0.3	2.66	9-8	Stable and unstable manifolds
0.3	2.66	9-8	Basin boundary Straddle trajectory
0.9	1.5	9-9	Stable and unstable manifolds
0.9	2.06	9-9	Stable and unstable manifolds
0.9	0.71	9-10a	Three basins of attraction (Wada basins)
0.9	0.71	9-10b	Blow-up of the three Wada basins
0.9	0.71	9-10c	Basin cell
0.3	1.4	10-3	A period 7 orbit on the Henon attractor

Ikeda map (I) with $z = x + iy$ $z \rightarrow$ rho $+ C2*z*\exp(i[C1 - C3/(1 + |z*z|)])$		
rho	Fig.	Abbreviated title
PRM	2-13a	A family of period 5 orbits
PRM	2-13b	Following family of period 5 orbits
1.0	3-1a	Trajectory and drawing of small box
1.0	3-1b	Blow-up of small box in Figure 3-1a
1.0	5-5	Chaotic trajectory
1.0	5-5	Approximate Lyapunov exponents
PRM	6-13	Bifurcation diagram
0.75	8-8	Saddle Straddle Trajectory in basin
1.0	9-6	Chaotic trajectory
1.0	9-6	Increasing pieces Unstable manifold
1.0	9-6	Unstable manifold surrounds attractor
1.0	10-2a	A period 10 orbit on an attractor
1.0	10-2b	Periodic points with period 10
PRM	11-1	Following family of period 3 orbits
PRM	11-2	Following family of period 2 orbits

For all the figures:
$C1 = 0.4$, $C2 = 0.9$, $C3 = 6.0$ (default values)

LOGistic map (LOG) $x \rightarrow$ rho $* x * (1 - x)$		
rho	Fig.	Abbreviated title
3.83	2-3a	Graph of the third iterate of the map
3.83	2-3b	Cobweb plot of a trajectory
3.83	2-3c	Trajectory versus time
3.83	2-3d	Different scalings in windows
PRM	6-1	Bifurcation diagram sent directly to printer

Lorenz system $- -$ Poincare return map (LPR)				
beta	rho	sigma	Fig.	Abbreviated title
8/3	PRM	10.0	6-15	Bifurcation diagram

OWN (Holmes map)			
$(x,y) \rightarrow (rho*x - x*x*x + C1*y, x)$			
C1	rho	Fig.	Abbreviated title
0.95	1.5	13-1a	Unstable manifold of (0,0)
0.95	1.5	13-1b	Saddle Straddle Trajectory

OWN (3PL map)

$x \rightarrow C1*x + rho$ if $x \le beta$

$x \rightarrow C2*x + (C1-C2)*beta + rho$ if $beta < x \le sigma$

$x \rightarrow C3*x + (C2-C3)*sigma + (C1-C2)*beta + rho$ if $x > sigma$

C1	C2	C3	beta	sigma	rho	Fig.	Abbreviated title
0.9	-7	0.5	0.2	0.8	PRM	13-2	Bifurcation diagram

forced damped Pendulum differential equation (P)

$x'' + C1*x' + C2*\sin x = rho*(C3+\cos(phi*t))$

C1	C3	rho	Fig.	Abbreviated title
0.2	0.0	2.5	2-12	Fixed points and period 2 points
0.1	0.1	1.5	2-25	Box-counting and Lyapunov dimension
0.2	0.0	2.0	7-7a	Basin (Low Resolution)
0.2	0.0	2.0	7-7b	Basin (High Resolution)
0.2	0.0	1.66	7-9a	Three Wada basins
0.2	0.0	1.66	7-9b	Blow-up of three Wada basins
0.2	0.0	2.0	8-11	Basin of attraction
0.2	0.0	2.0	8-11	Basin boundary Straddle Trajectory
0.2	0.0	2.0	8-11	Saddle Straddle Trajectories
0.2	0.0	2.5	9-7	Stable manifold of a fixed point
0.1	0.1	1.6	12-3	Chaotic saddle
For all the figures: C2 = 1.0 , phi = 1.0				

Piecewise Linear map (PL)

$(x,y) \rightarrow (C1*x + y + rho, C2*x)$ if $x \le 0$

$(x,y) \rightarrow (C3*x + y + rho, C4*x)$ if $x > 0$

C1	C2	C3	C4	rho	Fig.	Abbreviated title
-1.25	-0.0435	-2.0	-2.175	PRM	6-7	Bifurcation diagram
0.5	0.0	PRM	0.0	0.1	6-16	Bifurcation diagram
-1.25	0.18	2.0	-3.0	0.05	8-7	SST trajectory

Tinkerbell map (T)
x → x*x - y*y + C1*x + C2*y y → 2*x*y + C3*x + C4*y

Fig.	Abbreviated title
1-1	Attractor and consecutive blow-ups
2-22	Basin of infinity and attractor
2-22	Basin boundary Straddle Trajectory
2-26	Zooming in toward the chaotic attractor
3-2c	circle and its MI Iterates
5-4	Chaotic trajectory
5-4	Approximate Lyapunov exponents
7-4a	Basin of attraction (Low Resolution)
7-4b	Basin of attraction (Medium Resolution)
7-4c	Basin of attraction (High Resolution)
8-2	Chaotic attractor and ball in its basin
8-2	Basin of infinity and an attractor
8-2	Basin boundary Straddle Trajectory
10-1a	A period 10 orbit on an attractor
10-1b	Periodic points with period 10
10-5	Fixed points
10-5	Periodic points with period 3
10-5	Periodic points with period 5
10-5	Periodic points with period 7

For all the figures:
C1 = 0.9, C2 = - 0.601, C3 = 2.0, C4 = 0.5

Van der Pol equation (VP)						
x'' - C1*x'*(1-x*x) + C2*x + C3*x*x*x = rho*sin(phi*t)						
C1	C2	C3	rho	phi	Fig.	Abbreviated title
0.1	-2.5	2.5	5	1.0	1-5	Chaotic attractor
0.1	-2.5	2.5	5	1.0	3-4b	Zooming in toward a point on an attractor

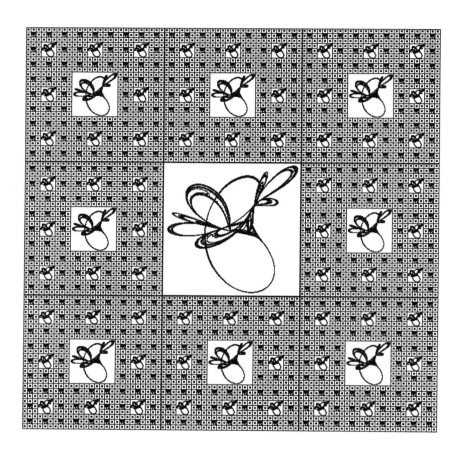

REFERENCES

In references of Springer-Verlag, "Berlin, New York, etc." means "Berlin, Heidelberg, New York, London, Paris, Tokyo, Hong Kong, Barcelona, Budapest".

Abraham, R.H. and C.D. Shaw (1992), *Dynamics: The Geometry of Behavior,* Addison-Wesley Publishing Company

Alexander, J.C. (1988, Ed.), *Dynamical Systems*, Proceedings of the Special Year at the University of Maryland 1986-87, Lecture Notes in Math. 1342, Springer-Verlag, Berlin, New York, etc.

Alexander, J.C., B.R. Hunt, I. Kan and J.A. Yorke (1996), Intermingled basins for the triangle map, *Ergodic Th. & Dynamical Systems* 16, 651-662

Alligood K.T., T.D. Sauer and J.A. Yorke (1997), *Chaos - An Introduction to Dynamical systems*, Springer-Verlag, New York

Alligood K.T. and J.A. Yorke (1992), Accessible saddles on fractal basin boundaries, *Ergodic Theory and Dynamical Systems* 12, 377-400

Anderson, P.W., K.J. Arrow, and D. Pines (1988), *The Economy as an Evolving Complex System*, Addison-Wesley Publishing Comp., Inc.

Aoki, N. and K. Hiraide (1994), *Topological Theory of Dynamical Systems*, North-Holland Math. Library, Vol. 52, North-Holland, Amsterdam

Arnold, V.I. (1983), *Geometric Methods in the Theory of Ordinary Differential Equations*, Grundlehren der mathematischen Wissenschaften 250, Springer-Verlag, Berlin, New York, etc.

Aronson, J. (1990), CHAOS: A SUN-based program for analyzing chaotic systems, *Computers in Physics* 4, 408-417

Arrowsmith, D.K. and C.M. Place (1990), *An Introduction to Dynamical Systems*, Cambridge University Press, Cambridge

Baker, G.L. and J.P. Gollub (1990), *Chaotic Dynamics, an introduction*, Cambridge University Press, Cambridge

Barnsley, M.F. (1988), *Fractals everywhere*, Academic Press, Inc. San Diego

Barnsley, M.F. and S.G. Demko (1986, Eds.) *Chaotic Dynamics and Fractals*, Academic Press, Inc.

Bedford, T. and J. Swift (1988, Eds.), *New Directions in Dynamical Systems*, London Math. Soc. Lect. Notes Series 127, Cambridge Univ. Press, Cambridge

Bélair, J. and S. Dubuc (1991, Eds.) *Fractal Geometry and Analysis*, Proceedings of the NATO Advanced Study Institute and Séminaire de mathématiques supérieures on Fractal Geometry and Analysis, Montréal, Canada 1989, Kluwer Academic Publishers, Dordrecht,Boston, and London

Benedicks, M. and L. Carleson (1991), The dynamics of the Hénon map, *Annals of Mathematics* 133, 73-169

Benettin, G., L. Galgani, and J.-M. Strelcyn (1976), Kolmogorov entropy and numerical experiments, *Physical Review A* 14, 2338-2345

Benettin, G., L. Galgani, A. Giorgilli, and J.-M. Strelcyn (1980), Lyapunov characteristic exponents for smooth dynamical systems and for Hamiltonian systems; a method for computing all of them, Part 1: Theory, *Meccanica* 15, 9-20; Part 2: Numerical application, Ibid., 21-29

Bergé, P., Y. Pomeau and C. Vidal (1986), *Order within Chaos. Towards a deterministic approach to turbulence*, (Translated from french by L. Tuckerman.) John Wiley & Sons, New York

Berry, M.V., I.C. Percival and N.O. Weiss (1987, Eds.), *Dynamical Chaos*, The Royal Society, London

Bowen, R. (1978), *On Axiom A Diffeomorphisms*, Reg. Conf. Ser. Math. Amer. Math. Soc. 35

Bowen, R. (1975), *Equilibrium States and the Ergodic Theory of Anosov Diffeomorphisms*, Lecture Notes in Mathematics 470, Springer-Verlag Berlin, New York, etc.

Broer, H.W., F. Dumortier, S.J. van Strien, and F. Takens (1991), *Structures in Dynamics, Finite Dimensional Deterministic Studies*, North Holland, Amsterdam

Butcher, J.C. (1996, Ed.), Special Issue Celebrating the Centenary of Runge-Kutta methods, *Applied Numerical Mathematics* 22 (Numbers 1-3)

Campbell, L., and W. Garnett (1969), *The life of James Clerk Maxwell*, The Sources of Science, No. 85, Johnson Reprint Corporation, New York and London

Campbell, D.K. (1990, Ed.), *Chaos - XAOC*, Soviet-American perspectives on nonlinear science. American Institute of Physics, New York

Campbell, D.K., R.E. Ecke, and J.M. Hyman (1992, Eds.) *Nonlinear Science, the Next Decade*, Special Issues of Physica D, M.I.T. Press, Cambridge

Casdagli, M. and S. Eubank (1992, Eds.) *Nonlinear Modeling and Forecasting*, Proceedings Volume XII, Santa Fe Institute Studies in the Sciences of Complexity, Addison-Wesley Publishing Company, Redwood City, California

Casti, J. (1989), *Alternate Realities, Mathematical Models of Nature and Man*, A wiley-Interscience Publication, John Wiley & Sons, New York

Cherbit, G. (1991, Ed.), *Fractals, Non-integral dimensions and applications* John Wiley & Sons, Chichester, New York, etc.

Collet, P. and J.-P. Eckmann (1980), *Iterated maps on the Interval as Dynamical Systems*, Birkhäuser, Basel

Crilly, A.J., R.A. Earnshaw, and H. Jones (1991, Eds.), *Fractals and Chaos*, Springer-Verlag, New York

Cvitanovich, P. (1984), *Universality in Chaos*, A reprint selection, Adam Hilger Ltd., Bristol

Dendrinos, D.S. and M. Sonis (1990), *Chaos and Socio-Spatial Dynamics*, Applied Mathematical Sciences 86, Springer-Verlag, Berlin, New York, etc

Devaney, R. and Z. Nitecki (1979), Shift automorphisms in the Henon mapping, *Commun. Math. Phys.* 67, 137-146

Devaney, R.L. (1989), *An introduction to Chaotic Dynamical Systems*, Second edition, Addison-Wesley Publishing Company, Inc.

Doedel, E.J. and J.P. Kernévez (1986), *AUTO: Software for continuation and bifurcation problems in ordinary differential equations*, Applied Mathematics, California Institute of Technology, Pasadena, CA 91125

Eckmann, J.-P. (1981), Roads to turbulence in dissipative dynamical systems, *Rev. Modern Phys.* 53, 643-654

Eckmann, J.-P. and D. Ruelle (1985), Ergodic theory of chaos and strange attractors, *Rev. Mod. Phys.* 57, 617-656

Ermentrout, B. (1990), *Phaseplane*, Version 3.0, Brooks-Cole

Falconer, K.J. (1985), *The Geometry of Fractal Sets*, Cambridge University Press, Cambridge

Falconer, K.J. (1990), *Fractal geometry*, John Wiley & Sons, Chichester, etc.

Fatou, P. (1919, 1920), Sur les equations fonctionelles, *Bull. Soc. Math. de France* 47, 161-271; 48, 33-94, 208-314

Feder, J. (1988), *Fractals*, Plenum Press, New York and London

Feigenbaum, M.J. (1978), Quantitative universality for a class of nonlinear transformations, *J. Stat. Phys.* 19, 25-52

Feigenbaum, M.J. (1979), The universal metric properties of nonlinear transformations, *J. Stat. Phys.* 21, 669-706

Feit, S.D. (1978), Characteristic exponents and strange attractors, *Commun. Math. Physics* 61, 249-260

Feudel, U., C. Grebogi, B. Hunt, and J.A. Yorke (1996), A Map with more than 100 coexisting low-period periodic attractors, *Phys. Rev. E* 54, 71-81

Fomenko, Y. (1994), *Visual Geometry and Topology*, Springer-Verlag, Berlin and Heidelberg

Franceschini, W. and L. Russo (1981), Stable and unstable manifolds of the Hénon mapping, *J. Stat. Phys.* 25, 757-769

Frank, M., G. Keller and R. Sporer (1996), Practical implementation of error estimation for the correlation dimension, *Phys. Rev. E* 53, 5831-5836

Georges, J. and D. Johnson (1992), *Dynamical Systems Software*, Software for the Apple Macintosh computer, Addison Wesley Publishing Company

Glass, L. and M.C. Mackey (1988), *From Clocks to Chaos, The Rhythms of Life*, Princeton University Press, Princeton

Gleick, J. (1990), *Chaos, The Software*, Autodesk

Grassberger, P and I. Procaccia (1983), Measuring the strangeness of strange attractors, *Physica D* 9, 189-208

Grassberger, P (1990), An optimized box-assisted algorithm for fractal dimensions, *Physics Letters A* 148, 63-68

Grebogi, C., E. Ott, and J.A. Yorke (1983), Fractal basin boundaries, long-lived chaotic transients, and unstable-unstable pair bifurcation, *Phys. Rev. Lett.* 50, 935-938, Erratum 51, 942

Grebogi, C., E. Ott, S. Pelikan, and J.A. Yorke (1984), Strange attractors that are not chaotic, *Physica D* 11, 261-268

Grebogi, C., E. Ott, and J.A. Yorke (1985), Attractors on an N-torus: Quasiperiodicity versus chaos, *Physica D* 15, 354-373

Grebogi, C., E. Ott, and J.A. Yorke (1987), Basin boundary metamorphoses: Changes in accessible boundary orbits, *Physica D* 24, 243-262

Grebogi, C., E. Kostelich, E. Ott, and J.A. Yorke (1987), Multi-dimensioned intertwined basin boundaries: basin structure of the kicked double rotor, *Physica D* 25, 347-360

Grebogi, C. and J.A. Yorke (1997, Eds.), *The Impact of Chaos on Science and Society*, United Nations University Press, New York - Tokyo - Paris

Guckenheimer, J., J. Moser and S.E. Newhouse (1980), *Dynamical Systems*, CIME Lectures Bressanone, Italy 1978, Birkhäuser, Boston

Guckenheimer, J. and P. Holmes (1983), *Nonlinear oscillations, dynamical systems and bifurcation of vector fields*, Applied Mathematical Sciences 42, Springer-Verlag, Berlin, New York, etc.

Guckenheimer, J. and S. Kim (1990), *KAOS*, Mathematical Sciences Institute Technical Report, Cornell University, Ithaca, NY, (A program for Sun workstations)

Gulick, D. (1992), *Encounters with Chaos*, McGraw-Hill, Inc., New York

Gumowski, I. and C. Mira (1980), *Dynamique Chaotique*, Cepadues Editions Toulouse

Gumowski, I. and C. Mira (1980), *Recurrences and Discrete Dynamic Systems*, Lecture Notes in Mathematics 809, Springer-Verlag, Berlin, etc.

Gurel, O. and O.E. Rössler (1979, Eds), *Bifurcation Theory and Applications in scientific disciplines*, Annals of the New York Academy of Sciences vol. 316, New York Academy of Sciences, New York

Hale, J. and H. Koçak (1991), *Dynamics and Bifurcations*, Texts in Applied Mathematics 3, Springer-Verlag, New York, Inc.

Hao, B.-L. (1990), *Chaos II*, World Scientific, Singapore

Hayashi, C. (1964) *Nonlinear Oscillations in Physical Systems*, McGraw-Hill Inc., (Princeton University Press, Princeton 1985)

Helleman, R.H.G. (1980, Ed.), *Nonlinear Dynamics*, Annals of the New York Academy of Sciences vol. 357, New York Academy of Sciences, New York

Hénon, M. (1976), A two-dimensional mapping with a strange attractor, *Commun. Math. Phys.* 50, 69-77

Hirsch, M., C. Pugh, and M. Shub (1977), *Invariant Manifolds*, Lecture Notes in Mathematics 583, Springer-Verlag, Berlin, New York, etc.

Hitzl, D.L. and F. Zele (1985), An exploration of the Henon quadratic map, Physica D 14, 305-326

Holden, A.V. (1986, Ed.), *Chaos*, Manchester University Press, Manchester

Holmes, P.J. (1980), *New Approaches to Nonlinear Problems in Dynamics*, SIAM, Philadelphia

Hubbard, J. and B. West (1990), *MacMath*, Springer-Verlag, Berlin, New York, etc. (A program for the Macintosh computer)

Hubbard, J. and B. West (1991), *Differential Equations: A Dynamical Systems Approach*, Part 1, Springer-Verlag, Berlin, New York, etc. ??

Ikegami, G. (1987, Ed.), *Dynamical Systems and Singular Phenomena*, World Scientific, Singapore

Iooss, G. (1979), *Bifurcations of maps and applications*, North-Holland Math. Studies 36, Amsterdam

Iooss, G., R.H.G. Helleman and R. Stora (1983, Eds), *Chaotic Behavior of Deterministic Systems*, North Holland 1983

Jackson, E.A. (1989), *Perspectives of Nonlinear Dynamics*, Volume 1, Cambridge Univ. Press, Cambridge etc.

Jackson, E.A. (1991), *Perspectives of Nonlinear Dynamics*, Volume 2, Cambridge Univ. Press, Cambridge etc.

Jakobson, M.V. (1981), Absolutely continuous invariant measures for one-parameter families of one-dimensional maps, *Commun. Math. Phys.* 81, 39-88

Julia, G. (1918), Memoire sur l'itération des fonctions rationelles, *J. Mathématiques Pures et Appliquées* 4, 47-245

Katok, A and B. Hasselblatt (1995), *Introduction to the Modern Theory of Dynamical Systems*, Encyclopedia of Mathematics and its Applications, Vol. 54, Cambridge Univ. Press, Cambridge

Kennedy, J. and J.A. Yorke (1991), Basins of Wada, *Physica D* 51, 213-225

Kifer, Y. (1988), *Random Perturbations of Dynamical Systems*, Progress in Probability and Statistics 16, Birkhäuser, Boston

Kloeden, P.E. and A.I. Mees (1985), Chaotic phenomena, *Bull. Mathematical Biology* 47, 697-738

Knuth D.E. (1981), *The Art of Computer Programming*, Volume 2: Seminumerical Algorithms, Addison-Wesley Publishing Company

Koçak, H. (1989), *Differential and Difference Equations through Computer Experiments*, With diskettes containing PHASER: An Animator/Simulator for Dynamical Systems for IBM Personal Computers, Second edition, Springer-Verlag, Berlin, New York, etc.

Kostelich, E.J. and D. Armbruster (1996), *Introductory Differential Equations*, Addison-Wesley, Reading, MA

Kuznetsov, Y.A. (1995), *Elements of Applied Bifurcation Theory*, Applied Mathematical Sciences, Vol. 112, Springer-Verlag, Berlin and New York

Lanford, O.E. (1982), A computer assisted proof of the Feigenbaum conjecture, *Bull. Amer. Math. Soc.* 6, 427-434

Lasota, A. and M.C. Mackey (1985), *Probabilistic properties of Deterministic Systems*, Cambridge University Press, Cambridge

Ledrappier, F. and L.-S. Young (1988), Dimension formula for random transformations, *Commun. Math. Phys.* 117, 529-548

Lichtenberg, A.J. and M.A. Lieberman (1983), *Regular and Stochastic Motion*, Applied Mathematical Sciences 38, Springer-Verlag, Berlin, New York etc.

Lorenz, E.N. (1963), Deterministic nonperiodic flow, *J.Atm. Sc.* 20, 130-141

Lorenz, H.-W. (1989), *Nonlinear Dynamical Economics and Chaotic Motion*, Lecture Notes in Economics and Mathematical Systems 334, Springer-Verlag, Berlin, New York, etc.

Lundqvist, S., N.H. March and M. Tosi (1988, Ed.), *Order and Chaos in Physical Systems*, Plenum press, New York and London

MacKay, R.S. and C. Tresser (1987), Some flesh on the skeleton: the bifurcation structure of bimodal maps, *Physica D* 27, 412-422

Mandelbrot, B.B. (1983), *The Fractal Geometry of Nature*, W.H. Freeman and Company, New York

Mané, R. (1987), *Ergodic Theory and Dynamical Systems*, (Translated from the Portuguese by S. Levy.) Ergebnisse der Mathematik und ihrer Grenzgebiete 3. Folge, Band 8. Springer-Verlag, Berlin, New York, etc.

Marek, M. and I. Schreiber (1991), *Chaotic Behaviour of Deterministic Dissipative Systems*, Cambridge University Press, Cambridge

Markley, N.G., J.C. Martin and W. Perrizo (1978, Eds.), *The Structure of Attractors in Dynamical Systems*, Proceedings North Dakota State University 1977, Lecture Notes in Mathematics 668. Springer-Verlag, Berlin, New York, etc.

Mattila, P. (1995), *Geometry of Sets and Measures in Euclidean Spaces – Fractals and rectifiability*, Cambridge studies in advanced mathematics 44, Cambridge University Press, Cambridge

May, R.M. (1975), Biological populations obeying difference equations: stable points, stable cycles and chaos. *J. Theor. Biol.* 51, 511-524

May, R.M. (1976), Simple mathematical models with very complicated dynamics, *Nature* 261, 459-467

McDonald, S.W., C. Grebogi, E. Ott, and J.A. Yorke (1985), Fractal basin boundaries, *Physica D* 17, 125-153

Melo, W. de, and S. van Strien (1993), *One-Dimensional Dynamics*, Ergebnisse der Mathematik und ihrer Grenzgebiete 3.Folge, Band 25, Springer-Verlag, Berlin, etc.

Milnor, J. (1985), On the concept of attractor, *Commun. Math. Phys.* 99 177-195; Comments On the concept of attractor: correction and remarks, *Commun. Math. Phys.* 102, 517-519

Mira, C. (1976), Structures de bifurcations "boîtes emboitées" dans les récurrences ou transformations ponctuelles du premier ordre, dont la fonction presente un seul extremum, Application d' un probleme de chaos en biologie, *Comp. Rend. Acad. Sc. Paris Ser. A* t282, 221-224

Molteno, T.C.A. (1993), Fast O(N) box-counting algorithm for estimating dimensions, *Phys. Rev. E* 48, 3263-3266

Moon, F.C. (1987), *Chaotic vibrations, An introduction for applied scientists and engineers*, John Wiley & Sons, Inc. New York, Chichester, Brisbane, Toronto and Singapore

Moser, J. (1973), *Stable and Random Motions in Dynamical Systems*, Annals Math. Studies 77, Princeton University Press, Princeton, NJ

Moss, F., L.A. Lugiato, and W. Schleich (1990, Eds.), *Noise and Chaos in Nonlinear Dynamical Systems*, Proceedings of the NATO Advanced Research Workshop on Noise and Chaos in Nonlinear Dynamical Systems, Cambridge University Press, Cambridge

Myrberg, P.J. (1958, 1959, 1963), Iteration der reellen Polynome zweiten Grades. I. *Ann. Acad. Sci. Fennicae* 256, 1-10; II. 268, 1-13; III. 336, 1-18

Nitecki, Z. and C. Robinson (1980, Eds.), *Global theory of dynamical systems*, Proceedings Northwestern University 1979, Lecture Notes in Mathematics 819, Springer-Verlag, Berlin, New York, etc.

Nusse, H.E. and J.A. Yorke (1989), A procedure for finding numerical trajectories on chaotic saddles, *Physica D* 36, 137-156

Nusse, H.E. and J.A. Yorke (1991), Analysis of a procedure for finding numerical trajectories close to chaotic saddle hyperbolic sets, *Ergodic Theory and Dynamical Systems* 11, 189-208

Nusse, H.E. and J.A. Yorke (1991), A numerical procedure for finding accessible trajectories on basin boundaries, *Nonlinearity* 4, 1183-1212

Nusse, H.E. and J.A. Yorke (1992), The equality of fractal dimension and uncertainty dimension for certain dynamical systems, *Commun. Math. Phys.* 150, 1-21

Nusse, H.E. and J.A. Yorke (1996), Basins of Attraction, *Science* 271, 1376-1380

Nusse, H.E. and J.A. Yorke (1996), Wada basin boundaries and basin cells, *Physica D* 90 (1996), 242-261

Oseledec, V.I. (1968), A multiplicative ergodic theorem. Ljapunov characteristic numbers for dynamical systems, *Trans. Moscow Math. Soc.* 19, 197-231

Ott, E. (1981), Strange attractors and chaotic motions of dynamical systems, *Rev. Modern Phys.* 53, 655-671

Ott, E. (1993), *Chaos in Dynamical Systems*, Cambridge University Press, Cambridge

Ott, E., T. Sauer and J.A. Yorke (1994, Eds.), *Coping with Chaos*, John Wiley & Sons, New York

Palis, J. and W. de Melo (1982), *Geometric Theory of Dynamical Systems. An introduction.* Springer-Verlag, Berlin, New York, etc.

Palis, J. (1983, Ed.), *Geometric Dynamics*, Proceedings Instituto de Matematica Pura e Aplicada, Rio de Janeiro, 1981, Lecture Notes in Mathematics 1007, Springer-Verlag, Berlin, New York, etc.

Palis, J. and F. Takens (1987), *Hyperbolicity & sensitive chaotic dynamics at homoclinic bifurcations*, Cambridge studies in advanced mathematics 35, Cambridge University Press, Cambridge

Parker, T.S. and L.O. Chua (1989), *Practical Numerical Algorithms for Chaotic Systems*, Springer-Verlag, Berlin, New York, etc.

Peitgen, H.-O. and P. Richter (1986), *The Beauty of Fractals*, Springer-Verlag, Berlin, New York, etc.

Peitgen, H.-O. and D. Saupe (Eds.) (1988), *The Science of Fractal Images*, Springer-Verlag, Berlin, New York, etc.

Poincaré, H. (1897), *Les méthodes nouvelles de la méchanique céleste*, Vols. I and II, Gauthier-Villars, Paris

Poincaré, H. (1899), *Les méthodes nouvelles de la méchanique céleste*, Vol. III, Gauthier-Villars, Paris

Pol, B. van der (1926), On relaxation oscillations, *Phil. Mag.* 2, 978-992

Rand, D. and L.-S. Young (1981, Eds.), *Dynamical Systems and Turbulence*, Proceedings Warwick 1979, Lect. Not. Math. 898, Springer-Verlag, Berlin, New York, etc.

Rasband, S.N. (1989), *chaotic Dynamical Systems of Nonlinear Systems*, John Wiley & Sons, New York

Robinson, C. (1995), *Dynamical Systems: Stability, Symbolic Dynamics, and Chaos*, CRC Press, Boca Raton, FL

Rössler, O. (1976), An equation for continuous chaos, Phys.Lett. A57, 397-398

Ruelle, D. and F. Takens (1971), On the nature of turbulence, *Commun. Math. Phys.* 20, 167-192

Ruelle, D. (1980), Strange attractors, *Math. Intelligencer* 2, 126-137

Ruelle, D. (1981), Small random perturbations of dynamical systems and the definition of attractors, *Commun. Math. Phys.* 82, 137-151

Ruelle, D. (1989), *Chaotic Evolution and Strange Attractors*, Cambridge University Press, Cambridge, New York, etc.

Ruelle, D. (1989), *Elements of Differentiable Dynamics and Bifurcation Theory*, Academic Press, Inc.

Ruelle, D. (1991), *Chance and Chaos*, Princeton University Press, Princeton

Saari, D. and J. Urenko (1984), Newton's method, circle maps and chaotic motion. *Amer. Math. Monthly* 91, 3-18

Shaw, R. (1981), Strange attractors, chaotic behaviour and information flow, *Z. Naturforschung* 36A, 80-112

Schuster, H.G. (1984), *Deterministic Chaos*, Physik-Verlag, Weinheim

Schroeder, M. (1991), *Fractals, Chaos, Power laws, Minutes from an infinite Paradise*, W.H. Freeman and Company, New York

Shiraiwa, K. (1985), *Bibliography for Dynamical Systems*, Department of Mathematics Preprint series No. 1, Nagoya University

Simó, C. (1979), On the Henon-Pomeau attractor, *J. Stat. Phys.* 21, 465-494

Sinai, Ya.G. (1989, Ed.) *Dynamical Systems II*, Encyclopaedia of Mathematical Sciences. Volume 2. *Ergodic Theory with Applications to Dynamical Systems and Statistical Mechanics*. Springer-Verlag, Berlin, New York, etc.

Sinai, Ya.G. (1991), *Dynamical Systems collection of papers*, Advanced Series in Nonlinear Dynamics, Vol. 1, World Scientific

Smale, S. (1980), *The Mathematics of Time. Essays on Dynamical Systems, Economic processes and related topics*, Springer-Verlag, Berlin, New York, etc.

Sparrow, C. (1982), *The Lorenz equations: bifurcations, chaos, and strange attractors*, Applied Mathematical Sciences 41, Springer-Verlag, Berlin, New York, etc.

Stein, P.R. and S. Ulam (1964), *Nonlinear transformation studies on electronic computers*, Rozprawy Matematyczne 39, 1-66

Stewart, I. (1989), *Does God Play Dice? The Mathematics of Chaos*, Basil Blackwell

Strogatz, S.H. (1994), *Nonlinear Dynamics and Chaos*, Addison-Wesley, Reading, MA

Takens, F (1983), Invariants related to dimension and entropy, In: Atas do 13° Coloquio Brasileiro de Matematica

Takens, F (1985) On the numerical determination of the dimension of an attractor, In: *Dynamical Systems and Bifurcations*, Lecture Notes in Math. 1125, 99-106, Springer-Verlag, Berlin etc.

Theiler, J (1987), Efficient algorithm for estimating the corelation dimension from a set of discrete points, *Phys. Rev. A* 36, 4456-4462

Theiler, J (1990), Estimating fractal dimension, *J. Optical Society of America A - Optics and Image Science* - 7, 1055-1073

Thompson, J.M.T. and H.B. Stewart (1986), *Nonlinear Dynamics and Chaos. Geometric methods for engineers and scientists*. John Wiley & Sons, Ltd.

Tufillaro, N.B., T. Abbott, and J. Reilly (1992), *An experimental approach to Nonlinear Dynamics and Chaos*, Addison-Wesley Publishing Company, Redwood City

Verner, J.H. (1978), Explicit Runge-Kutta methods with estimates of the local truncation error, *SIAM J. Num. Anal.* 15, 772-790

Vincent, T.L., A.I. Mees, and L.S. Jennings (1990, Eds.), *Dynamics of Complex Interconnected Biological Systems*, Birkhauser, Boston

West, B.J. (1985), *An Essay on the Importancy of being Nonlinear*, Lecture Notes in Biomathematics 62, Springer-Verlag, Berlin, New York, etc.

Wiggins, S. (1988), *Global Bifurcations and Chaos*, Applied Mathematical Sciences 73. Springer-Verlag, Berlin, New York, etc.

Wiggins, S. (1990), *Introduction to Applied Nonlinear Dynamical Systems and Chaos*. Text in Applied Math. 2, Springer-Verlag, Berlin, New York, etc.

Wolf, A. (1986), Quantifying chaos with Lyapunov exponents, In "*Chaos*" (Ed. A.V. Holden), 273-290, Manchester University Press, Manchester

You, Z., E.J. Kostelich and J.A. Yorke (1991), Calculating Stable and Unstable Manifolds, Intern. Journal of Bifurcation and Chaos 1, 605-624

Young, L.-S. (1982), Dimension, entropy and Lyapunov exponents, *Ergod. Th. & Dynam. Sys.* 2, 109-124

Zaslavsky, G.M. (1985), *Chaos in Dynamic Systems*, (Translated from the Russian by V.I. Kisin.) Harwood Academic Publ. Chur, etc.

Zhang, S.-Y. (1991), *Bibliography on Chaos*, Directions in Chaos - Vol. 5, World Scientific Publishing Co. Pte. Ltd., Singapore

INDEX

Applied Mathematical Sciences

(continued from page ii)

(continued on next page)

Applied Mathematical Sciences

(continued from previous page)